Iroquois Medical Botany

THE **Iroquois** AND THEIR NEIGHBORS

Laurence M. Hauptman, *Series Editor*

Iroquois
Medical
Botany

JAMES W. HERRICK

Edited and with a Foreword by
DEAN R. SNOW

SYRACUSE UNIVERSITY PRESS

Copyright © 1995 by Syracuse University Press
Syracuse, New York 13244-5290

All Rights Reserved

First Edition 1995

21 22 23 24 25   11 10 9 8 7

This book is published with the assistance of a grant from the John Ben Snow Foundation.

For a listing of books published and distributed by Syracuse University Press,
visit https://press.syr.edu.

ISBN: 978-08156-0464-8

Library of Congress Cataloging-in-Publication Data

Herrick, James W.
    Iroquois medical botany / James W. Herrick ; edited and with a
Foreword by Dean R. Snow. — 1st ed.
        p.   cm.—(The Iroquois and their neighbors)
    Includes bibliographical references and index.
    ISBN 0-8156-0295-2 (cloth : alk. paper) :
    1. Iroquois Indians —Ethnobotany. 2. Iroquois Indians—Medicine.
3. Botany, Medical—New York (State) 4. Medicinal plants—New York
(State) I. Snow, Dean R., 1940–   . II. Title. III. Series.
E99.I7H47   1994
615'.321'097—dc20   94-32760

It is with much love and affection that I dedicate this book to my little cosmos:

Laurie,
Daniel,
Kelly

JWH

JAMES W. HERRICK lectures in anthropology, sociology, psychology, and gerontology, and studied under William N. Fenton. His publications include articles in *The Psychological Record, The New York State Journal of Medicine,* and *The International Journal of Aging and Human Development*, and chapters in *The Anthropology of Medicine: From Culture to Method* and *Reassessment in Psychology: The Interbehavioral Alternative.* He was a contributing author to *The Academic American Encyclopedia.* He and his family live in Chittenango, New York.

DEAN R. SNOW is Professor of Anthropology at the University at Albany, SUNY, where he has directed the Mohawk Valley Project. He organizes the annual Conference on Iroquois Research and is the author of many books and articles. These include *The Archaeology of North America* and *The Iroquois.* He is a co-author of *Atlas of Ancient America.*

Contents

TABLES

FIGURE

Foreword

James Herrick completed his doctoral dissertation on the subject of Iroquois medical botany in
1977. He was guided through several collections of
unpublished notes by William N. Fenton, then
Distinguished Research Professor of Anthropology
at the University at Albany, State University of New
York. On the botanical side, he was guided by
Stanley J. Smith, who was then Curator of Botany at
the New York State Museum. Smith helped him to
organize plant lists according to the scheme then current at the museum, and Fenton helped him to
understand Iroquois uses of the plants. The result
was a dissertation that ran over 560 pages long. It has
been cited repeatedly by ethnobotanists in the years
since its completion.

When William Starna and I started the Mohawk
Valley Project in the early 1980s, we were both aware
of Herrick's work and its potential significance for
the project. Our initial intention was to clarify the
Mohawk Iroquois archaeological sequence. It soon
became clear that the sequence had the additional
potential to clarify regional demographic changes
that took place after exogenous epidemics, colonial
warfare, and other benefits of European colonization
began to wash over the region. Our sites began to
produce botanical remains that appeared relevant to
the study of both subsistence and health care, and we
saw the importance of Herrick's dissertation with
growing appreciation.

By 1986 the Mohawk Valley Project was enjoying
the support of several funding agencies. The largest
of these was the National Endowment for the
Humanities, but the project had also received sub-
stantial funding from the National Geographic
Society, Earthwatch, Arkell Hall, the Rochester
Museum, and the Evans Foundation. William Starna
had by this time traded his codirectorship of the

project for new commitments in ethnohistorical research, and I had turned my attention to raising new
funding to carry the project through to 1989. The
trajectory of the project was by now moving toward
completion, and I judged that it was time to begin
publishing more than annual reports and technical
articles on it.

My 1986 application to NEH contained provisions
for the publication of a revised version of *Iroquois
Medical Botany*, as well as a compilation of a volume
of historical narratives relating directly to the
Mohawk Valley up to the early nineteenth century. I
saw both as essential foundation volumes for later
project volumes. It seemed likely that two more volumes, one containing site reports and demographic
analysis and the other covering public and private
artifact collections, would come next. However, both
of these core project volumes would have had to also
contain large passages on ethnobotany and historical
documentation, so large in the aggregate that the
projected volumes would have been prohibitively
long. Thus it made sense to publish botanical and
historical volumes first, so that their contents could
be cited without repetition in later volumes. I was
pleased that NEH and other contributors agreed.
NEH funds were matched by the Wenner-Gren
Foundation, Arkell Hall, the J.M. McDonald
Foundation, the Lucius Littauer Foundation, and
Hartgen Associates.

I edited both the narrative chapters and the plant
entries during 1989. In this I was assisted by William
Starna and, of course, James Herrick. The plant list
was rearranged according to the new ordering found
in Richard Mitchell's (1986) *Checklist of New York
State Plants*. He later generously read and amended
the plant entries. John Haynes generously did the
same for the lower species found near the

beginning of the list. Susan Bamann culled sources for illustrations of 181 of the plants, none of which were illustrated in the original dissertation. She also oversaw their reproduction at scales suitable for this volume. Janet Snow set the pages and kept track of cross-references, no small task, especially in the plant list chapter.

I have tried to provide illustrations of as many of the relatively obscure weedy plants as possible. Illustration of many other plants was not undertaken for a variety of reasons. Many fungi and other lower plants defy meaningful illustration. Many plants have yet to be accurately drawn. In many other cases, I judged that better illustrations already existed in readily available field guides. This was especially the case for well-known trees and wildflowers. Readers are advised to use such guides in conjunction with this volume and Mitchell's (1986) *Checklist of New York State Plants*. Many plants not yet illustrated here or elsewhere will eventually be illustrated in forthcoming New York State Museum bulletins. In the meantime, serious students of Iroquois medical botany will have to consult directly with that institution.

Multiple (sometimes contradictory) orthographies and the limitations of typesetting options forced the deletion of most native Iroquoian terms for plants. Perhaps a qualified linguist will eventually become interested in publishing these terms, which are not necessary to an understanding of either botany or archaeology. I experimented with alternative ways to present the data, but in the end concluded that Herrick had found the best system in the first place. We added many alternative common plant names, supplied new numbers to the species list, and found ways to make variations in fonts help readers find their way through the thicket of native Iroquoia.

This version of Herrick's book upgrades the plant list from an unillustrated appendix to an illustrated chapter. However, the significance of his analytical chapters is not diminished. Readers who take the time to read the early chapters carefully will be rewarded by what will be to most of them a new perspective on the plant world around us. Traditional Iroquois practitioners had (and have) a sophisticated perspective on the plant world that contrasts strikingly with that of modern medical science. The practitioners whose initials appear in the plant lists have been the authoritative sources for information recorded by Herrick, Fenton, Waugh, and others. To understand them is to understand part of the Iroquois cosmos. Modern archaeologists may choose to use this volume as a handy means to understand the presence of certain plant remains on Iroquois sites, but if they let it, it can take them much deeper into Iroquois culture than that.

DEAN R. SNOW

Preface

The primary goal of this study is to take the topic of Iroquois medical botany from the descriptive level of investigation to the levels of understanding and explanation. That is, I want to go from a strictly "emic" (culture from the viewpoint of the native) perspective to an "etic" (scientific) one. By doing this it is hoped that medical behaviors in *any* culture will be better understood. Inherent in such a goal are the problems of cross-cultural communication faced by all cultural anthropologists. It will be seen that it is not enough to simply match European or Western cultural concepts and categories with those of the Iroquois people in the name of discovering how they have traditionally conceived of health, illness, and plant medicines. At the same time, concepts and categories familiar to members of Western culture must be used so as to provide a point of reference for those interested in the beliefs and behaviors of others.

Superficially, it may appear that I have relied upon the inner workings of my own mind more than the testimonies of the various sources of cultural information made available to me. Actually, what I have done is construct models that I believe account for the presence of often contradictory and confusing statements made by both native sources and various ethnographers. Rather than assume that one native source or ethnographer was correct while another was wrong or confused, attempts have been made to take into account the free-floating, highly flexible, seemingly contradictory nature of Iroquoian conceptions of health and medicine. In using this approach, I feel that I have come extremely close to understanding plant medicines as they were and are conceived by traditional Iroquois people.

It is possible to use one's own cultural concepts in describing, understanding, and explaining the cultural behavior of others only after attempts have been made to point out fundamental, cross-cultural differences in perceptions of the cosmos. Attempts to do this have been made in the first two chapters of this study. Once the differences between the Western or European and Iroquoian views of the things and events of the universe are understood, a cautious attempt is made at analyzing the traditional beliefs of the Iroquois regarding health, illness, and the use of medicinal plants in terms of both native and Western concepts. It is seen that any culturally imposed, European concepts necessarily become altered as they are used in relation to a perception of the universe that is quite unlike that of the Westerner. Discrete categories and concepts become replaced by overlapping categories and concepts, with questions of degree dominating over questions of kind. Causes and effects become supplanted by models of mutual interactions. A self-generating, ever-expanding universe replaces a static one.

Chapter 3 undertakes an analysis of traditional Iroquoian conceptions regarding illness and disease etiologies. Details of four overlapping etiological categories are discussed. Borrowed notions and their relationships to native theories of disease causation are also dealt with in this chapter. Classes of symptoms are paired with native etiologies, the establishment of these relationships being critical to determining the relative degrees of power thought to be possessed by various modes of treatment in general and by particular plant medicines specifically.

Treatment categories and health actors are discussed in chapter 4. Special emphasis is given to the health actor categories "herbalists," "clairvoyants," and "witches" as they are involved in practicing counterwitchcraft. The use of plant medicines by these health actors is seen to play a significant role in the overall conception of plant medicines in traditional Iroquois culture. Collection techniques and the use of various plants in acts of divination are discussed, as is the topic of preventive medicines. The role of "white" doctors (i.e., health actors using a European-based, scientific medical belief system and technology) as alternatives to traditional forms of healing is treated in conjunction with a discussion of what have been termed the "confidence" or "peace of mind" medicines of the traditional Iroquois. Communal medical practices are dealt with as they relate to traditional conceptions and applications of plant medicines.

Chapter 5 of this study builds on the previous discussions of traditional Iroquoian medical-religious behaviors. From these discussions an attempt is made to isolate from the 454 species contained in chapter 6 those plants that were and are thought to possess greater amounts of power. The result of this analysis is a checklist of powerful medicinal plants in traditional Iroquois culture.

Throughout the text, there is frequent mention of various plants. A number that appears before each plant refers to its position in chapter 6. The reader is referred to the introduction of chapter 6 for information regarding languages spoken by the various sources of cultural information also mentioned throughout the text.

Most of the data contained in chapter 6 are those collected by F.W. Waugh between 1912 and 1914 and by W.N. Fenton between 1933 and 1942. Since their principal sources of information were speakers of the Seneca and Mohawk languages (the two most mutually unintelligible languages of the Five Nations), I concentrated on herbalists at the St. Regis Reservation (Mohawk) in Canada and the Allegany and Tonawanda (Seneca) reservations in western New York. Along the way, sources of native herbalism were sought at the Onondaga, Cattaraugus, and Six Nations reservations.

It should be noted that it was never my intent to expand upon what was already an abundance of ethnographic data provided by native herbalists who were both less acculturated and less prone to antianthropologist sentiments than anyone I might have encountered in the early 1970s. Some of Waugh's contacts, for example, were in their eighties when he worked with them in the early 1900s. Any new ethnobotanical information I was able to gather as a byproduct of my attempt merely to verify Waugh's and Fenton's field notes pales in comparison to their work.

Despite the wealth of information presented in chapter 6, the reader should realize that it does not constitute all that is known about Iroquois use of medicinal herbs. That is, it is not a compendium of the work of Parker, Beauchamp, Waugh, and others even though these ethnographers are referred to in the main text. At the same time, the information contained therein provides more than ample data for achieving the goal of this study: namely, to understand the use of herbal medicines within the broader cosmological, ceremonial, and ritual context of traditional Iroquois culture. If this goal has been achieved, one should discover that use of medicinal herbs by the Tuscarora, Huron, Cherokee, or any other Iroquoian people (and many non-Iroquoian people, as well) will follow the same basic cultural patterns and principles that have guided and continue to guide the people of the Longhouse.

I am very aware of how fortunate I was to have known and worked with a person of Professor William N. Fenton's stature. My sincerest thanks go to him for his guidance, patience, and generosity. Thanks also go to my botanical mentor and friend, the late Stanley Jay Rexford Smith, former Curator of Botany at the New York State Museum and Science Service. And my gratitude is expressed to those Native Americans with whom I worked; for their appreciation of the idea that a foreigner might be interested in their cultural ways. Finally, the support and encouragement of Dean R. Snow and those enlisted by him to help in this project is appreciated and gratefully acknowledged.

Financial support for my field research was provided by the Wenner-Gren Foundation and the National Museum of Man, Ottawa.

Iroquois Medical Botany

Introduction

In aboriginal times the tribes of the Iroquois League or Confederacy (the Ho-de-no-sau-nee or "People of the Longhouse," as they called themselves) occupied most of what is now New York State. The "keepers of the eastern door" were the Mohawks, with their territorial neighbor to the west being the Oneida nation or tribe. The capital of the Confederacy was located in central New York, its territory being inhabited by the Onondaga tribe. To the west of these "firekeepers" were the Cayugas, and farther to the west were the "keepers of the western door," the Seneca nation. Deganawida, a Huron, was the legendary founder of the League. He and his spokesman, Hiawatha (an Onondaga later adopted by the Mohawks), are said to have traveled amongst the then five feuding tribes in an attempt to promote friendship and peace. Archaeologists estimate that this occurred sometime between A.D. 1400 and A.D. 1600. A sixth nation, the Tuscarora, was adopted into the Iroquoian kinship state in the early eighteenth century. They occupied lands south of the Oneida and Mohawk territories, extending southerly to the Susquehanna River region.

Languages of all five original Confederacy members are considered part of the Northern Iroquoian group, with Cherokee being a temporally and spatially distant relative of the Southern Iroquoian group. Each nation, however, had its own distinctive language. As geographic separation increased, as in the case of the westernmost Seneca and the easternmost Mohawk, mutual intelligibility decreased. Tuscarora, although considered a member of the Northern Iroquoian group, is believed to be an older branch of the Iroquoian languages. It is, however, closer in structure to the languages of the tribes of the Five Confederacy Nations than it is to Cherokee.

Iroquoian speech is characterized by the combination of what English speakers would consider several words into one. A single word, for example, might include a verb root, a noun root, a pronoun, and various monosyllabic prefixes and suffixes that communicate to the listener such things as tense or time limitations, the circumstances under which action is taking place, the intent of the speaker and of the person or persons carrying out the action, the nature of the action, the sex and number of persons acting, whether or not an object is being acted upon and in what manner, quality, quantity, possession, and so forth.

The traditional means of food production of the Iroquois people consisted of slash-and-burn horticulture. Women, with help from the men, prepared the fields for planting, but women alone did the planting, tending, and harvesting. Hunting, gathering, and fishing rounded out their subsistence patterns. Corn, bean, and squash were cultivated. Deer, moose, elk, beaver, and bear were hunted, and a variety of wild nuts, seeds, berries, greens, fungi, and roots were gathered seasonally. Digging sticks, hoes, stone knives, snares, hooks, nets, traps, drags, spears, blow guns and darts, packs, snowshoes and toboggans, clay pipes and pottery, and bows and arrows comprised part of their traditional technology.

A dozen or more hilltop palisaded villages ranged in population from fewer than 100 to 1000 or more persons. Rights to the communally owned cleared lands surrounding their villages were determined on the basis of one's mother's ancestry, or the ancestry of one's adoptive matrilineage. Elm bark-covered longhouses, some of which were expanded to reach lengths of over 300 ft, housed matrilocal, matrilineal "fireside" families consisting of monogamous parents and their offspring. At this level of family organization, both mother's and father's (especially father's sister) relatives had special kinship obligations and expectations.

The Bear, Turtle, Heron, Hawk, Snipe, Deer, Eel, Duck, Ball, and Beaver clans were the symbolic and temporal extensions of the matrilineages. They constituted the underpinnings of Iroquois social and political organization. The cultural rule of exogamy, necessitating marrying outside a specified group of clans (i.e., outside of one's moiety), served as an underlying framework for the economic, political, and ceremonial interdependence of families, villages, and tribes. It served as a basis for the unification of the Confederacy as a whole.

The exogamous clans functioned in the economic realm by facilitating and promoting mutual aid, hospitality, friendship, reciprocal gift-giving, food sharing, condoling, feasting, and reciprocal ritualism and ceremonialism at the local level. Political unity was accomplished by the existence of councils of clan chiefs who were both appointed and deposed by clan matrons. Ranking clan chiefs, in turn, were represented on village, tribal, and Confederation councils. All had great prestige, but little power; unanimity was required in the making of policy. Certain lineages of each clan possessed chiefly titles. Should a suitable candidate be unavailable, elder lineage matrons could loan a title to a lineage not possessing one. War chiefs, unlike the inherited and appointed peace chiefs or sachems, achieved their positions through warfare or oratorial skills at councils.

The death of a Confederacy sachem was followed by a ceremony designed to install a new chief. This Condolence Council involved a bilateral division of all the clans of the Confederacy, with the clear-minded moiety ritualistically mourning, then feasting and socializing with, the grief-stricken moiety.

The yearly ecologically based cycle of ceremonies of the Iroquois centered on important animals and plant spirit forces and the hunting, gathering, and agricultural activities that comprised their traditional mode of subsistence. Included were Midwinter (a New Year's ceremony concerned with the ritual renewal, thanksgiving, begging, honoring, remembering, and fulfilling the various spirit forces of the cosmos—all in the hope that the earth would again awaken in the spring), Maple (giving thanks for its sweet water), Seed Planting (thanks for the arrival of planting season and seed blessing rites), Strawberry (first fruits), Thunder (thanks and begging for rain), Sun (thanks to sun), Bean (thanks for the mature bean), Green and Little Corn (thanks for the ripening corn, beans, and squash), and Harvest ceremonies. Confession over wampum in order to achieve a pure mind, an opening address of thanks involving the burning of tobacco (its smoke conveying messages to the sky world), closing speeches, ritual games (peach stone, ball, hoop and javelin, snowsnake, and bone games), sacred and social dances, and feasts featuring soups generally characterized these ceremonies. Their content and structure could be altered by dream revelations. Certain ceremonial functionaries and participants in the various rites were determined on the basis of moiety membership. Faithkeepers were the ceremonial counterparts of peace chiefs. Representing their respective moieties, they set dates and generally organized the calendrical ceremonies. They were either named at birth or appointed as adults at the Midwinter or Green Corn ceremonies by the matrons of clans possessing their titles.

Rituals deriving from the calendrical ceremonies, as well as those extracted from associations concerned with the powers of certain animals (otter, eagle, buffalo, bear), spirits of the dead, mythical creatures (little people, magic animals, False and Husk Faces), and charms (hunting charms fashioned from powerful animals, Little Water Medicine bundles) went into making up the various Medicine societies. Their functions were to cure sick or unfortunate persons on demand or at regularly scheduled ceremonies, and to renew cures and the powers of the Medicine societies themselves. Membership was achieved by having been cured by a particular society, by having dreamed of a society's tutelary or totem, or upon the advice of a clairvoyant.

Autumn was the time of councils and the time for the formation of hunting and war parties. The latter activities typically took men far from home, thus bolstering the domain of the "clearing" for the women. Warfare could be massive in scale, but could also be no more than an attempt at seeking revenge at the raiding and feuding levels. Scalping, headhunting, torture, human sacrifice, cannibalism, or, ideally, the capture and adoption of prisoners, were the byproducts of the warrior's quest for glory and prestige.

After 1600 A.D., at its height, the Iroquois Confederacy dominated tribes from Hudson Bay to the Carolinas, and from the Atlantic to the Mississippi River. A strategic geographic position allowed them to play the Dutch, English, and French off against one another in the seventeenth and eighteenth centuries, to their own economic and political advantage. Geography also forced the western tribes of the New World to deal with the Iroquois before connecting with European trading centers on the east coast.

The Proclamation of 1763 made the Appalachian Mountains the boundary line between European and Native American settlements. This line was pushed farther west and made more clearly definable by the Treaty of Fort Stanwix in 1768, but the ensuing struggle between the Crown and the colonists soon made even this line of demarcation meaningless.

Caught up in the struggle occurring between King George and his rebellious American children, most of the tribes of the Confederacy chose to side with their historic ally, the British. Consequently, many were forced to relocate in western New York and Canada after their villages and crops were destroyed by American Generals Sullivan and Clinton. Those allied with the Americans and those that stayed behind and survived ultimately found themselves restricted on lands reserved for their use. This was the beginning of the reservation system.

After the American Revolution, reservation life had undermined the traditional status of Iroquois men as free-ranging hunters and warriors. Drunkenness, murder, brawling, adultery, and accusations of witchcraft replaced pride, prestige, social order, political stability, and economic well-being. It was at this time that the charismatic Seneca prophet and chief Handsome Lake appeared on the scene. He initially warned against any association with the powerful animal and other spirit forces of the Medicine societies, against performing the White Dog Sacrifice (said to be a symbolic remnant of a human sacrifice rite), and preached against gossiping, the use of alcohol, and the practice of witchcraft (including the use of love medicines that contributed to infidelity). He advocated plow agriculture by men, a valuing of nuclear families residing in single family dwellings (with the traditional communal dwelling becoming the Longhouse or church of his followers), and marital fidelity and stability. The old beliefs and ceremonies were revitalized, and new values that brought about the economic, technological, social, and political changes necessary for the cultural survival of his people were instituted. Today, a conservative minority of Longhouse adherents may be found at the Tonawanda, Allegany, Cattaraugus, St. Regis, and Onondaga reservations in New York, and at the St. Regis, Six Nations, Oneida, and Caughnawaga reservations in Canada.

1

Plants in the Cosmos
of the Iroquois

Like other Native American peoples, the Iroquois have traditionally adhered to certain beliefs and values that are quite different from those held by members of modern-day urban-industrial society. These beliefs and values and the cosmology or world view within which they are expressed are manifest in every detail of Iroquois life. They are especially relevant to a discussion of traditional medical practices. Here an attempt is made to relate the Iroquoian cosmology to its basic cultural themes and values. In doing this, a first step will have been taken in understanding how the Iroquois have traditionally used plants as medicines. We begin with a synopsis of the Iroquois creation story.

THE STORY OF CREATION

Before humans inhabited the earth there existed only the sky world or the "world beyond the sky." The same plants and animals found on earth also existed in this world, as did humans or humanlike beings with human emotions, a matrilineal social structure, and bark-covered lodges. Since the sky world was *beyond* the sky, it did not have a moon, sun, or stars—all of which existed *in* the sky and were created at a later time. A Tree of Light located in the center of the sky world next to a chief's lodge provided the only source of light.

In an older version of the creation story an adolescent boy and girl found the opportunity to be alone together. The girl combed her companion's hair and was somehow impregnated. She refused to divulge the boy's name to her family. The boy eventually died during a couvade for the mother of his child. The child was female.

Another version has the boy and girl being siblings under the control of their mother's brother. Before their uncle died he gave his niece instructions to marry the distant chief who owned and lived near the Tree of Light. The girl went to the chief, underwent a cruel ritual so as to earn the respect of her husband's people, then, as Fenton (1962:290) has put it, was "persuaded to engage in the most intriguing . . . gambit of aboriginal literature." It was during this game that the breaths of the two newlyweds intermingled, resulting in the young woman's impregnation.

Jealous over his wife's (by now referred to as "Sky Woman") pregnancy, Sky Chief commenced to brood. To relieve his depression, village folk and all the supernatural forces (except Wind) assembled and tried to guess the "wishes of his soul" as revealed in his dreams. Sky Chief's dream centered around an uprooting of the Tree of Light and the casting down into the hole thus formed all living things, including the Tree. After the dream was guessed, an attempt was made to fulfill it. The Tree was uprooted. Sky Chief and his wife sat down to eat, dangling their legs over the edge of the hole that had once encased the Tree's roots. Suddenly, in an apparent attempt to avoid the destruction of all growing things in the sky world, Sky Chief sprang to his feet and shoved Sky Woman into the hole. The surprised woman slowly drifted downward to what was a landless world of oceans. The Tree of Light was repositioned to plug the hole in the sky dome, and life continued as usual in the sky world.

Before considering what happened to Sky Woman after she fell through the hole in the sky, one other version of the events in the sky world that resulted in her fall deserves attention. This one, presented in A.F.C. Wallace's now classic *The Death and Rebirth of the Seneca* (1972), is reported to have been the one with which the late eighteenth-century Seneca prophet Handsome Lake was familiar. In this version, the youngest of five sons was pining over his unrequited love for a young woman. He enlisted his parents' help in persuading the woman to join him and end his illness. She consented, and he regained his health, but suffered a relapse. The cause of his weakness this time, however, was a dream that yearned to be fulfilled. He informed his brothers of his dream and they proceeded with an attempt to fulfill it. They had to uproot the Tree of Light lest their brother die. The Tree was uprooted and cast into the hole. It now became a source of light for the earth below. Light in the sky world was henceforth supplied by a smaller Tree of Light that had been growing alongside the now discarded one. The young man summoned his wife to the edge of the pit, they sat down together, and she was impregnated by a gust of air from the South Wind. Just before he pushed her through the hole in the sky dome he informed her that she was to become the mother of earth beings.

Depending on which version is consulted, Sky Woman either took to earth with her three ears of corn, meat, firewood, and a child, or arrived empty-handed. In the latter case she was presented with corn, meat, firewood, a miniature mortar and pestle, a small pot, and a soup bone by Meteour.

A multitude of birds, especially water birds such as the loon, bittern, and duck, guided Sky Woman to earth and cushioned her fall onto the back of a huge mud turtle floating on the primal sea. In some versions she brought with her soil scraped from the hole through which she was pushed; in others, Muskrat and Beaver immediately dove for mud from the depths of the sea, plastering it to Turtle's shell. Sky Woman, by now referred to as "Earth Mother" (and, ultimately, the "Old Woman" and the "Wicked Grandmother") began walking about. As she walked, the earth expanded under her feet and vegetation sprouted in her wake. Turtle then prophesied that a great race of beings would grow upon the earth, and that he, in accordance with a dream that had been dreamed in the sky world, would play a role in the creation of these beings.

Some versions have Earth Mother giving birth to

a female child after her descent to Turtle's back. This child matured, then against her mother's constant warnings, went into the water to play and catch young water birds. As a result of this she became pregnant. According to Fenton (1962), the Seneca version has the young woman becoming impregnated by the West Wind while swinging in a tree. Still other versions involve Turtle leaving a pointless magical arrow beside her as she sleeps. The result, however, is that the young woman becomes pregnant.

She soon discovered that she was going to give birth to twins. Even before they were born they were heard arguing in their mother's womb. Their final prenatal argument centered around how they would leave their mother's body. One argued for the conventional way, the other for a shorter, but deadlier, route through his mother's armpit. Each went his separate way and the mother eventually died. Before dying, however, she called her mother (Sky Woman or Earth Mother) to her side and informed her of what she could expect from her grandsons. The Good-Minded One would create and regulate all that was good and beneficial on the earth, while the other (the Evil-Minded One that caused his mother's death) would create and rule over all that was evil, destructive, and harmful. She then instructed her mother to bury her with her feet toward the east, with a fire at her feet for ten days. Through time, two cornstalks would begin to grow out of her grave over the location of her breasts. In some versions the body of the dead woman was mutilated by the evil son, who then convinced his grandmother that this was the work of his brother. The grandmother and evil twin then put the woman's head up in a tree and used it as a source of light. Later, the good son removed his mother's head and propelled it into the sky, thus creating the sun.

The bases to a fundamental theme in traditional Iroquois culture are found by considering the personalities and behaviors of the Good Twin and the Evil Twin. The Good Twin is also known as the Creator, Good Spirit, Master of Life, Sky-Holder, Thrown-Away, Good-Minded, the Light One, or Sapling. The Evil Twin is also called the Toad-Like One, Dark One, Flint, Ice, or the Great World Rim Dweller. The notion that there exist in the universe things and events that are intrinsically good and things that are intrinsically evil pervades Iroquoian theories of health and medicine.

Flint became the favorite of Sky Woman or Earth Mother, his grandmother. She repudiated the Good Twin and threw him into the brush. The Creator, having been rejected, sought out his father—who was Turtle or Wind, depending on which version is consulted. In the Turtle version, the Creator, with the help of a magical arrow left behind by Turtle, parted the magically imbued waters of a lake at the bottom of which resided his father. Turtle then taught his son about such things as husbandry, hunting, house building, the use of fire, and the roasting of maize. The Wind version has the Creator in a running competition with three of his brothers—the winner gained access to a flute and the contents of a bag holding game animals. In both versions the Creator was taught about the more practical aspects of living on earth. Flint, the Evil Twin, did not undergo such training.

The version of the creation reported by Wallace (1972) does not include a discussion of the Creator's visit to his father and the subsequent impartation of certain cultural ways. Instead, an account of the various acts of creation by the Good Spirit or Good Twin, and the countering, destructive deeds of his jealous twin brother, Flint, is given. Included among the former are: creating the sun, moon, man, woman, vegetables and herbs, the rain, rivers and streams with two-way currents, improved corn, and game animals such as deer, bear, beaver, and elk. In the version analyzed by Fenton (1962), the Creator's efforts were also being constantly thwarted by his evil grandmother—the now figuratively and literally fallen Sky Woman. The contributions of the resentful and envious Flint to the earth world included obnoxious animals such as frogs, bats, owls, meat-eating monsters, snakes, worms, spiders, insects, and the like. He redesigned streams with but one-way currents garnished with rapids, rocks, whirlpools and falls. He created a corn-retarding blight, caves, diseases, ice, storms, and death itself. The baleful, harmful, depressing, troublesome, wicked, stinking, burning, irritating, dank, rotting, painful, repulsive things and events of this world, then, were the works of Flint, the Evil-Minded One. At the same time, he busied himself with the destruction or "spoiling" of his brother's beneficial creations.

The ongoing struggle between the Creator and Flint began with a preliminary bout between the former and his evil grandmother. They engaged in a game of peach pits, with the winner gaining control of the earth and its food sources. The Creator won, then took on his brother in a battle to the death. In some versions the two carried on their battle throughout the entire earth, with the end result being the destruction of everything. In others, the struggle between the two brothers developed into a contest or demonstration of power—each claiming that he was the true, all-powerful Creator. Some versions suggest that Flint had at this time taken the form of a giant, grotesque, hunch-backed monster known as the Great World Rim Dweller.

To prove their respective claims, they each attempted to move a mountain while holding their breath. Flint went first, but managed only to shake and move the mountain a very short distance. The Creator then succeeded in moving the mountain so that it was close to his brother's back. Then, depending on the version, Flint was either instructed by the Creator to turn around or, out of Flint's curiosity and impatience, he suddenly reeled around to discover his brother's power. In both cases his nose was broken and permanently bent as it collided with the side of the mountain (see origin of False Faces, below). The Creator won the contest.

With Flint's defeat came certain concessions. He was to help man-beings or the people of the earth as a grandfather helps his grandchildren. He would help the people by giving those who wore masks imitating him the power to handle hot coals, withstand the cold, drive away high winds, obstruct and undo witches, and cure disease. In addition to wearing these masks in his likeness, the masks were to be fed corn mush. It was also the case that the smoke created from the burning of Sacred Tobacco (*Nicotiana rustica*) would not only please him, but would in addition act as a vehicle for the transmission of messages to him in the sky or spirit world. The origins of the False Face Society lie in the practice of wearing, nurturing, and burning tobacco for these grotesque masks. Since the behaviors centering around the masks have as their goal the propitiation of the transformed Evil-Minded Twin, it is not surprising to find them being used in dealing with a wide variety of the dangerous and harmful things and events created by him. These include such things as winter, disease, monsters, and windstorms. The masks were to become, as Wallace (1972:92) points out, "the faces of many gods."

This phase of the creation story ends with the Creator and Flint returning to the sky world by way of two divergent paths in the Milky Way. Neither, however, had by any means finished his work on earth. Their heavenly pathways would come to symbolize their diametrically opposed attitudes toward the earth and the beings that dwelt upon it. It will be seen that these paths were to become well worn by the departed souls of humans who had chosen one way of life over the other. They were the paths to heaven and hell.

The next phase of the creation story tells of the various social and cultural crises that faced human beings, and begins with the Creator (in the guise of *Odendonniha* or "Sapling") and *Awenhaniyonda* ("Hanging Flower") being the first husband and wife. They produced many offspring which, in turn, also produced many offspring. But there was silence in the world. No one performed any ceremonies, they merely wandered about and stood. It was then that the Creator returned to earth from the sky world.

He began by giving the people the four (sacred) ceremonies or rites which were patterned after those performed by beings in the sky world. These were the Great Feather Dance, the Skin Dance, the Men's Personal Chants, and the Bowl Game or Peach Stone Game. The performance of these rites would please the beings living beyond the sky. He told them when they were to be performed: at first fruits, when foods mature in the fall, and at midwinter. He also told them how: with mutual congratulations, much rejoicing, and repeated thanks to the Creator. The people were also instructed to perform the Stirring Ashes rite and the Sacrifice of the White Dog at the Midwinter Ceremonial. The first was an act symbolizing a call to assemble and conduct ceremonies, while the second symbolized the Creator's attire. It was during the Midwinter Ceremony that the people should beg the Creator for the return of all the plants and animals that had diminished in number or died during the winter. Human infants would also be begged for. The things to be thanked in one united voice or mind during the Thanksgiving Address were reviewed. These included everything from the grasses and other things *on* the earth, up through those spirit forces *in* the sky, such as the moon and sun, and finally, the spirit forces *beyond* the sky, including the Creator. Responsibility for seeing that repeated daily thanks were carried out between ceremonial

occasions was given to the elder women of each family. The Creator also took the opportunity to instruct the people in the proper etiquette for greeting one another. They were instructed to do this with kindness, stroking, and repeated thanks to the Creator for the fact that there are others of one's kind on earth.

Finally, he cautioned the people that they must remember to thank him, and also be thankful for each other. Failure to do so could result in their coming under the influence of his evil brother, Flint, and all his evil creations. He warned of the dire consequences of neglect: internal dissension, self-destruction, the dissolution of families, and the beginnings of divided minds that would ultimately take divergent paths to the sky world. Flint, it was warned, was relentless in his attempt to undermine the Creator's good works and to control the world.

The people followed the ceremonial ways for a time. Eventually, however, they could not agree on anything and found it impossible to love one another or to unite their minds in ceremony. Peculiar happenings begin to occur. People disappeared, there was a solar eclipse, family conflicts increased, there was a period of continual rain and high winds. A rainbow formed and the people turned to Sapling for its explanation. He said that it was a sign that they had neglected the ceremonies and that the Creator had returned.

The Creator reminded the people of the two minds and paths they could take, then told them that the rainbow was their reminder that he was still using his power for their benefit and protection. However, when the rainbow stretched across the entire sky from east to west, and the Thunderers came from the east, life on earth would end. Apparently, some life was washed away with this most recent long rainfall, and the Creator began to renew it again and also supplement the initial four ceremonies he had mandated. It was at this time that he gave the people two berry ceremonies (Strawberry and Mulberry) and the Sap or Maple Ceremony. The importance of the gifts of love and peace was reiterated. Without these behaviors observance of the Creator's cherished ceremonies would be impossible.

Time passed, disagreements again cropped up, ceremonies were neglected, and the people began to see monstrous animals in the forest, evidence that Flint was gaining even more control of the earth.

People were killed by these monsters. They also began to die for no apparent reason.

The Creator then reappeared briefly. He warned of Flint's growing influence and told the people that they must not forget love and peace before departing.

Social decay continued. Murders, lack of respect among families, insanity, lying, fear, and illness proliferated. Women and children could only weep. Again, the Creator appeared and called for an assembly.

He began with the pronouncement that people must look out for themselves individually after each has grown up. Then he informed them that the anxiety and sorrow they were experiencing was the result of not loving one another. Because of this, Flint was spoiling and destroying all that was good. Two clearly divided paths had now been defined.

The people of the earth world had been neglectful to the point that the Creator could no longer thwart his evil brother's influence. Disease, the faceless one, traveled the earth unrestrained. In response, stopgap measures were undertaken by the Creator. He gave them herbal medicines, instructed them as to how they should go about collecting them, and informed them that only a few persons would ever know all their names and uses.

An outcome of the people's neglect of the ceremonies was the destruction and diminution of the gifts of wild plant and animal food sources. To compensate for these losses and ensure the survival of man-beings, the Creator provided the gifts of domesticated corn, beans, and squash, all of which would require much hard labor when compared to that needed to harvest the once abundant wild foods. Accompanying these three gifts were the ceremonies that would allow thanks and beseechment, the Planting, Green Corn, Harvesting, and other ceremonies.

The Creator then reminded the people that at the time of the end of the world he would send another person. Furthermore, all things growing on the earth would at first grow less in quantity and power, then would cease to grow altogether. Evil things in the earth that had been placed there after Flint lost the contest of power would emerge. There would be earthquakes, and Flint would succeed in seducing the minds of all human beings. Once this was done, all of the earth would become spoiled. He then told them that they would have to share among themselves the new gifts of corn, beans, and squash, and that these foods, the Three Sisters, were to be regarded as mothers to all the people. He informed the people of what they would have to do to receive these gifts. An adolescent boy and girl would have to be chosen to make a journey following the path of the Creator leading to the grave of Earth Mother. There, they would find the Three Sisters growing. They would also learn of the practice of interring the dead. Again, the people were reminded that they had to think of him and thank him repeatedly.

For a time, after the people had sent the two youths to collect the Three Sisters, things proceeded smoothly. But deaths continued to occur, and many people became so disturbed in mind that they could not perform their ceremonies properly, especially the rites of the four sacred ceremonies.

A solution to their problem was sought. A good, quiet boy, the Creator in disguise emerged on the scene with a suggestion. He pointed out to the people that the Creator provided diversity among other living things, and that humans, too, should become diverse. This diversity would allow them to assist one another, to lift the minds of those who are worried or sorrowed. Thus the boy proposed that humans imitate plants and animals and break up into groups of different kinds of humans, thus leading to the formation of moieties and clans. The importance of women as producers of life was restated, and a means for naming and determining membership in clans was disclosed. The families, led by elder women, were to cross a river by clinging to a grapevine. At dusk the vine would become detached, leaving half the people on one side, half on the other. In the morning the elder women were to go to the river to draw water and take note of the first animal they saw. These animals would determine their clan names.

Finally, the question of establishing a regular place of council and the seating arrangement of clans at these councils was addressed by the young man. After solving these problems, he was given the name His-Mind-Is-Great and the eldest man of the Wolf clan repeatedly thanked him for his accomplishments. Then, the eldest man of the Deer clan (cousins of the Wolf clan) did the same. Now, the whole matter was finished, and there would always be tribes on either side of the river. The creation story ends.

Table 1 summarizes events in the creation myth. Cultural themes are abstracted from these events.

IROQUOIS VALUE ORIENTATIONS

Kluckhohn and Strodtbeck (1961) have delineated what they believe are the five basic areas around which cultural values are centered. The first of these deals with *time* or *temporal focus*, its major dimensions involving *past, present,* or *future* orientations. *Mankind's relationship to nature* comprises a second value area. In this case, *below, in harmony with,* and *above* nature make up the range of possible value orientations. Third, the *human nature* value orientation concerns itself with whether the innate character of mankind is basically *evil* (mutable or immutable), *neutral* (and mutable), both *good and evil* (and immutable), or *good* (mutable or immutable). Fourth, mankind's relationship with other human beings, or *relational* orientation, is broken down into the *lineal, collateral,* and *individualistic* modes. Finally, the *activity* orientation deals with "the problem of the nature of man's mode of self-expression in activity" (Kluckhohn and Strodtbeck 1961:16). Its modes include *being* (behavior that is spontaneously emitted), *being-in-becoming* (the goal of personal development dominating), and *doing* (which accentuates personal accomplishments and the material).

In all cases Kluckhohn and Strodtbeck (1961) stress the idea that all of the various modes of each of the five value areas are seen to operate in every culture. It is the dominance of one mode or modes that makes such an inquiry meaningful. The rank ordering of value orientation alternatives was accomplished through a series of structured questionnaires.

In the present study, no attempts have been made to validate empirically the dominant value orientations existing in Iroquois culture. This does not, however, prevent one from attempting to distill from the literature general statements regarding their traditional, dominant values.

Using Opler's (1946) thematic approach to the study of culture, Fenton (1962:297-298) also formulated what was intended as a provisional list of Iroquoian cultural themes. His themes were derived from the various accounts of Iroquoian cosmology as discussed earlier, and include the following:

1. The native earth, our mother, is living and expanding constantly, imparting its life-giving force to all growing things on which our lives depend.
2. Renewal—the alternation of seasons and tasks is attuned to ecological time as are the lives of plants and animals that rest when the earth sleeps.
3. It is women that count.
4. Paternity is a secondary consideration— *agadoni* ("father's people") comes after *ohwachira* ("mother's people").
5. Twins are lucky and creative, but siblings fight—only the pure who have been secluded succeed.
6. Hospitality is a right and a duty to share— throwing ashes is the negation of sharing, hospitality, friendship, peace, and harmony.
7. The bench or bed is the place of reflection and counsel.
8. Do not oppose the forces of nature—going to water, traveling, making medicine, participating in ceremonial circuits.
9. There is a regular way—life and customary procedures are fairly established.
10. *Orenda* ("supernatural power") adheres to animate and inanimate things, to aspects of the environment, and to sequences of behavior.
11. Restraint is important—one must not exert too much power and "spoil it."
12. Acting the role by impersonation is a means of taking on power.
13. By continual and repeated greeting and thanks, the hierarchy of spirit forces between the earth and sky that the Creator appointed to assist man in the enjoyment of the earth must be remembered and balanced in the fulfillment of their tasks.
14. Dreams or words that the soul desires, compel fulfillment.
15. Smoke symbolizes thought, desire, community, and government—it is a vehicle for sending up words to the gods in the sky world.
16. Earth shakers are friendly spirits to be awed and propitiated.
17. A man never refuses when asked and never shows fear.

Table 1. Themes Deriving from Events in the Creation Myth

Events in the Sky World

	Sky World is prototype of earth life
	Dreams compel fulfillment
1. Sky Woman falls to earth	Basic cultural ways brought from Sky World

Events on Earth

2. Sky Woman gives birth to a female child	
3. Daughter then gives birth to twins	The earth is to be influenced by good and evil
4. Wicked grandmother repudiates Good Twin (Creator)	
5. Father Turtle teaches Good Twin domestic activities	
6. Grandmother and Good Twin bet for control of food supply	Good triumps over evil, but requires good luck
7. Struggle between Flint and Creator; Flint loses contest of power	Good triumphs over evil; evil can be made to do good through imitation and appeasement; smoke is vehicle to Sky World
8. Creator assigns duties to Moon, Sun, Thunderers, etc.	Spirits in sky to assist humans
9. Creator and Flint return to Sky World via different paths	Good-minded humans to follow the Creator: Evil-minded to follow Flint

Earthly Human Events and Visits by the Creator

10. Humans multiply, but wander aimlessly; there are no ceremonies	Ceremonial life needed for meaningful existence
11. Creator gives Four Sacred Rites	Four Sacred Rites please supernaturals; imitation of activities in the Sky World pleases supernatural beings
12. Gift of Stirring Ashes Rite	People must be roused to unite their minds in repeated thanks; they must rejoice at life

---

Table 1. *Continued*

Earthly Human Events and Visits by the Creator

13. Gift of the White Dog Sacrifice	Purity needed to communicate with Creator; Tobacco and smoke are vehicles for messages to the Sky World
14. Gift of Thanksgiving Address	All spirit forces require repeated thanks
15. Gift of love/peace	Peace and love needed for uniting minds in repeated thanks
16. People become negligent; disease, social decay	Without ceremonies, life is endangered
17. Rainbow appears; Creator gives Berry and Maple ceremonies; value of love and peace restated	Creator will continue to protect humans, signs given
18. Continued social decay; ceremonies neglected	Flint's influence increases; two divergent paths well established
19. Disease and death travel earth unrestricted	Disease and death are faceless entities created by Flint to destroy humans
20. Creator losing control; provides medicinal herbs as stopgaps; gifts of corn, beans, and squash	Herbs used on individual bases to counter Flint's disease and death; easily acquired wild food sources dwindle, humans must work harder to cultivate their sustenance
21. Deaths increase, people become distraught and neglectful	Unclear minds of bereaved interfere with uniting minds in ceremonies; ceremony given to clear minds of bereaved
22. Quiet youth proposes division of humans into groups	Imitation, mutual aid, clans, moieties, love, peace, and clear minds allow humans to perform ceremonies

---

18. Things go by twos and fours—forked path, divided mind, sex, seasons, moieties, life-death, balance of forces.
19. A pattern of reciprocity obtains between moieties for quickening life and faculties, for restoring society, and for all ceremonial associations of renewal.
20. Culture is an affair of the mind.

Table 2 attempts to rank order each of the five value orientations found in traditional Iroquois culture. For each value orientation, Fenton's (1962) thematic statements are used to support the order presented. The following paragraphs offer a discussion of certain of these cultural themes as they justify the proposed dominant value orientations in Iroquois culture.

**Table 2. Dominant Iroquoian Value Orientations as Supported
by Fenton's (1962) Discussion of Cultural Themes**

Value Orientation	V-A Dominating in Traditional Iroquois Culture	Supporting Thematic Statements
I. Human Nature	Good/Evil (mutable) *over* Good = Evil	11, 17
II. Time	Past = Present *over* Future	2, 3, 8, 11, 13
III. Man/Nature	In-Harmony *over* Below *over* Above	5, 8, 9, 11, 18
IV. Activity	Being-in-Becoming *over* Being *over* Doing	1, 6, 13, 17, 19
V. Relational	Lineal = Collateral *over* Individual	1, 3, 4, 6, 11, 17, 19

Thematic statement 1 may be interpreted as relating to the activity value orientation. The earth is thought to be ever-expanding, imparting its life-giving force to all growing things on which our lives depend. Although this statement suggests that life force was given only to sustenance-producing plants and nonhuman animals, it must be assumed that human animals were also seen as being growing things. This would certainly be the case in folk societies that are dominated by a sharing and cooperation ethic. Thematic statements 6, 17, and 19 further support the idea that a being-in-becoming activity value orientation dominated. They also tell us a great deal about the relational value orientation in traditional Iroquois society. Never refusing when asked, the conception of hospitality as a right and a duty, and the system of moiety reciprocity sustain the importance of collateral relationships. The life force imparted to humans, then, was given with the idea that it must be utilized by individuals in such a way that it resulted not only in the social responsibility of sharing, but in the social responsibility of restraint (see theme 11) as well. It was a power given with the

stipulation that individuals use it for personally developing themselves into responsible citizens. Because of this, the traditional Iroquois activity value orientation could be said to be dominated by the being-in-becoming mode. This is followed in importance by the being mode. The least important activity value orientation in traditional Iroquois society was that of doing. In the area of medicine or the use of traditional treatments and explanations for disease, one is more likely to be dealing with the being-in-becoming world view stressing personal development. In chapter 3 it will be seen that taboo violation played an important role in native Iroquoian theories of disease causation. In the broadest sense, violation of the established ways (cf. statement 9) is tantamount to a failure on the part of individuals to *develop* themselves as responsible members of society.

The lineal and collateral value orientations dominated over the individualistic value orientation in traditional Iroquois society, and it was the lineal relationship through females that formed the backbone of their traditional social organization. Iroquois

society now has many individual members who
subscribe to the European-based tradition of valuing
individualistic relations over both lineal and
collateral ones. However, traditional Iroquois culture
evolved in such a way that females and an awareness
of female ancestry became dominant cultural
concerns. Females were the caretakers of gardens,
collectors of wild foods, and female ancestry was the
mechanism by which past, present, and future access
to property was determined. Thematic statements 1
and 3 point up this cultural conception of female
importance.

Thematic statement 4 reinforces the importance
of females in traditional Iroquois culture, while at the
same time supporting the idea that lineal and collateral relationships co-dominated. The *agadoni*
("father's people" or "clan") principle served as a cultural safety valve to ensure proper male socialization
should maternal rejection occur, or should there be a
lack of adult males in the mother's lineage. This
same principle is seen to extend to the entirety of the
social and political makeup of Iroquois society, it
being a mechanism to ensure the reciprocal exchange
of demands and obligations needed for survival in
traditional societies (cf. Fenton 1975).

Thematic statement 11 is especially relevant with
regard to both the human nature and man-nature
value orientations. That the equable or tranquil person succeeds may be interpreted as supporting the
mutable, good and bad inherent qualities of individuals. Similarly, individuals must not spoil or upset the
harmonic or balanced aspect of the universe. The
admonition "never refuse when asked" (theme 17) is
a more subtle expression of the notion that both
good and bad (i.e., the potential for refusing when
asked) reside in mankind's basic human nature. This
bivalent nature of human beings was, however, but
one small element in what the Iroquois traditionally
believed to be a bivalent cosmos. More will be said
about this in the discussion of the Iroquoian cosmos
that follows.

A past temporal focus may be inferred from thematic statements 2, 3, and 13. A concern for past kinship running through the female line and all of the
social and political factors that depended on such a
concern, made an awareness of or sensitivity to individual, familial, social, and cultural events in the past
of paramount interest to the traditional Iroquois. At
the same time, themes 3 and 13 make it clear that

renewal and repeated greeting and thanks were valued individual and social behaviors. Both involved
the operation of present behaviors upon past events
or principles established by the Creator, and both
further reinforced the dominant position of a being-
in-becoming activity value orientation. When it is
considered that "the Iroquois do not serve the supernatural through the ceremony; they participate in the
activities of the ceremony" (Tooker 1970:27), further
support for the dominance of a being-in-becoming
value is found. The concepts of renewal and repeated
thanks played a great role in traditional Iroquoian
notions of preventive medicine and further linked the
realms of medicine and religion.

Certain aspects of the Iroquoian languages also
reflect a past-present *over* future time value orientation. The *aorist* (Lounsbury 1953) or *indicative*
(Chafe 1967) prepronominal prefix is used as a tense
marker. This tense "may have a present- or a past-
tense meaning in a given context so long as it is definite; i.e., so long as it refers to actions that have
taken or are taking place at some particular point in
time; broadly, it is a non-future!" (Foster 1974:272).

A connecting link between conceptions of disease
causation and the use of plant and animal medicines
in traditional Iroquois culture is found in an analysis
of the man-nature value orientation. In order to
understand this relationship, it will be necessary to
relate the man-nature value orientation of the
Iroquois to the concept of *orenda*, then point out why
various things, events, people, and places were
thought to possess varying degrees of this power
(Foster 1974:101).

The Iroquois traditionally believed that all things,
events, persons, and places possessed inherent life
force, power, spirit, soul, or *orenda* (see thematic
statement 10). These animistic (belief in souls) beliefs were accompanied by the animatistic (belief in
the existence of a raw, impersonal power) notion that
all things and events were capable of influencing
other things and events of the cosmos through the
exertion of their inherent power. This exertion of
power might then disrupt the inherent power of
those other things or events.

We are told in thematic statement 13 that the
hierarchy of spirit forces between earth and sky that
the Creator appointed to assist man in the enjoyment
of the earth must be remembered and balanced in
the fulfillment of their tasks (Fenton 1962:298). Seen

in this statement is the already noted emphasis on remembering, this being an indication of the Iroquois concern for personal and communal development, while at the same time suggesting a past temporal focus. From this act of remembering through continual and repeated greeting and thanks, it was believed that a cosmic balance could be realized. Man was thought to be very much a part of this balanced universe and was further thought to possess the ability to increase his supply of inherent power (see theme 12). In keeping with his bivalent human nature, however, he had to restrain the use of any such acquired power lest he spoil it and upset the dictates of the Creator. At the same time, an individual was not to rebel against desires expressed or experienced as he walked in the spirit world (i.e., while he dreamed). Thwarted desires of the soul would lead to an imbalanced life force or spirit of the individual. It was the responsibility of both individuals and groups of individuals to see to it that such imbalances were rectified.

So important was this idea that man is part of a balanced cosmos (especially manifested in thematic statements 8, 9, and 18), that it is very likely that there was little, if any, intellectualizing on the part of the Iroquois about what has been for European or Western peoples a long-standing philosophical problem, namely mankind's place in nature. This concern over man's place in nature is most likely experienced only by those who are convinced that they are somehow apart from it. At the base of such a world view are techno-environmental circumstances that allowed the cultural ancestors of modern Westerners to rely less on other-worldly explanations (involving religion) and more on worldly explanations (involving science) for things and events in their surroundings. This increased accumulation of and reliance upon worldly explanations demanded a system of discrete, fixed categories that would facilitate the transmission of cultural knowledge to future generations. A world view that contrasts sharply with that of the traditional Iroquois people resulted.

Aspects of the Western or European-based world view that are *not* typical of most traditional or folk cultures include such things as (1) a reliance upon hierarchical principles and causes-and-effects, (2) authoritarianism and individualism, (3) competitiveness, (4) unity by similarity and repetition, (5) a reliance upon categorization and taxonomic

methods, (6) a belief in one truth, and (7) a belief that past and future information is inferable from the present (after Maruyama 1974). The traditional Iroquois and many non-Western peoples of the world hold a world view that stresses just the opposite. Appreciating these differences is essential to understanding how the Iroquois traditionally conceived of illness and medicine. For example, it is contended that they (1) did not think in terms of hierarchies, but rather stressed interactional relationships, (2) stressed cooperation, (3) thought in terms of symbiotic relationships, (4) saw harmony in diversity, (5) relied upon contextual factors and contextual analyses, (6) believed that one must learn different views and take them into consideration, and (7) believed that information could be nonredundantly generated (Maruyama 1974). Many of Fenton's (1962) Iroquoian cultural themes are seen to substantiate these perceptions of the universe.

In traditional Iroquois culture, the key unifying principle that underlies all living and nonliving things and events of the universe was that of harmony or balance. By stressing this very simple principle, a philosophical place was provided for the acceptance and tolerance of heterogeneity and harmony in diversity. The communal concerns of coordination, cooperation, symbiosis, and mutual interaction and adjustment became secondary manifestations of it. Individually and culturally, such a world view cultivated an openness or receptivity to new or different ways of viewing things and events of the universe. This, again, leads to a great deal of adaptability and flexibility. Perhaps it was just such a philosophy that facilitated the formation and success of the Iroquois Confederacy. Fenton (1965:263), in fact, has pointed to these same cultural characteristics as being responsible for the perpetuation of much of Iroquois culture into modern times. These include elaboration, projection of symbols, transfer of functions, and obligatory reciprocity.

The flexible nature of traditional Iroquois culture further manifests itself in the area of medicine. It will be seen to confound attempts at trying to find precise prescriptions or the application of precise medical techniques for specific diseases. The same holds for attempts at discovering plant names and their uses. Fenton (1940b:793) observed that "botanical knowledge is uncodified and . . . plant names and medicinal usage can differ between families living in the same local group, to say nothing of group differences existing between localities. The ethnologist is left hoping that his revealed information constitutes a fair sample of what is generally known." Through time, not only did the traditional Iroquoian world view encourage intrasocietal experimentation and innovation, it was such that it readily accepted newly introduced medical beliefs and techniques as well (cf. Fenton 1941a). This will be seen to confound efforts directed at finding truly native, vis-à-vis acculturated, beliefs concerning medical matters.

THE IROQUOIS COSMOS

Fenton (1936, 1962), Speck (1949), Chafe (1961), Shimony (1961), Tooker (1970), Wallace (1972), and Foster (1974) have attempted to represent the structure of the Iroquoian cosmos through a content analysis of the *kanõ:nyõk* (Seneca), "Let it be used for showing appreciation or gratitude," commonly called the Thanksgiving Address. The *kanõ:nyõk* (cf. Fenton's theme 13) is an opening address that accompanies all calendrical ceremonies.

The version of the Iroquois cosmos given in Table 3 is borrowed from the authors cited above. As is true of the story of creation, there is no one correct or complete list of things and events thanked in the Thanksgiving Address. Depending on the ethnographer being consulted, different labels are seen to be used for each subdivision. If, however, one feels the need to categorize the Iroquois cosmos, the simplest division might be the sky world versus the earth world. On the other hand, we might want to be less concerned with specific categories and more concerned with things and events in the universe of the Iroquois as they exist along a cosmological continuum.

Perhaps the most realistic way of conceptualizing the cosmos of the Iroquois would be with the phrase "structured, yet flexible." Foster expressed a similar thought when he stated that "the hierarchy rule applies to the arrangement of divisions [i.e., spirit forces of the earth, spirit forces in the sky, and spirit forces beyond the sky] rather than to the absolute arrangement of sections" (Foster 1974:101). He further added that although the precise arrangement of the things, events, people, and places within each of the three divisions might vary from speaker to speaker, there existed an underlying conceptual basis

to each division. These conceptual bases include: (1) things, events, persons, and places having functional *use* to mankind (the earth spirit forces), and (2) things, events, persons, and places having *duties* to mankind (the spirit forces in and beyond the sky). Even these distinctions, however, may give a false impression as to what the critical feature is that may account for a thing's relative position in the cosmos.

In the traditional Iroquois world, everything had a use and a duty as intended by the Creator. In reference to plants, it is often heard that everything is good for something. Everything had a use, and this use was divinely imparted. Likewise, collectors of various medicinal herbs were required to "remind" a plant of the divinely imparted duty it had to human beings. Foster's (1974) "use vs. duty" distinction, then, may not be as meaningful culturally as it appears. What may be more meaningful is the relative amounts of duty and use possessed by each thing and event in the universe. Some items of the universe clearly have duties and uses that are seen to affect the welfare of the entire nation. Others might affect the welfare of much smaller groups or, perhaps, individuals. It is the individually oriented uses and duties of medicinal plants that may account for their low position in the cosmological hierarchy. Plants used as both medicines and food would have much more general uses and duties than plants used for the specific purpose of maintaining, restoring, or upsetting individuals. If there is a uniting principle that accounts for the relative position of various things and events in the Thanksgiving Address, it may be that things and events with more specific or individualized uses and duties are thanked before those things and events having widespread duties or uses.

THE RELATIVE DEGREE OF POWER POSSESSED BY VARIOUS THINGS AND EVENTS IN THE IROQUOIS COSMOS

The question of why different items of the Iroquois cosmos are assigned different positions at different times in the Thanksgiving Address may be answered through a consideration of a cultural-ecological rule of thumb. The more beneficial or harmful a thing, event, person, or place is to a people, the more power the thing, event, person, or place is thought to possess, and the higher its relative position in the cosmological hierarchy. Combine with this

Table 3. A Comparison of Various Representations of the Iroquois Cosmos

The Creator
Handsome Lake
The Major Deities
    (the Four Beings,
    the Four Angels,
    Winds of the Four
    Quarters)
	Upper Pantheon	or	Spirit Forces
			Beyond the Sky
				or	Sky World		

Stars Moon Sun Thunders (Winds)	Lower Pantheon	or	Middle Pantheon	or	Spirit Forces in the Sky	or	Sky World

Fire
Our Sustenance
    (Corn, Beans,
    Squash)
Birds
Animals
Water
Trees (Saplings
    Poles, Bushes)
Medicine Plants
Grasses (Herbs)
Fruit
Earth
	Spirit Forces
	on the Earth
		or	Lower
			Pantheon
				or	Earth Bound		

People							
					Beneath the Earth		

principle a world view dominated by contextual, interactional, and symbiotic concerns, and the result is a series of things and events on a cosmological continuum with a potential for being in a constant state of flux. It is further seen that those things and events having the least amount of power will have the greatest amount of flexibility within the cosmological hierarchy. These least powerful things and events would be those with the potential for affecting the lives of only a relatively few individuals (e.g., medicinal plants).

Fenton's (1971) discussion of ecological time as related to the yearly cycle of activities and ceremonies points up the fact that the Iroquois were aware of celestial and earthly correlates of the four seasons. The four seasons and the prevailing winds of each are personified through the four beings in the hierarchy. There are questions in the minds of some (e.g., Foster 1974:75-79) as to whether or not these beings are recent additions to the cosmology, or are aboriginal to Iroquois belief. Their recognition may, in fact, have been a Christian borrowing by way of Handsome Lake, and could represent a modified Trinity. It seems more likely, however, that the idea of there being four beings whose job it was to assist the Creator would be very ancient to a people who were finely attuned to their environment. These beings, along with the Creator and the Christ-like Handsome Lake, possess great amounts of power. They appear later in the Thanksgiving Address because they were and are thought to have far-reaching influence over the Iroquois people. The Creator may also be thanked after the thanking of certain terrestrial entities. That is, he may be thanked more than once in the Thanksgiving Address. This may be seen as further support for the specific-to-general sequence of events outlined above. Add to this Tooker's (1970:32-33) statements that "in Iroquois ritualism there is no constant association between a rite and the being so honored by the performance of the rite" and that "any rite may be used in a ceremony addressed to any being, though certain rites are more often used than others," and the more or less fluid position of all beings or items in the Thanksgiving Address is sustained.

Moving down the cosmological continuum to the spirit forces in the sky, one would expect to find greater amounts of variation in the relative positions of these entities. This would be owing to the fact that each is intimately connected with a particular phase of the ecological cycle. The stars are indicators of the proper time for commencing and terminating fall and winter hunting activities. The moon and sun are further time markers upon which cyclical activities depend, as well as being sources of heat and light. The Thunderers are the beings responsible for bringing water to corn, beans, and squash, and all other growing things. These were all clearly thought to possess much power. Their power, which was exerted through the activities of their respective deities, would vary from season to season according to the duty or duties desired by the Iroquois.

One finds other such contextually based, adjustmental recitations of entities in the Thanksgiving Address dominating, such as the order of spirit forces thanked in the category of spirit forces on the earth. For example, when periodically performed ceremonials are carried out and there is thought to be a great need for ceremonies that are more directly concerned with medical matters than with matters of sustenance, priorities may lean in the direction of herbs and animals as opposed to corn, beans, and squash. Therefore, animals, birds, poles, herbs, or winds might be thanked before Our Sustenance. This would be due to (1) the mythological importance of animals in the area of medicine, (2) the notion that the wind is a carrier of certain diseases, and (3) the heavy reliance upon plants in the treatment of individual illnesses. If medical matters were of critical importance to a particular community at a given point in time, one might even find mention of the specific category of medicinal plants. The important point to be remembered if one is to understand Iroquois conceptions of things and events in the universe is that the context within which the Thanksgiving Address is recited will determine the relative position of the things and events recognized and thanked by the speaker. The context might be a need for plant medicines, a need for wild plants or animal food sources, a need for domesticated food sources, a need for morality, or a need for water. Because of this, variations are to be expected both within and between local groups.

THE CLASSIFICATION OF PLANTS IN TRADITIONAL IROQUOIS CULTURE

Much of what is known about Iroquois plant taxonomy stems from the work of Smith (1885), Beauchamp (1888, 1902), Waugh (1916), Parker (1910), and Fenton (1936, 1940a, 1940b, 1942, 1949). Their findings are summarized below. In general, plants are named on the following bases:

1. Structural features
2. Functional features
3. Resemblances to other plants, especially in cases involving introduced species
4. Assumed kinship relations (e.g., "it is the sister of . . .")
5. Utility (e.g., "people pound the wood" for *Fraxinus nigra* or black ash, a tree from which baskets are made)
6. Metaphorical statements (e.g., "whippoorwill's shoe")
7. Habitat (e.g., "vegetable matter floating on water")
8. Onomatopoeic devices (e.g., the sound a berry makes when it is picked)
9. Noun roots that stand for nothing but a particular plant (e.g., the Seneca word for American elm, *kaõkœ:?*)

It is important to realize that a single species (as defined in the Linnaean system) may have several different names and that, conversely, different species may share the same name. The context within which the plant is used is seen to dictate the name that will be applied. As might be expected, those plants having particular noun roots are consistently called by one name both within and between communities and tribes. The critical feature of these plants is their widespread or common utilitarian or sustentative value to the Iroquois people. Individual herbalists who are experts in the use of a wide variety of plant remedies are seen to use plant names not known or shared by others.

Regarding the classification of plants, the following general principles are seen to operate in traditional Iroquois culture. (1) There is a male and female of every plant. These, in some cases, correspond to what are considered different species and genera in the Linnaean system. There are, however, contradictory statements regarding this principle. (2) The plant kingdom, in general, is considered to be feminine (adult, active). (3) Some plants that closely resemble each other in structure are considered "sisters." Examples include *Chimaphila umbellata* or Pipsissewa, and *Gaultheria procumbens* or wintergreen. The same is true for plants sharing a function, such as corn, beans, and squash as food plants. (4) Herbal remedies, as a whole, are thought of as "grandfathers." (5) Plants of the same species growing near one another are thought to be "married."

In addition to the above-mentioned general classificatory principles, Fenton (1940b:790) has made the following observation concerning the arrangement of medicinal plants in the Iroquois cosmos:

> Native taxonomy is consistent at one point. All the medicinal plants fall into one hierarchy of form that parallels the sequential pattern for addressing spirit-forces in prayers. The criteria are relative stature and function. Proceeding from the earth upward, classification begins with the grasses and herbs closest to the earth, and steps upward through weeds, shrubs, and poles to the huge timbers of the standing forests. The Seneca assign plants to these classes: low grass or anything close to the earth; the weeds and stalks that die away in autumn, for instance, the various goldenrods; third, shrubs or bushes, then the growing poles, the dogwoods, for example; and lastly, growing timber. There are, besides these, noxious weeds, such as poison ivy and nettles, and the cultivated staples: corn, beans and squashes.

Although Fenton's (1940b) statement regarding the relative stature of plants is true, it may be true

only in a correlative way. Plants thought to be less critical to the survival of the Iroquois people as a whole are those that are thanked before the more powerful plants that possess more widespread uses. Grasses, herbs, and weeds are used primarily in a medical way. They do lie close to the earth, but it may be their individually oriented function that accounts for their being dealt with first in the Thanksgiving Address. Weeds and stalks are also used in medical ways, but more of these types of plants are seen to provide nutritive functions as well. The shrubs, bushes, and growing poles similarly provide nutritive, medicinal, and, in some cases, utilitarian functions. The growing timbers also provide food and medicines, but, because of their far-reaching utilitarian value, they are thanked at a later point in the Thanksgiving Address. The same holds true for the Three Sisters as regards their nutritive value. They are certainly of less stature than the standing timbers, but are thanked at even a later point than the trees. Relative stature, then, appears not to be the central feature in native Iroquoian plant taxonomy. What may be of more importance is the relative amount of power thought to be possessed by each, with the power being a function of a plant's relative nutritive, utilitarian, and medicinal worth.

THE POWER OF PLANTS IN TRADITIONAL IROQUOIS CULTURE

Keeping in mind the importance of contextual factors as they relate to the structured, yet highly flexible cosmos of the Iroquois, it is possible to construct a rough picture of the relative power or spiritual force possessed by various plants. This discussion will be of further value when topics centering around the application of specific plant medicines are considered in chapters 4 and 5 of this study.

In terms of a generalized conception of power, those plants having the greatest benefits or detriments to human beings are those that are thought to have the greatest power. Benefits include utilitarian value, sustentative value, or value as restorative or preventive medicines. Detrimental plants include poisons, those possessing obnoxious structural or functional characteristics, those that bring about violent physiological reactions, or those thought to be of use in practicing witchcraft. In terms of those plants benefitting humans, the following general arrangement of plants is seen to exist:

Our Sustenance (corn, beans, and squash)
The Trees
Other Food Sources (berries, fruits, and nuts)
Poles, Saplings, and Bushes
Herbs (medicinal plants)
Grasses (also medicinal, but closer to earth)

These positions may shift in relation to the ceremonial/ecological context within which the Thanksgiving Address is recited. Overall, as far as the assignment of power to plants is concerned, the Three Sisters (corn, beans, and squash) are clearly the most important to the Iroquois because of their vital, nutritive value. Other important sources of food are the fruits, especially strawberries and raspberries. Nut and sap producing plants would also possess much power owing to their nutritive value. The reliability of the Three Sisters as sources of food, and the reliability of certain utilitarian trees would make them of greater general value to the Iroquois people than the less reliable nut, berry, or fruit producing plants. This latter class of plants would have a more limited use and, like plants that are used on an individual basis (such as herbal medicines), are seen to occupy a lower position in the Thanksgiving Address.

An often neglected plant in traditional Iroquois culture is the sunflower. At the time of Waugh (1916) and Parker (1910) the role of the sunflower as a source of food had clearly diminished. It is likely, however, that it was at one time an important food source for the Iroquois people. Both the seeds of *Helianthus* sp. (sunflowers) and the tubers of *Helianthus tuberosus* (Jerusalem artichoke) are reported to have been important sources of food to native North Americans (Fernald and Kinsey 1958).

The importance of the sunflower to the Iroquois is revealed in their creation stories. In relating the Myth of the Earth Grasper for example, Hewitt (1925-1926:491) recorded that "the first thing was the sunflower which he ... planted there beside his lodge, and then said 'That will continue to be a sign for the coming uterine families when soon human beings will establish themselves here on earth.'" Furthermore, the sunflower is clearly represented in

what Parker (1912:620) has depicted as being the "tree of light" in the symbolic art of the Iroquois. In the area of medicine, it is seen that sunflower oil was used to anoint or "feed" the powerful False Face masks.

Generally speaking, any wild or cultivated food source having its own ceremony would be one that was of great value. It follows that they would be thought to possess great amounts of power. The Raspberry, Strawberry, Maple, Green Bean, Little Corn, Green Corn, and Our Sustenance ceremonies are examples of these.

There does not exist a specific ceremony for squash even though it, along with corn and beans, is included in the biannually performed ceremony of Our Sustenance. The reason for this may be that unlike corn and beans, which require more specific environmental conditions for their proper development, members of the Cucurbitaceae family (pumpkins, squashes, and melons) grow with little care and are readily insect pollinated (Hill 1952:310, 338, 382).

It should be noted that in addition to the nutritive value of corn and the gourds, both plants played important utilitarian and medicinal roles in the lives of the Iroquois. The former was used in the making of mats, moccasins, baskets, toys, bottles, masks, and beads. The latter was used in making containers, floats, and medicine rattles (Waugh 1916; Parker 1912). Waugh (1916:45-46) reports that when the kernels from abnormal ears of corn were planted, the ears they produced were then wrapped in the clothing of a sick person and placed under a medicine man's pillow. He then dreamed of the appropriate remedies. And it is the mythical wild cornstalk and wild squash that are the main ingredients in the powerful substitute Little Water Medicine (Fenton 1979). This overlapping of the utilitarian, nutritive, and medicinal values of various plants would have to be taken into account in all cases. One way of determining the relative amounts of power possessed by certain plants based on these three qualities would be to make a checklist. In doing this, however, the fluid, contextually based nature of the Iroquois world view would be sacrificed in the name of discovering precise categories that do not really exist. With this in mind, we might end up with the following representation of the relative worth of certain plants in the Iroquois cosmos:

Corn, Beans, and Squash
Maple and Sunflower
Pine and Hemlock
Strawberry and Raspberry
Elm and Basswood
Other Berry, Fruit, and Nut Producing Plants
Other Utilitarian Trees and Poles
Herbs and Grasses used in what we would
    consider a medicinal way

Regarding the placement of the pine and hem-lock, it appears that any plant that remained green throughout the year was thought to be very powerful. This notion is supported mythologically. Hewitt and Curtin (1918:81-82) relate a story involving a journey to the dwelling place of the Seven Sisters and their mother. Next to their dwelling place stood a large pine tree. It was said that these people were immune from the effects of normal *orenda* or magic power. From this it may be inferred that the pine tree was included in this story in a metaphorical way since it too was immune from normal amounts of *orenda*. Hemlocks are frequently mentioned in various other Iroquois tales as well. Nonhereditary chiefs who earned their status through warfare or oratorial skills were called Pine Tree chiefs. Replacements for dead or deposed hereditary chiefs were "raised up" like fallen trees during the Condolence Council. The immunity of the *Pinaceae* to the effects of winter is clearly the rationale behind the use of the great pine, the "Tree of Peace," as a symbol of the invulnerability of the Iroquois Confederacy. It also accounts for the relatively high position of pines in the cosmos. It will be seen in chapter 5 of this study that this evergreen or ever-growing principle also operates in the use of certain medicinal plants.

The elm and basswood were highly valued for their utility. It was the bark of the bottomland American elm (*Ulmus americana*) that provided the Iroquois with housing, vessels, and boat-building material. The soft, easily sculptured wood of the basswood (*Tilia americana*) was used for making the False Faces (Fenton 1942).

In support of the significance of the pines and the elm to the Iroquois are the various representations of the World Tree (the pine) and the Celestial Tree (an elmlike figure) in decorative art. Figure 1 (after Parker 1912) is typical of such symbolic representations. The parallel lines represent the earth, while the curved line represents the sky dome (see also Fenton 1963).

Missing in Parker's (1912) treatment of Iroquois tree myths and symbols is a discussion of the significance of finding unusual combinations of various structural parts of plants on a single plant. For example, trees are found at the ends of branches of larger trees. In other instances, trees are found with a single large flower at their tops (Parker 1912:617). These attempts to represent many diverse plants in a single plant may be a reflection of the generalized belief that every plant is good for something. Highly abstract representations would, therefore, avoid the possibility of neglecting or not remembering even the most insignificant plants put on earth by the Creator (cf. Fenton 1940b). This broad recognition of the usefulness of all plants also manifests itself in the performance of the Bush Ceremony.

It is in dealing with those plants that are honored in the Bush Ceremony that further problems arise with regard to determining relative values and powers. An analysis of less valuable plants according to their nutritive, utilitarian, and medicinal properties would yield only a very rough approximation of the relative value assigned to them by the Iroquois. Generally, most plants that had a nutritive value and/or a utilitarian value were also used in what we would call a medicinal way. These included the barks of pines and elms, various nut-bearing plants such as the chestnut, hazel nut, and oak, and various fruit-bearing plants such as mandrake or ground lemon, black cherry, wild plums, raspberries, and strawberries. Other plants were used *only* as "medicines." Attempts to focus on the relative values of plants that do not have either far-reaching utilitarian or nutritive functions would, therefore, result in a further distortion of the free-floating structures found in traditional Iroquois culture. What is important is that the medicinal plants, as a whole, were of significantly less value to the Iroquois people than plants having nutritive and/or utilitarian value. The reason for this is that medicinal herbs were used to ensure individual, as opposed to communal, survival.

If we are to be concerned with the relative powers possessed by various medicinal plants, then the exis-tence of the following subclasses of plants must also be taken into account:

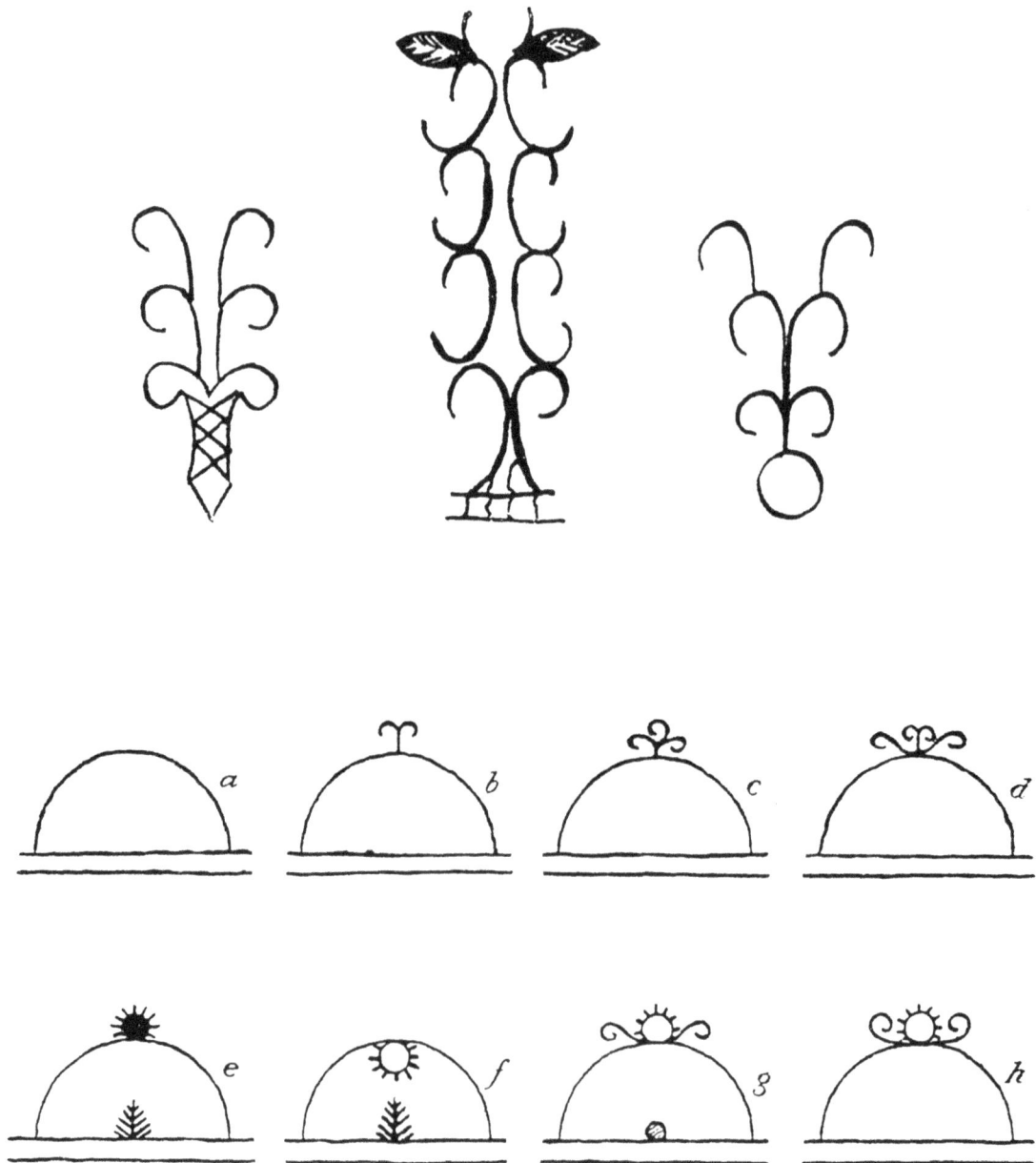

Figure 1. The World and Celestial Trees as Represented in Decorative Art (Parker 1912:615, 617).

1. Plants having nutritive, utilitarian, and medicinal functions.
2. Plants having utilitarian and medicinal functions.
3. Plants having nutritive and medicinal functions.
4. Plants having medicinal functions alone.

Once these differences are recognized and understood, one may deal with more limited questions centering around the criteria by which medicinal powers are determined. To do this successfully, however, an analysis of the entire health, medical, and religious complex of Iroquois culture must be undertaken. Chapters 2, 3, and 4 of this study attempt such an analysis. In these chapters Iroquoian conceptions of good health, illness, and medicine are treated in relation to the general statements that have been made about Iroquoian culture up to this point. Chapter 5 returns to the specific topic of powerful medicinal plants in traditional Iroquois culture.

2

Folk Conceptions of Health and Medicine

In order to understand the logic behind traditional or folk conceptions of health and medicine, it is again necessary to put aside our cultural notions of discrete categories, causes, and effects. More precisely, it is necessary (1) to forget about specific amoebae, bacteria, viruses, and the like as being the causes of specific diseases or symptoms; and (2) to forget about our distinctions between illnesses of the physical variety versus those of the mental variety. It will be seen in the following discussion that such distinctions or notions are often not relevant when analyzing traditional or folk conceptions of health and medicine.

The notion that every thing and event of the universe lives and has life force, soul, or power is the foundation principle upon which many traditional medical beliefs are built. Furthermore, the notion of a balanced universe or cosmos is at the base of what we would consider good health. It follows that ill health in traditional cultures simply involves the idea that the inherent balanced life force or power of some thing, event, person, or place has become upset. Rubel (1960), Price-Williams (1962), Bidney (1963), and Gonzales (1966) are just a few of the medical anthropologists who have observed these notions in non-Western cultures.

Because of the balance of power thought to pervade the universe, and because such balance is conceived as being a delicate one, it follows that this delicate balance may be upset in almost an unlimited number of ways. Furthermore, it is believed that the exertion of life force or power by any person, place, thing, or event upon any other person, place, thing, or event may either upset, maintain, or restore the balance of the latter. It is in this sense that a traditional conception of medicine is used. This accounts

for what appears to be the contradictory concept of "bad medicine" reportedly used by some English-speaking North American Indians. The idea that medicine equals the mysterious or unexplainable exertion of power by some things and events upon other things and events in either a good or a bad way may be inferred from the writings of Clements (1932), Grinnell (1935), Greenlee (1944), Corlett (1935), Parker (1928), and, more recently, Vogel (1970). An understanding of this conception of medicine is necessary to understand the significance of love, weather, corn, war, and hunting medicines as used by the Iroquois.

It is important to remember that although we tend to connote goodness with the concept of medicine, in many non-Western cultures (and especially in North American Indian cultures, which have frequently been analyzed in a Western conceptual mold) any concept that approximates our concept of medicine will have as part of its meaning the inclusion of badness. A medicine man, that is, an individual thought to possess what might be termed an above average ability at controlling the exertion of various spirit forces, would therefore be an individual with the ability not only to *cure* illness or misfortune, but to *cause* misfortune or illness as well. He or she may be a witch.

In reviewing the various folk theories of disease or illness causation (i.e., the causes of spiritual imbalances in human beings or any other thing or event), we are again faced with the trappings of our culture as reflected in our language. A search for causes of illnesses may leave the ethnographer with the impression that he or she is being humored or perhaps intentionally misled by the various sources of cultural information. It is only after a philosophical break is made from an insistence upon specific causes for specific symptoms or syndromes that the frustrations of studying non-Western medical beliefs are relieved.

Even when very general notions of disease causation in traditional societies are discussed, there is a problem of inappropriate categorizing. The result of this is a lack of complete appreciation of and sensitivity to the ways that members of other cultures see the universe and their place in it. For example, reports on traditional theories of disease causation have included the following factors: (1) taboo violation, (2) disease object intrusion, (3) spirit intrusion, (4) soul loss or dreaming, (5) the wrath of God or other spirit forces, and (6) sorcery or witchcraft (Rivers 1924; Clements 1932; Corlett 1935; Roberts 1944; Ackerknecht 1945; Montagu 1946). That any of these supposed categories may be combined with any other or others makes such distinctions meaningless both to the student of cross-cultural medical beliefs and to those being studied.

Fegos (1959) has approached this problem from a somewhat different point of view. He too tends to give the impression that his proposed categories represent native categories as opposed to those demanded by his culture's scientific community. To him, folk illnesses may be divided into two broad classes. The first of these addresses how illnesses are thought to be manifested and includes (1) object intrusion, (2) spirit or demon intrusion, and (3) soul capture or straying. The second category deals with how the experienced imbalance or illness is thought to be brought about, these being (1) sorcery or black magic, (2) sleep and dreams, and (3) ritually bad states and moral delinquency. Polgar (1962) relies on similar categories when he makes a distinction between "etiological notions" and "incidence notions"; while Frake's (1964) distinction between "diagnostic criteria" and "etiology" serves the same purpose. It is upon the second of Fegos's (1959) two broad categories that the present study will focus. By concentrating on how an illness is thought to be brought about, we will be getting at the heart of how traditional people see personal illness, namely, as if being just one manifestation of a delicately balanced, volatile cosmos.

In many non-Western societies, animistic and animatistic beliefs are accompanied by the idea that the universe and every thing and event in it are thought to be in a delicate state of balance. When the balance of any particular person, place, thing, or event is upset to some degree by the spirit force of any other person, place, thing, or event, that imbalanced thing or event will be observed to be in some way malfunctioning. Earthquakes, volcanos, tornadoes, hurricanes, too much or too little rain, an insufficient amount of wild or domesticated plants or animals, the loss of a battle, bad luck at hunting, and so forth, may all be thought of as being manifestations of spiritual imbalance. In such cases, the earth or any thing or event on earth would be considered "ill" in the same way that an improperly functioning or imbalanced person would be considered ill.

In terms of this conception of illness, the question that must be asked in order to discover the causes of personal illness in traditional cultures is this: What conditions or circumstances are thought to bring about disruption in an individual's inherently balanced spirit force? From the numerous studies dealing with conceptions of disease causation in non-Western societies, the following answer is seen: An individual's inherent, balanced life force or spirit force may become upset when that individual violates the rules and regulations established by the Creator or other supernaturals to govern mankind's behavior in relation to all other things and events of the universe. Further, the individual may not only violate these rules during his waking hours, but may also violate them while his or her spirit is in the spirit world (i.e., while he or she is dreaming). In such a conception of the causes of disease in traditional societies, the important principle of faulty interactional behaviors with various things and events of the universe emerges as the cornerstone of disease etiological notions. From this, Fegos's (1959) second broad category is sustained in importance, while his somewhat culturally biased distinctions between (1) people and other things and events (this being reflected in the distinct category "witchcraft" or "sorcery") and (2) the waking world and the spirit or dream world are appropriately minimized.

There are two additional features of traditional theories of disease causation that are seen to be of critical importance when considering Iroquois medical beliefs. These are: (1) the notion that illnesses in children, up to a certain age, are induced by the offensive, improper, or immoral behavior of adult kinsmen; and (2) the notion that the negative power responsible for the death of an individual resides in or near the corpse for a period of time after death. The latter of these beliefs results in the frequently observed cross-cultural phenomenon of "death sickness," to which children are especially vulnerable (cf. Downs 1971; Faron 1963; Hogbin 1947). Both of these etiological notions are clearly subsets of the broader belief that illness is the result of offensive, abusive, or improper behavioral interactions. In the case of death sickness, for example, the improper behavior would involve failure to take proper precautionary measures when coming into contact with the dead.

What we would term "medical treatment" would, in many traditional cultures of the world, best be described as an attempt at restoring spiritual balance to an improperly functioning individual or individuals. The sick individual(s) or some other societal member(s) will, at some point in time, perceive or observe signs of spiritual imbalance and then attempt to restore balance to the ill individual. Restoration of balance to a malfunctioning individual (or any other thing or event) would necessarily involve the concept of good medicine. Bad medicine would be involved when power was exerted for the purpose of disrupting the spiritual balance of any person, place, thing, or event.

The behaviors and roles of the sick individual as that individual interacts with various health actors in non-Western contexts have been detailed by Polgar (1963). It is seen that what we call "symptoms" are thought to be the manifestation of the individual's spirit force becoming imbalanced through abusive, improper, offensive, or immoral behavioral acts carried out in interaction with other persons, places, things, or events. Once these symptoms appear, certain variables determine the action of the sick individual. These variables (after Mechanic 1966) include: (1) the number and persistence of symptoms; (2) the individual's ability to recognize symptoms; (3) the perceived seriousness of symptoms; (4) the extent of social and physical disability resulting from the symptoms; (5) the cultural background of the defining person, group, or agency in terms of an emphasis on tolerance, stoicism, etc.; (6) available information and medical knowledge; and (7) the availability of sources of help and their social and physical accessibility. Of these, the last is least applicable to traditional societies. A culturally recognized healer is

usually quite accessible. An addition to Mechanic's (1966) list would be the seriousness of the offensive behavioral act believed to be responsible for his or her experienced discomfort or pain.

Any one or all of these variables determine what sort of treatment behavior will be carried out or sought by the sick individual. If the experienced or observed symptoms are few in number, mild, or do not require any drastic alterations in the everyday routine of the sick person, then that individual is most likely to enter into self-treatment. This is also true if symptoms are experienced by a person who (due to his or her personality makeup) does not typically overreact to slight imbalances or is familiar with the possible outcomes of certain symptomatic experiences. Perhaps no treatment whatsoever will be attempted, this being quite possible in a society whose individuals see occasional imbalances in a delicately balanced universe as normal and expected as opposed to pathological. All of these variables would, of course, vary from culture to culture and from individual to individual within a culture. Frake (1964), for example, has reported that the Subanum patient, regardless of how minor an ailment is, rarely relies on introspection and usually seeks out another lay or professional person. Frake's use of the word "rarely," however, suggests that the Subanum do not always turn to others for help and may, as theorized above, avoid any sort of treatment altogether in some cases of mild, expected imbalances.

An understanding of the relationship between health actor categories proposed by Polgar (1963) and symptomatic severity is essential to understanding the various types of treatments used by the Iroquois. We begin by juxtaposing his categories of self-treatment, lay health actor, and professional health actor against the symptomatic categories of mild, serious, and severe. "Mild" symptoms occur when the experienced or observed imbalance is characterized by one or a few symptoms that give the sick individual a relatively low degree of discomfort for a relatively short period of time. No treatment or self-treatment will occur. In the "serious" category, the number, intensity, and/or duration of symptoms experienced or observed increases. When the sick person becomes immobilized, when there is great discomfort experienced as the result of a degenerative illness, or when death is thought to be imminent, the "severe" stage is realized.

As symptomatic discomfort or pain intensifies past a certain unspecified point, or as the number of symptoms increases, or the longer the symptom or symptoms persist, the more likely it is that the sick individual will turn from self-treatment to others for help in restoring his or her spirit to a balanced state. The biological and behavioral history of the sick individual, along with sociocultural factors are the determiners of this point. Again, depending on the above factors, the sick individual would either seek help from a friend or a relative who is not a culturally recognized professional, or seek help from a professional or culturally recognized individual thought to possess special abilities at manipulating or influencing spirit forces.

Depending on symptomatic factors and individual biological and behavioral considerations, the sick individual may seek medical treatment at any one of these stages without necessarily passing through any other or others. For example, in some cases a relative may also be a professional health actor. Also, when a professional treats himself or herself, both self-treatment and professional health actor treatment are involved. As with all other categories used to describe traditional medical behaviors, it must be remembered that such constructions are the conceptual tools of the anthropologist and not necessarily those of the native.

HOW AND WHY FOLK MEDICINES WORK

Part of the problem with accounts of the medicines and medical techniques of traditional people is that readers are usually left with one of two general impressions: (1) that they are a novelty of sorts—a collection of bizarre beliefs and customs that are, at most, amusing and, at least, foolish, or (2) that they are intriguing in that they leave the reader with the notion that the people being considered are somehow privileged to otherwise undiscovered or long forgotten secrets of the universe. Although the present study is certainly not intended to leave the reader with either of these impressions, the forgotten secret impression may be created as a discussion is undertaken of how and why folk medicines may actually work.

One "secret" forgotten by modern-day Westerners is the importance of being relatively free of stress or anxiety behaviors so as to enhance mechanisms of

defense and recovery. Another is to be found in Ackerknecht's (1945:33) statement that "man is a fairly solid animal, and many diseases are self-limited anyway." This idea is also reflected in Saunders's (1965:191) comment that "folk medicine, like scientific medicine, undoubtedly derives much of its prestige and authority from the fact that the majority of such persons get well regardless of what is done" (see also Skinner 1965:86). Both technologically simple and technologically complex societies are guilty of assuming too much responsibility in the curing or healing process. The primary difference between the two groups, however, rests in the folk conception of healing through restoring spiritual balance versus the Western or scientific notion of healing through human mastery. The former lends itself quite easily to religious or spiritual interpretations of the healing process and could be characterized as involving faith as opposed to reason. The latter, however, is seen to be equally based on faith. The faith in this case is merely placed in worldly as opposed to otherworldly powers. In both approaches to healing, an organism that has evolved in such a way that it is able to recover from a wide variety of ailments is at the base of medicine's success (cf. Dubos 1961). In both approaches to healing, the element of faith plays a critical role in reducing anxiety and stress.

Contributing to an understanding of how and why primitive medical practices actually work are developments in the area of physiological psychology as they relate to psychosomatic and psychogenic illnesses and the related phenomenon of voodoo death (Barber 1961; Lex 1974). Both psychosomatic and psychogenic illnesses fall under the general heading of behaviorally induced illnesses. Psychosomatic illnesses are said to involve structural somatic changes in the organism that are brought about through faulty interpersonal relations variously described as being stressful, frustrating, aggressive, hostile, or anxiety laden. Psychogenic illnesses involve symptoms with no apparent structural somatic alterations (e.g., hysterical blindness, paralysis, etc., also known as conversion disorders). An attempt at tracing the cross-cultural development of the Western concepts of placebos, psychosomatic, and psychogenic illnesses, as well as psychotherapeutic techniques, can be found in Herrick (1976). It is felt that a consideration of the historical and cross-cultural development of these concepts, along with an appreciation for the "primitive logic" underlying etiological notions in many non-Western cultures, is necessary to understanding how and why primitive medicines work. The reader is also referred to Bidney (1963), Horton (1967), Laughlin (1963), and Northrop (1950) for a treatment of the similarities between "primitive" and "scientific" logic. In the discussion that follows, the reader should keep in mind the idea that the psychophysiologic responses of the autonomic/involuntary and somatic/voluntary components of the peripheral nervous system are not as mutually exclusive as scientists in our culture once thought. Intentional psychological control over physiological activities once believed to be uncontrollable is at the base of today's behavioral medicine, as well as Zen, yoga, hypnosis, biofeedback, meditation, autogenic training, and sentic cycles. All humans are capable of doing this as they interact with their respective cultural belief systems regarding the nature of illnesses and their treatment (cf. Moerman 1979).

It will be remembered that folk conceptions of disease causation in human beings often involve the notion that an individual has somehow upset his or her inherent spiritual balance via an offensive, abusive, immoral, or improper behavioral act. Once this idea became culturally instituted, an individual who knowingly violated or offended a particular person, place, thing, or event could expect to have his or her balanced spirit force upset in some way. The symptomatic manifestations of this imbalance would be expected to follow. If symptoms were experienced when no apparent violations of the established ways had occurred, it could be assumed that the offending individual did so unknowingly, or unconsciously while in the spirit or dream world.

If the offensive behavioral act were knowingly committed, it is not difficult to understand why apprehension, anxiety, fear, or stress might follow. Depending on the seriousness of the offensive behavioral act, and depending on what counteractive or restorative acts of supplication were initially performed to appease the offended person, place, thing, or event, the offending individual might then experience various degrees of anxiety, stress, or fear. Once this condition of fear or anxiety was established, a series of behavioral and/or physiological changes would take place. These changes would involve what may be considered the behavioral and

neurophysiological participants in stress, fear, or anxiety. All of these would contribute to the vulnerability of the organism in three regards. First, under the highly stressful or anxious conditions experienced by an individual who is anticipating punishment (the degree of punishment, again, being related to the supposed seriousness of the offense), the organism may be experiencing any of the classic physiological responses that characterize what we would term "psychosomatic illnesses." These might include such things as hypertension, increased heart rate, increased secretion of gastric juices, anorexia nervosa, or urticaria. Secondly, under such conditions of stress and anxiety, the normally disease-resistant organism with its built-in mechanisms of recovery and defense would be unable to function effectively, thus allowing normally mild disruptions to overwhelm the organism. Finally, the organism experiencing stress or anxiety behaviors in response to a chronically threatening situation is one that is highly field-dependent or suggestible. The verbal suggestion operating in this case would be as follows: Offenses to a particular person, place, thing, or events will result in that person, place, thing, or event causing you physical and/or spiritual harm at a distance, with the physical harm being a manifestation of the spiritual imbalance incurred. The seriousness of the punishment for committing the offensive act will be thought to vary with the power of the person, place, thing, or event that has been violated (cf. Hallowell 1941).

This verbal suggestion or autosuggestion will operate to some degree whenever a societal rule is broken. Parker (1928), Ackerknecht (1945), and Montagu (1946) have made similar observations, although they do not detail the behavioral dynamics that underlie this very important cause of illness in both traditional and industrial societies (cf. Herrick 1976). Since many of the illnesses experienced in traditional societies may be the result of suggestion or autosuggestion deriving from the belief that illness is a punishment for going against the established ways (cf. Ackerknecht 1945:28), it is not surprising to find a wide variety of ceremonial techniques or procedures that successfully reduce the degree of symptomatic discomfort or remove symptoms altogether. Suggestively acquired or augmented symptoms or discomfort would be readily removed through countersuggestive measures. These countersuggestive therapeutic techniques might be performed individually or communally. The neurophysiological phenomenon of tuning (see Gellhorn and Kiely 1973) and the increased suggestibility which accompanies it may be heightened through the use of psychoactive drugs, rhythmic dancing, singing or chanting, and the monotony of beating drums or rattles.

Yet another factor contributing to the success of folk medical techniques involves the practical application of various plant and animal drug materials. In general, these drugs work on a symptomatic basis. They work to counteract or intensify built-in biological mechanisms of defense, resistance, and recovery, such as coughing, diarrhea, vomiting, sleep, sweating, and diuresis. By labeling these drug materials "practical," it should in no way be inferred that other medicines and techniques are impractical, or that what is in operation is a distinction between material and spiritual medicines. When plant- or animal-derived medicines are used, it is always assumed by the native that it is the spirit or power of the plant or animal that is effecting the cure.

Combining the last two factors discussed above, another emerges that may account for the efficacy of traditional or folk medical practices. Simply stated, this factor involves the fact that the use of all drug materials is accompanied by the placebo effect (cf. Lehmann and Knight 1960). That is, all medicines enhance the expectations and hopes of the patient, reduce anxiety and stress, and ultimately enable built-in mechanisms of recovery and resistance to function effectively. Recognition of the role played by the placebo effect in all treatment behavior is crucial to understanding both Western and non-Western medical beliefs and practices. It further allows the ethnographer to appreciate what, at times, seems to be the nonsensical use of pharmacologically inappropriate medicines. As regards Iroquois medicines in particular, this point was recognized and appreciated by Fenton. He has stated that "some of their [the Iroquois] medicine has merit, and it does work cures even when suggestion is the sole ingredient having therapeutic power" (Fenton 1949:237). It is this very elementary, yet highly significant, principle that plays perhaps the largest role in the success of many medical endeavors, folk or otherwise.

There also exist various other practical medical techniques that add to the success of traditional

medicines. Surgical techniques, sweat baths, and the avoidance of certain foods would be included here. Accompanying these "practical" measures would be the ever present placebo effect. This would be true even in cases of preventive techniques such as sweat baths, the idea being that a person who believes that he or she has undertaken measures to ensure proper balance will be one who is less anxious or fearful than one who has not done this. Being less anxious, he or she is now in a better position to benefit from built-in mechanisms of defense and/or recovery. The less anxious person would also be less vulnerable to suggestion overall.

In light of the above discussion, Isaac's (1976-1977) research on the use of plant medicines among the Iroquois and the reported decline in their "pragmatic" application over a 56 year period (1915 to 1971) may not be as significant as it appears. This type of research focus may, in fact, say more about Western medical science and its concern for the practical and pragmatic than it does about the medical beliefs and behaviors of those in traditional societies. Such concerns become even less relevant when we consider that the pharmacological action of drugs can be brought about by suggestion alone. For example, a drug substance that behaves biophysiologically as a gastric activator may behave as a gastric inhibitor when the subject taking the drug has been told that its effect is the latter (Wolf 1950). Clearly, psychological and cultural factors may override reality and pragmatism. It is the recognition and appreciation of this fact that makes the study of the culture concept so intriguing to those who understand it.

A summary of the factors that account for how and why traditional or folk medicines work would include the following:

1. Human beings are fairly disease-resistant organisms, but anxiety, fear, and stress behaviors are counterproductive to built-in mechanisms of defense and recovery, and also heighten suggestibility. Treatment reduces anxiety, fear, and stress.

2. The always subjective discomfort or pain accompanying illness (and, in some cases, the actual acquisition of symptoms) may be suggestively or autosuggestively intensified. Such discomfort (or symptoms) may be easily

allayed or removed through individual or communal techniques of countersuggestion.

3. Drug materials work to counteract or intensify built-in, biological mechanisms of resistance and recovery.

4. Both active and inactive drug materials involve the placebo effect in that they relieve anxiety and stress.

5. There are the additional benefits of other "practical" therapeutic techniques.

6. There is a placebo effect accompanying all therapeutic techniques, whether considered practical or not.

3

Conceptions of Illness in
Traditional Iroquois Culture

As discussed in the previous chapter, the concept of medicine in non-Western cultures is used in a much broader sense than used in Western culture. It is thought of as involving the exertion of power by any person, place, thing, or event upon another person, place, thing, or event in such a way that the inherent power or life force of the latter is either maintained, restored to a state of balance, or disturbed to some degree. Thus the separation of medical matters from religious or nonmedical matters may be seen to be a culturally biased distinction. Furthermore, the separation of the personal health of individuals from the health or balanced state of all other things and events of the cosmos can be viewed as a Western cultural bias that makes human beings the center of worldly concerns.

Rioux would probably disagree with this broad conception of medicine. He has stated that "today [at Six Nations] medicine has gathered around itself other traits which formerly were not so intimately linked with it" (Rioux 1951:154), thus suggesting that medical matters have expanded in recent times. A restricted definition of medicine may account for this conclusion.

Similarly, Shimony (1961:261) has stated that "'health,' as it is currently conceived at Six Nations, includes not only physical well-being, but also the maintenance of life, mental ease, and good luck"; and further, that the modern Longhouse adherent sees the world as being "panmedical" (1961:263). From these statements, we get the impression that cognitive distinctions *were* made in the past between ceremonies and rituals dealing with individual people versus ceremonies and rituals dealing with communal or political matters. The latter include those dealing with the agricultural/sustenance cycle or the welfare of the society as a whole. However, I am not convinced that this was the case. It will be seen in chapter 4 that communal or religious ceremonies are merely medical behaviors involving groups of persons acting on behalf of society or some segment of society, or on behalf of a spiritual force or forces. Tooker's (1960:70) contention that the calendrical ceremonies of the Iroquois were built out of rites that were once part of their curing rites is especially relevant to this thesis. Using this line of thinking, the supposed transformation of the Dew Eagle Ceremony from one centering on war, external politics, or communal interests to one dealing with both communal *and* individual well-being (cf. Fenton 1953:102) may not have been a transformation at all. The breakdown of the practical functions of the League during post-Revolutionary times resulted in this once politically oriented ceremony being concerned simply with communal well-being in general, as well as with individual well-being. Whatever the ritual focus of the Dew Eagle Ceremony was, its common purpose involved a group of persons acting together in an attempt to restore, maintain, or upset the spiritual balance of a person or persons. It was always "medical" as the concept is employed herein, and it was always "religious" as the concept is used in Western culture.

Although matters of personal health may dominate among today's conservative Iroquois, the larger, more traditional and inclusive concepts of health and medicine must be kept in mind if the discussion of Iroquois medical practices that follows is to be fully understood. Similarly, bad health, illness, or disease must be understood in a general sense as a disturbance of balance in the power or life force of some person, place, thing, or event. When this is done, the comment that "if x is not done, there will be damage to health" (Shimony 1961:261) becomes a statement that may be used in understanding the etiologies of *all* cosmic imbalances, not just imbalances in individual human beings. If, for example, the "x" in this statement were "the Green Corn Ceremony," the damage to health (i.e., a harmonic universe) resulting would be the failure of the yearly corn crop.

Conforming to the broader conception of medicine used in this study, Parker (1928:9) characterized the Iroquois conception of medicine as simply involving "the magic needed to secure certain results." This magical power was, and is, thought to be bivalent in nature. Further understanding of Iroquoian conceptions of health and medicine may be obtained from a consideration of the Seneca verb root -*atō?*-, meaning "cause, provide, change into, or become." With a repetitive allomorph, the meaning of this word becomes "recover from an illness." Recovery from an illness or a cure, then, may simply involve *changing* the organism back into one that is experiencing spiritual balance. "Their bodies are transformed" (*honōtya?ta:tō?s*) is also based on the above-mentioned verb root, but is combined with the reflexive and the noun root -*ya?t(a)*-. This has the meaning "form, shape, body, or doll," all of which connote some sort of sympathetic, magical associations (Chafe 1967:44, 86). It is highly probable that *honōtya?ta:tō?s* is a concept used to express illness brought about through magical means. It may also refer to the witch or transformed agent thought to be responsible for illness.

There would be many ways that the life force of a human being might become upset in Iroquois culture, and there would also be many ways that this imbalance might manifest itself. For example, a person might knowingly or unknowingly upset his or her balanced state by violating one of the established ways. She/he might then exhibit what we would term various physical or behavioral mental disturbances. In the case of traditional people in general (cf. Ackerknecht 1945), and in the case of the Seneca specifically, however, one should understand that "no dichotomy is recognized between physical and mental disease" (Snyderman 1949:220).

Iroquoian conceptions of health and medicine, along with the general principles of renewal, thanksgiving, and balance, form the backbone of Iroquois

ceremonialism (cf. Morgan 1962; Tooker 1970; Shimony 1961; Wallace 1972; Fenton 1936). With this in mind, it is now possible to take a closer look at more specific native etiological notions.

IROQUOIS NOTIONS REGARDING CAUSES OF IMBALANCES IN HUMAN BEINGS

Wallace (1972) divides Iroquois theories of disease into (1) natural causes, (2) witchcraft, and (3) unfulfilled desires of the soul. Rioux (1951:153) distinguishes between "empirical" and "magical" aspects of Iroquois medicine. However, the terms "natural" and "empirical" connote a lack of otherworldly influence. This is not likely for holders of animistic and animatistic beliefs. It may be that the secular/nonsecular dichotomy of Westerners has been imposed upon the Iroquois in this case. Accidents, battle wounds, and the like may have been ethnocentrically credited to the secular category. Even if the terms "good luck" and "bad luck" are applied to so-called natural occurrences, we are still dealing with a mysterious energy force. The concept of luck has become a secularized force in Western culture, but we may not assume the same has occurred for other cultures (cf. Lefly 1975).

A look at the Seneca concept of accident further supports the idea that we may be guilty of imposing our world view on the Iroquois. Chafe (1967:45) reports that the verb root -*atyeō*- translates as "happen unexpectedly or accidentally to." Although it is true that in our culture an accident is by definition an unexpected chance occurrence with secular connotations, it is a mistake to assume that the unexpected in traditional cultures necessarily has a similar secular connotation. Experiencing an unexpected illness that is eventually diagnosed as being the result of witchcraft does not make the illness an accident in our sense of the word. Neglecting an unknowingly possessed hunting charm similarly does not make the illness a secularized chance occurrence.

One way to determine what the Iroquois have traditionally considered to be the causes of imbalances in human beings is to consider the etiological statements made by various native medical practitioners as collected by Waugh, Fenton, and myself. To do this successfully we must always consider the behavioral history of the sick individual, what Frake (1964:204) has called the "etiological criteria." It refers to the response received from the question: How did the patient encounter his illness? We must also consider the "pathogenic criteria," or the agent or mechanism that produces or aggravates an illness, as well as the "symptomatic criteria," which consist of responses to inquiries regarding the "attributes of an illness as perceived by the patient or an observer (Frake 1964:201, 202). Polgar's (1962) "incidence notions" and "etiological notions" approximate the same subject matter.

Beginning with the premise that illness is thought to occur when the organism is in a state of spiritual disequilibrium, it is seen that there appear to be at least four overlapping causes of imbalance in humans as conceived of by the traditional Iroquois. Because each causal agent is thought to be influencing the balanced, yet constantly fluctuating, life force of the individual, there is necessarily the element of spirituality involved in the treatment and diagnosis of *all* illnesses. This is true even when the ethnographic data suggests a pragmatic cause or cure.

As noted earlier, separating the general categories of natural and supernatural causes may have been the result of Western ethnographers imposing their cultural distinctions upon the Iroquois. This misunderstanding may have stemmed from the fact that when herbalists or other medical practitioners were questioned, they may not always have included a discussion of the spiritual component of imbalances in humans. Instead, they may only have discussed symptoms, thus creating an impression of pragmatism. This would be especially true in cases of mild illnesses or imbalances. For the traditionally minded Iroquois, however, this seemingly pragmatic discussion of symptoms would necessarily have been accompanied by the implicit assumption that one of the four overlapping causal principles was in operation. These occasional, expected imbalances would be thought to have a wide variety of potential causes. Only after time had passed with no recovery would the processes of verbalizing and delineating specific causes occur.

So-called nonmagical or empirical causes of imbalance, such as those included under our cultural categories of accident or bad luck, would also be thought to involve the exertion or manipulation of power in the universe. This, of course, would only be true for those present-day Iroquois who have not yet fully adopted a Western scientific world view. Henry

Red Eye, for example, recommends the use of 164, *Sedum telephium,* "when someone has witched you and you get hurt as a result, maybe sprain your ankle" (Fenton, field notes n.d.). Similarly, David Jack has reported that 158, *Pyrola elliptica,* is good for sore legs "if a man is a good worker and someone wants to spoil him" (Waugh, field notes 1914).

It should be noted that as time has passed and the Iroquois have become more and more acculturated, while at the same time remaining alienated from the dominant European culture, the medical information shared with anthropologists has become concerned less with traditional explanations for illness and more with explanations dealing simply with biological interpretations of illness. Perhaps this has occurred partly out of resentment toward white anthropologists who have been perceived as exploiters of the Indian and partly out of fear of derision. Resentment and fear of derision have combined with the already well-established belief that the gathering of such information by nonbelievers constitutes a breach of the established ways and may result in the destruction of the power of the medicine and the administrators of it. Fear of losing social status and prestige on the part of the herbalist by sharing arcane medical knowledge must also be considered (Fenton 1940b). Waugh (field notes 1912, verified by Fenton [field notes 1933]), for example, received the following bit of cultural information from Herb Johnson: "Plants can make themselves invisible when they wish, or when they are being sought for improper purposes, such as telling a white man what they are good for. The plants are then believed to be 'laughing' at the medicine man." This notion of plants hiding and laughing at their collectors when accompanied by whites was also noted by botanist Stanley Smith as he and others tried to learn native names for various plants (Smith, personal communication 1974). Such beliefs are indicative of the problems encountered in attempting to learn traditional medical beliefs in general. Even Parker (1928:12) recognized that "it is not an easy matter to get the real remedies of the Indians, for their herb doctors are cautious about imparting much information." More specifically, Fenton (field notes 1938) has recorded Tonawanda Seneca Jesse Hill as saying:

> I will not divulge any of the medicines which I use because of an old belief that if one tells the medicines, he loses the power to help himself. If I buy the root from

someone he will not tell me the name of it. I have to pay in addition to learn the various medicines. The old Indians would pay cash to each other to learn the medicines. My mother said that if I sold my knowledge of the plants and the medicines, I would lose that medicine. The old people say that if I sell my recipes, I cannot use them again.

In light of this, it is significant to note that when I compare the data collected by Waugh during the years 1912-1915 with those gathered by Fenton (1933-1939) and, ultimately, with my own information, I find a steady decrease in the degree of emphasis placed on the four etiological categories described in this study. With Waugh, a whole series of events is frequently found. These events range from the initial behavioral interaction thought to have precipitated the spiritual imbalance, to a discussion of the internal object thought to be interfering with the balanced vital principle contained within the blood, to a discussion of symptoms or external manifestations of imbalances. Fenton's data contain fewer complete discussions of the entire sequence of events thought to result in illness. My data primarily contain symptomatic features, with occasional mention of the action or role of blood as a causal factor.

In summary, there are at least five factors responsible for receiving the false impression that the Iroquois traditionally used their medicines in both magical/spiritual and empirical/pragmatic ways. These include: (1) the gradual adoption of European conceptions of health and medicine, and perhaps the desire on the part of recent herbalists to gain credibility in the eyes of the scientific establishment, (2) fear of derision by outsiders for giving traditional etiological notions, (3) the recent practice of avoiding discussions of *any* traditional ways with those who are not Native Americans, especially anthropologists, (4) the long-standing cultural belief that medicine or the ingredients of various medicines should not be revealed to anyone lest they and their users lose their effectiveness, and (5) the fear of losing social status and prestige by making public secret medical knowledge. Of these, the first is probably least important simply because medicine and medical matters form the nucleus of conservatism among the Iroquois of the recent past (cf. Shimony 1961). Traditional healers are the least acculturated of all Native Americans.

In order fully to understand how the Iroquois traditionally conceived of illness, it is necessary to

discuss each interrelated causal principle with Iroquoian conceptions of physiology as well as with the logic thought to underlie notions of imitative and contagious magic. Crucial to native conceptions of physiology is the idea that the blood is the seat of the vital principle or life force of the individual. Vitality is thought to reside or exist within the blood in a state of equilibrium, and a state of disequilibrium (illness) is thought to occur when it is disturbed to some degree. Linguistic support may be found for this belief by considering the Seneca concept meaning "catching a disease." This concept (e.g., *?kæ?t*, "I caught it") is based on the verb root *-(C)(æ)-* (with the objective) meaning "put in" or "incorporate" (Chafe 1967:47). Many references to this idea of the blood being the primary thing that is affected in illness are found in the ethnographic data. A good example of this notion is seen in Sarah Snow's use of 370, *Eupatorium rugosum* (see FndB). This medicine is thought to separate the blood from the disease.

What we would term "symptoms" would, in the traditional Iroquois way of conceiving of illness, be considered external manifestations of a disturbed, internally contained, vital principle. This internal principle could become upset through four interrelated ways. These include: (1) violating the established ways set down by the Creator for the purposes of maintaining a balanced, harmonic universe; (2) denying oneself things that are desired in either the earth world or the spirit world; (3) interacting with things, events, places, the dead, or people that are thought to exude or radiate negative power, evil, or *?otkō?* (see *-atkō-* in Chafe 1967:43); and (4) offending an individual who is thought to have access to great knowledge regarding the manipulation of spirit forces. In all cases, the person who has upset his or her inherent spiritual balance as a result of one or more of these causes may have done so either knowingly or unknowingly.

The notions of imitative and contagious magic serve to interrelate these four causal principles. The two of them in combination are sometimes called sympathetic magic. With imitative magic, like produces like, while with contagious magic objects once in contact with each other are believed to continue to influence each other at a distance. These principles of magic, as well as the idea that blood is the seat of a delicately balanced vital principle, will now be discussed as they relate to the four overlapping causal notions proposed earlier.

OFFENSIVE BEHAVIORAL ACTS OR TABOO VIOLATION

Various etiological statements suggest that some sort of offensive behavioral act or violation of the rules established by the Creator has been carried out by the sick individual. In most of these statements, the behavior that is to be avoided is stated along with the symptoms associated with it. Probably the best examples of this are found in behaviors relating to menstruating women. Illness may result from having sexual relations with one, eating food prepared by one, or simply by coming into contact with one. From this, it may be inferred that the blood of the menstruating woman is somehow thought to be contaminated, evil, or *?otkō?*. The close relationship thought to exist between the concept of taboo and menstruation is also evidenced linguistically in the Seneca concepts for each. The verb root *-ahtyawee-* means "to be taboo," while the verb root *-ahtyaweet-* means "to menstruate." When the latter concept is combined with the reflexive morpheme, it takes on the meaning "to avoid something taboo" (Chafe 1967:38-39).

Like other things and events deemed *?otkō?*, menstrual blood has the ability to spoil medicines. Lightning bugs and witches may have similar effects on good medicine. There exists a specific Seneca concept that describes this strongly held belief. The verb root *-ke^n-*, meaning "see" or "find," when combined with the noun root *-yá?t(a)-*, meaning "form, shape, or body," forms the concept of *waōwōya?take^n?*, "a menstruating woman spoiled the Little Water cure by looking at the patient" (Chafe 1967:65). Contact with such women either directly or indirectly may, therefore, result in a spoilage of the blood. This notion is a good example of the principle of contagious magic. The contamination remains on food even when the woman who has prepared it is no longer present. Similarly, cases involving unborn children being contaminated by a menstruating woman illustrate this principle.

In certain instances there exists the idea that some sort of internal object may be involved in sicknesses resulting from not avoiding menstruating women. For example, if a man "goes with" a woman during her menses, her breath will cause his face to

break out and this may eventually "turn to worms." In other cases of taboo violation, rich foods may cause a fever which, in turn, may cause vomiting. If the solid food taboo is broken, worms will exist within the body and result in *godjigwe^{n}s*—a term used in reference to a variety of venereal diseases (see *-jikwe^n-*, Chafe 1967:62).

Several symptomatic characteristics are mentioned in conjunction with offensive behavioral acts of one type or another. These include shortness of breath, stricture, swelling of the body, suppressed menses, facial eruptions, diarrhea, sterility, piles, sore mouth, bad luck at hunting, vomiting, fever, swelling and stiffening of the joints, insanity, rotting of a limb, bloody urine, stomachache, pains in the sides, and intensified soreness of pregnancy.

Other offensive behavioral acts or taboo violations include such things as (1) mistakes in ritual (especially those performed by members of medicine societies); (2) cheating or using unfair magical tactics for the purpose of winning a game; (3) neglecting False Faces by not "feeding" them corn mush or tobacco, or by not guessing their desires, neglecting spirits of the dead, neglecting powerful medicine charms or the tutelaries of charms (such as the Pygmies) that may be held knowingly or unknowingly by a family; (4) improper acts of supplication before hunting; (5) not periodically participating in a medicine society after having been cured by its medicine; or (6) not properly making up or completing a previously performed "hurry up" ceremony that was designed to expedite matters in certain critical cases of illness. In this last case, the offensive behavioral act of neglecting proper ritual may manifest itself through illness affecting the offender's immediate family in the present, or in future generations. One could also view an act such as not behaving as a friend toward a fellow societal member as being an offensive behavioral act. This act might not only lead to illness in the person who needs a friend, but also in the members of the medicine society who have the responsibility of providing friends for others. The society in this case is usually the Eagle Society.

On one occasion at St. Regis, I was told that not all illnesses were acts of punishment. The respondent may have been thinking of a case when the cause of misfortune was believed to be an "unwarranted" act of witchcraft. But it is doubtful that any act of witchcraft could be considered unwarranted. The violation of certain established ways may have prompted the act of witchcraft. Inciting envy or not exhibiting proper restraint might, for example, easily make another individual jealous. The jealous person could be a witch or, more likely, a person willing to enlist the help of one.

Other misfortunes are clearly punishments for immoral behavior, that is, taboo violation. I was once told the following:

> A man once pulled up someone else's corn and potatoes. Later, he was dragged by a horse and suffered a year and a half. There was another case which involved a man who threatened to burn down the Longhouse. He suddenly died, and his father accused the Longhouse members of killing his son and he, too, suddenly died [Herrick, field notes 1973].

In regard to this statement, I then asked: "Could your medicine have helped these people?" The answer received was: "No, it was too late for them." From this, it may be inferred that some taboos, once broken, cannot be set right again through the use of medicine. Some acts that are particularly offensive to the Creator, such as threatening to burn down a longhouse, are therefore beyond the control of human intervention.

Seneca concepts that relate to the etiological category of taboo violation include the following: -$ye^{n'}$?hi-/-$ye^{n'}$?hi?- ("to make a mistake"), which, when combined with the causative, dative, and duplicative morphemes, means "bring trouble to." A related concept is that of -$ke^{n'}e^{n'}ht$-, which, with the reflexive, translates "let spoil" or "miss an opportunity" (Chafe 1967:65). A last example is the concept meaning "to have indications that a ceremony is overdue," which is built on the noun root -$we^{n}n(\bar{o})$-, "voice, word, language" and the verb root -$\bar{o}ne^{n'}ht$-, "to swallow" (Chafe 1967:85).

Also relevant to a discussion of improper or delinquent behavioral acts as causes of imbalances is the Seneca concept of -$asha$-/-$asha(\alpha)$-, which means "to look out for" or "to take care of." When combined with the inchoative morpheme signifying "coming into being" and the objective, this comes to mean "remember" (Chafe 1967:39). To remember someone, then, is to take care of them. The rituals and ceremonies designed to remember an individual or individuals, or the spirits of various things and events of the universe, abound in Iroquois culture. These concerns are, of course, subsets of the broader cultural themes that stress renewal, hospitality, and reciprocity.

UNFULFILLED DESIRES

Wallace (1972) has written extensively about the Iroquois interpretation of dreams and how it may be seen as a type of psychoanalysis. Perhaps it would be more correct to say that psychoanalysis itself is only another cultural variation on human dream analysis. Freud's psychoanalytic principles may be characterized as his attempt to make an already well-established folk theory regarding one of the possible causes of illness acceptable to the medical science establishment existing in Europe in the late nineteenth century. Freud simply transformed the concept of spirit into the concept of the unconscious and put the burden of responsibility for psychosomatic and psychogenic illnesses on the individual (unsuccessful resolution of biologically based conflicts) as opposed to the immoral acts of kinsmen, although kinsmen still played a role in his theory.

The following passage from Wallace (1972:61) relates the thoughts of Father Raguemeau (1647) regarding the subject of Iroquois dreams:

> Now they believe that our soul makes these natural desires known by means of dreams, which are its language. Accordingly, when these desires are accomplished, it is satisfied; but on the contrary, if it be not granted what it desires, it becomes angry, and not only does not give its body the good and the happiness what it wished to procure for it, but often it also revolts against the body, causing various diseases, and even death.

It is as if the dream or spirit world was the original source of events in the earth world, and that the latter was thought of as being merely a reflection of the already transpired happenings of the former. This type of thinking is very similar to the type of idealism that is often credited to Plato. One major point of difference, however, would be the apparent lengthening of the time span thought to occur between the events in the spirit world and the actualized events in the earth world. For the Iroquois, this time span was greatly lengthened. A dream or desire could be actualized in the earth world through a sort of imitative magic, especially in cases involving unpleasant experiences. As Wallace (1972:71) has expressed it, "a wish, although irrational and destructive toward self or friends, was fateful, and the only way of forestalling the realization of an evil-fated wish was to fulfill it symbolically." In the case of desiring or dreaming about pleasant experiences, the immediate gratification of these wishes would serve the purpose of avoiding any future conflicts that might ultimately lead to spiritual imbalance (illness). This could, therefore, be considered a preventive medicine. Presumably, the same logic would exist in the event that an unpleasant experience was dreamed about. An unpleasant experience such as torture, loss of property or possessions, or the eating of human flesh would have been considered secondary to the imbalance that would occur were these desires to be thwarted.

An important distinction that should be made in a discussion of Iroquois etiological notions is that made by Wallace (1972) between "symptomatic" and "visitation" dreams. The latter of these two types of dreams are sometimes called "particular" dreams by the Iroquois (Fenton, personal communication 1977). They are said to involve direct contact with an individual by a powerful, otherworldly being. The symptomatic dream, on the other hand, is said to involve the guessing or revelation of dream content by a clairvoyant. They "displayed in their manifest content relatively humble and mundane matters: wanted objects, like dogs, hatchets, knives, clothing, etc.; familiar dances and rituals, and their ceremonial equipment; and familiar animals, birds, and plants" (Wallace 1972:72). In general, the desire to return home, lonesomeness for a living or deceased loved one, or the desires of a dead person residing in the spirit world are seen as being possible causes of imbalances. In some cases, specific symptoms such as restlessness, craziness, melancholia, diarrhea, or pain in the heart are given. In other cases, it is suggested that unfulfilled desires may be treated simply by freeing the body of the internal object or objects thought responsible for the experienced imbalance. In these instances, vomiting and defecation are included in the treatment. Sometimes, the patient needs a friend and the clairvoyant will prescribe acquiring a ritual friend at the Dew Eagle Ceremony.

There is a conceptual link between desires of the soul and witchcraft. It will be seen in the discussion of witchcraft to follow that what is called an onōhwhit (see -noō'hkw-, Chafe 1967:70) by many of Waugh's and Fenton's informants is a love medicine that is thought to have the power either to bring people together or to make people attracted to material objects such as baskets. The principle upon which the application of love medicines is built is that of sympathetic magic, and the use of such magic for the purpose of affecting another individual is considered an act of bewitchment. Handsome Lake specifically warned against their use. Onōhwhit is the medicine for lonesomeness, a condition that may be inflicted upon another through the manipulation of a love medicine. It may be inferred that it is "like a love medicine" in the sense that it involves manipulation of a force or power, in this case, unrealistic or unfulfillable desires for another living or dead individual. These desires, in turn, are responsible for the feelings of lonesomeness. Again, the idea of a balanced organism becoming imbalanced through some sort of behavioral act is revealed. Further, it is seen that even in cases involving what we might consider a mental or behavioral illness, things such as melancholia, depression, or love sickness, a model incorporating both physiological and magical/spiritual elements is employed.

Related to the etiological notion of unfulfilled desires is the belief that the grief, loneliness, and depression brought about by the death of a loved one or chief was the cause of profound mental derangement. Mental derangement would, in turn, interfere with a family's, moiety's, community's, or the League's ability to unite their minds in ceremony. In the case of a family member's death, members of the opposite moiety would attempt to relieve the suffering of the bereaved family by performing the Ten-days' Feast. When a chief died, performance of the Condolence Ceremony by the chief's opposite moiety served the same purpose.

With a little flexibility in thought it is possible to see connecting links between the etiological category "unfulfilled desires" and the categories of "things and events considered ʔotkō?" and "taboo violation." A case in point is seen when an individual dreams of performing a feast for a powerful charm but does not perform it. Certain animal charms are considered evil because of the great powers they possess. This failure to act could also be considered a violation of a taboo because it goes against the established ways set down by the Creator. Another example of the interrelatedness of taboo violation and unfulfilled desires is found in the belief that one's family members will have "their heads stuck to the ceremony" (resulting in illness) if one dreams of joining a medicine society *with* family members, but excludes family members when the dream is acted out (Fenton 1936:11).

THINGS, EVENTS, PEOPLE, AND PLACES THAT RADIATE OR EXUDE EVIL

It is clear from the Iroquois creation story that there exist certain things, events, people, and places that are *?otkō?*. Many of these evil forces were forcibly contained within the earth by Flint, the personification of evil, and it is believed that they are periodically released to harm human beings living on the earth. A passage from Hewitt (1904:196-197) relates that: "They [the animals] were changed, becoming *utgo^n* [*?otkō?*], and the reason that it thus came to pass is that some customarily put forth their *orenda* [magical power] for the purpose of ending the days of man-beings; and, moreover, they still haunt the inside of the earth."

Although only animals are mentioned in this particular passage, it may be assumed that *any* nuisance, threatening, or generally obnoxious thing, event, person, or place was also thought of as being the work of Flint, thereby being a possible source of *?otkō?* or evil. Working on this assumption, it is seen that the etiological statements cited below involve things or events that are thought to be *?otkō?*. Some are by their nature thought to radiate or exude evil power, while others are considered evil because of their hindrance to humans. These things and events include mice, rats, squirrels, bloodsuckers, snakes, salt, mustard, pepper, soda, partridges, toads, bats, foxes, worms, ants, thousand-legs, spiders, moths, certain birds, and "legs" or *ohna?tsa?*. With a few possible exceptions, most of the animals considered *?otkō?* are nuisance animals that feed on wild or stored foods also eaten by humans. Bloodsuckers have an obvious connection with vital blood that necessarily makes them evil. It is probably the acrid qualities of soda, salt, pepper, and mustard that account for their being included in this category.

Menstrual blood is in itself considered *?otkō?* in that it is conceived of as being spoiled or contaminated blood that is periodically discharged by life-producing women. This notion could account for the fact that contact with menstruating women is considered a taboo violation. It certainly accounts for the practice of poisoning or bewitching another by secretly mixing menstrual blood with foods.

The fact that an aura of *?otkō?* surrounds the menstruating woman is also revealed by certain linguistic evidence. The Seneca verb root *-a-* (with the duplicative) means "to defile" or "to affect adversely by touching" and is used in some instances with the noun root *-ksá?t(a)-* ("child") to indicate that a menstruating woman touched a child and caused it to behave badly (Chafe 1967:37, 67). Such a notion provides an additional conceptual link between the etiological categories of taboo violation and coming into contact with things or events deemed *?otkō?*, as well as that of witchcraft.

The idea that death may be viewed as not only a cause of illness (death sickness), but as an extreme case of illness or imbalance in itself is suggested in many statements (Herrick 1977:211-238). The negative power or evil responsible for driving the life force out of an individual's body is thought to reside in or near the corpse for a period of time after death. From this, the idea follows that if a person has been contaminated by the *?otkō?* of death, that person is thought to be capable of transporting evil with him and should, therefore, avoid those already in a state of imbalance. This would be done out of the fear that a contaminated person might intensify the amount of imbalance already being experienced by a sick individual. Failure to do so would constitute a breach of the established ways. Relating to this notion of contamination by the dead are Waugh's (1916:131) observations that "animals inhabiting graveyards should not be killed for food, or 'bad luck' will result," and that "the spirits of dead people were ... in these animals." The same reasoning is applied to plants that grow in or near graves.

The belief that "when paralysis occurs, it is thought that part of the body is dead" (Herrick 1977:213) is especially important and is supportive of the idea that death may be thought of as being an extreme case of spiritual imbalance or illness. Because paralysis is considered "dead on the body," the idea is suggested that only part of the life force of the individual has been driven from the body. This would conform to Wallace's (1972:63) observation that the Iroquois soul "occupied all parts of the body, and so had head, arms, legs, trunk, and all the rest of the anatomy (in ethereal counterpart) of the corporeal body." What remains a mystery, however, is whether or not paralyzed (i.e., dead) limbs or other bodily parts were thought to be sources of evil in the same way that a completely dead individual was. That is, were victims of paralysis thought to radiate *?otkō?*, and were they barred from seeing or visiting those already imbalanced? Evidence does exist that an already paralyzed individual may be worsened by seeing someone who has been to a funeral. In keeping with the principle that good health equals a spiritually balanced organism, those who have become contaminated by the evil power surrounding a corpse may also render ineffective or spoil the effects of good medicine.

Other things, events, and places that are considered *?otkō?* include the powerful hunting or medicine charms. These charms are composed of various anatomical parts of certain magical or powerful animals. When magically combined with water, they go into making up *nika:neká?a:h* (after Chafe 1967:54) or Little Water Medicine (cf. Fenton 1979). The guardians of these charms, the "Little Folk" or "pygmies" as they are called, are also powerful and *?otkō?* (Parker 1909, in Fenton 1968:119). The False Faces are also considered *?otkō?* (Wallace 1972:84). Undoubtedly, other paraphernalia, ritual events, and places associated with various ceremonies designed to manipulate spiritual power would also be considered *?otkō?*. In fact, many of the things and events of the Iroquois cosmos that are considered *?otkō?* are simply those that are thought for one reason or another to be very powerful. This power has the potential for producing both good and evil consequences. When uncontrolled, these things and events are considered evil because of their upsetting influences upon normally balanced things.

An animal or a human that is pregnant is also thought to possess great amounts of power, thereby being a potential source of *?otkō?*. I was told at St. Regis that a pregnant woman has two minds. This apparently makes her more knowledgeable and powerful. Along these same lines, Peter John (Waugh, field notes 1912) was recorded as saying that "in making medicine by a society, a man is chosen whose wife is with child. . . . this is favorable to the idea of the medicine doing its work of healing." In keeping with this, we find pregnant women being kept from handling a hunter's rifle lest he be unable to hit anything. Also, Waugh (1916:131) recorded John Jamieson, Jr. as saying that "the flesh of pregnant animals was no good and produced diarrhea."

Similarly, Mrs. Carpenter, an herbalist at Onondaga, informed me that one should never have a pregnant or menstruating woman with him or her when picking medicine. "She will somehow remove the strength from the plants. Even after picked, they should not be allowed to see the medicines."

There also seems to be a relationship between the concepts of *?otkō?*, witch, great power, or poison and the color white. John Jamieson, Jr. (Waugh, field notes 1912) reported that albino animals were never killed, and added that the Creator had commanded not to kill them. When the game was just about gone he promised to send two pure white animals to breed some more of this kind. And it was a white dog adorned with white wampum that was periodically sacrificed and burned in an attempt to thank and honor the sun: the God of War in earlier times. The performance of certain medicine rituals in the dark could also be interpreted as symbolically involving whiteness. In this case whiteness is avoided so as not to contaminate powerful medicines. For example, the Dark Dance is performed by the Pygmy Society members during the Little Water Medicine Ceremony.

A further possible expression of the significance of whiteness in Iroquois culture is seen in the use of salt. John Echo (Waugh, field notes 1914) reported that there was no restriction on salt, except that menstruating women should not eat it. If, however, the "salt is put in a cloth and then in the fire and heated (scorched a bit), when the cloth burns off it can then be used." The whiteness removed, salt would then become safe to use by the already dangerous menstruating women.

The idea that whiteness is associated with evil and/or witchcraft extends to the plant world as well. John Jamieson, Jr. (Waugh, field notes 1914), for example, reported that "white leaves on milkweed or other plants indicate one which should not be used . . . they are *?otkō?* or witch." This principle would also extend to the "whitish feather," 260, *Cicuta maculata*, the highly poisonous suicide medicine of the Iroquois (see Fenton 1941b).

The special bivalent position of whiteness in Iroquois medical practices is also seen as it enters into the use of very powerful, good medicines. In discussing *nika:neká?a:h*, the Little Water or Great Bird Medicine, Hewitt and Curtin (1918:491) have stated that:

It is so powerful in *orenda,* or magical potency, that when it was given to the sick by the charm holders the patient was forbidden to eat anything that was colored; he could eat, however, pure white beans and pure white cob corn. If anything black or in any manner colored was eaten, the taboo was broken and the man or woman would die, as the medicine's virtue was destroyed.

Yet another case of the symbolic importance of whiteness in Iroquois culture is seen in Beauchamp's recounting of "The Good Hunter and the Great Medicine." In this instance, whiteness is associated with wisdom or accumulated knowledge. These qualities are thought of as potential sources of evil and/or witchcraft in the minds of the Iroquois.

The eagle's head had become white in his long and wise life, and from his lofty eyrie he had looked down, and knew every force of nature and every event of life. This white-headed sage said that the dead would not revive unless the scalp was restored [Beauchamp 1922:32].

WITCHCRAFT

In some instances, statements of disease etiology involve the malicious acts of one individual against another. Many specific statements clearly illustrate this causal principle (Herrick 1977:211-238). As in the case of taboo violation, most discussions of witchcraft include symptomatic features. About 25% of the time an internal object is also mentioned. Snyderman (1949:219-220) substantiates this observation.

It may be seen that the Seneca regarded certain types of illness as due to the failure of the individual's personal power to protect him from witches who place material objects in his body. Accordingly, the witches destroyed his physical and mental health. The treatment accorded was therefore geared to remove these foreign objects and to act as a positive path for physical and mental health.

From the statements made in reference to acts of witchcraft, it is seen that these inserted objects might include such things as broken pieces of bone, bloodsuckers, snakes, poison roots, insects, worms, thousand-legs, and broom splints. In a few cases menstrual blood is mentioned, sustaining the theory that spoiled blood is in itself *?oktō?.* This in turn may be used as a poison. Of course many of the objects that are discussed here are also things existing in nature that are thought to radiate or exude *?oktō?* with or without human intervention.

Contagious magic is clearly involved in the causal notion that "if a man chews . . . (60. *Acorus calamus*)

and spits on place where woman's menses have fallen, it will cause profuse menstruation at once" (Herrick 1977:214). Again, the principle of sympathetic magic is seen. This principle forms the foundation upon which all acts of witchcraft are based.

Specific symptoms associated with witchcraft-caused imbalances include the following: (1) consumption and internal hemorrhage, (2) abscesses in the body, legs, or neck, (3) dizziness or craziness, (4) a rotten stomach, (5) suicidal wishes, (6) profuse menstruation, (7) skin or muscle twitching, (8) excessive drinking, (9) breaking out like cancer, (10) swelling after copulation, (11) pain in a particular place, (12) generalized pain, (13) foul language, (14) diarrhea, (15) fever and headache, (16) coughs and colds, and (17) piles.

It might be hypothesized that the concept of "witch" is a culturally imposed category of disease causation that reflects a European world view. What a European would consider a witch, a *person* who manipulates supernatural power for evil purposes, could easily be subsumed under the more general Iroquoian etiological category of things and events considered *?otkō?.* The noun for "witch" in Seneca is derived from the verb root *-atkō-* that translates "to possess evil power." This is the verb root upon which *?otkō?* is based (Chafe 1967:43). Because the traditional Iroquois did not rigidly separate human beings from all other things and events of the cosmos, it is not surprising to find the word "witch" being used in reference to *any* powerful and/or potentially dangerous thing or event.

For the Iroquois, all Iroquois witches, human or otherwise, are thought to be capable of transforming themselves into something or someone else. Shimony (1970:248) makes this point, but limits her discussion to human witches. That this ability of transformation is not limited to humans is evidenced in John Jamieson, Jr.'s report that hummingbirds are witches that have the power to fly into holes in logs and change into snakes, mice, or other animals (Waugh, field notes 1912). There even exists a specific Seneca concept that refers to this magical ability possessed by witches. *Kaháih* translates as "magical transformation, usually for an evil purpose" (Chafe 1967:64).

A conceptual link that bridges the gap between the etiological categories "witchcraft" and "unfulfilled desires" is found in the Seneca verb root meaning "to poison." To begin with, poisons in general are referred to as *?otkō?*; while the verb root *-atyanō?-* ("to poison"), when combined with the distributive *-?syō-* (which may connote the ubiquitous quality of the action taking place), comes to mean "cause to dream." An evil thing, event, place, or person may all consequently cause an individual to dream. One might speculate that these evil causes of dreams would, in turn, necessitate distasteful, harmful, or dangerous acts to effect their fulfillment (Chafe 1967:44-45).

A similar interpretation may be made of the Seneca concept *-ye^{n}ōni-/-ye^{n}ōny(a)-*, a verb root meaning "to show, signal, or indicate." With the distributive, duplicative, and translocative, this word comes to mean "touch" or "bewitch." An example is *ho?thakye^{n}ōnyō:?*, "he touched me, bewitched me." With the reflexive and dative, this word translates as "occur to one that one should do something"; and *?ōkatye^{n}ōnye^n?* means "it came to me (e.g., in a dream) that I should do it" (Chafe 1967:88).

The topic of love medicines or *onōhwhit* is relevant in discussing the overlapping nature of the etiological categories of witchcraft and unfulfilled desires. This concept is based on the verb root *-noō'hkw-*, meaning "to love" or "to have affection for." There also exists a concept meaning "love potion," which is built on the verb roots *-nō'e-* ("like" or "be content with") and *-ye^{n}tei-* ("to know" or "be aware of"), forming *yenō'ehtaye^{n}te:ih* (Chafe 1967:70-71). The traditional Iroquois believe that it is possible for an individual to exert or manipulate power in such a way that he or she is capable of making one individual desire another individual. Although the notion of unrequited love is, in Western culture, associated with an individual's love for another individual, the Iroquois do not limit the concept in this way. The key to understanding the Iroquois notion of love medicine is, therefore, to think of it as being simply the exertion of power by any person, place, thing, or event for the purpose of bringing together two people, two objects, a person with an object, a person with a place, and so forth. Presumably, when evil or malevolence are associated with the concept of *onōhwhit*, the people or objects involved *do not* choose to be together. Love medicines may also be used by an individual for the purpose of compelling another individual to become attracted to him or her. The evil connotations of

witchcraft would either not apply or be minimized in such cases. The term "medicine" would suffice.

Two Seneca concepts that might refer to this ability or power to compel others to desire people, places, or objects are *thakōye^{n}nōhtō:ni:h*, meaning "he makes them do it," and *shōkwe^n?nikōe^{n}tíyōhkōh*, which translates as "he has gained control over our minds" (Chafe 1967:50, 83). The first of these concepts is based on the verb root *-(e^n)nō'ht-* ("to know"), while the second is based on the verb root *-tiyō'hkw-* ("to gain control") and the noun root *-?nikō'e^n-* ("mind" or "spirit").

Some plants, animals, or other things and events deemed "witch" are also thought capable of compelling other objects or people to perform acts that they do not want to perform. For example, any plant or animal that "makes you crazy" would fall into this category. It is not surprising, therefore, to find the plant 168, *Agrimonia gryposepala*, being considered a "basket medicine" or *ye?ásyohka:tha?*. People apply it to baskets to attract customers. Nor is it surprising to find certain plants (e.g., 123, *Sarracenia purpurea*) being considered both a "basket medicine" and a "love medicine," their common characteristic being the power to compel attraction. The critical feature separating the basket medicine from the love medicine would be the motive behind the use of each. In the former case people are attracted to objects, while people are attracted to people in the latter case. Principles of sympathetic magic as they determine the potential uses of various plants are at the base of the use of both love and basket medicines. For example, the hooked seed receptacles of *Agrimonia* and the insect-trapping abilities of *Sarracenia* are key attributes. These same principles are seen to be at the base of the European notion of the doctrine of signatures, the idea that each plant gives a clue to its possible medicinal use through various structural or functional characteristics.

The ingenuity of the Iroquois in applying witching medicines believed to have the power to compel attraction is seen in the following accounts of the use of 123, *Sarracenia purpurea*, or pitcher plant:

> This is a medicine for peddling. Grandmother said was not for love medicine. But some of them, if you want to talk to some nice girl or marry her, use that medicine, she will come right over. If you don't want to talk she will get down in front of you on hands and knees and cry. That's what they say. I never used that kind [A. Jonas, as recorded by Fenton, field notes 1938].

The pitcher plant is supposed to be good as an *onōhwhit*. An old woman near here used to take this medicine and put it on herself using tobacco, telling it what to do, and then go into the crowd so that a man would chase her. She also used to use tobacco and this medicine before she went to town to sell her baskets. Then people all flocked around her to buy them [Sarah Snow, as recorded by Fenton, field notes 1933].

When a witch, through magical means, compels a person to be attracted to another person, place, thing, or event, the illness that results could also be thought of as falling into the category of unfulfilled desires. To want to be close to someone, some place, or something while at the same time being unable to realize these desires (because of a lack of mutual interest on the part of the other person, a lack of a means for obtaining an object, or the great distance a person is from a particular place) necessarily leads to a spiritual imbalance in the "witched" person.

There exist several Seneca concepts that refer to the illnesses brought about by unrealized desires for people or places and the feelings of depression that accompany them. The Seneca word meaning "to be melancholy" or "to be depressed," for example, is the verb root -*we^{n}ne$^{n'}$?t*-. Chafe (1967:85) reports that this concept is built upon the verb root -*e^n*- ("to leave something") or -*á?se^n*-/-*e^n*- ("to drop" or "to knock down") and the noun root -*we^{n}n(ō)*- ("voice, utterance, word, language"). Similarly, according to Michelson (1973:62, 130) the Mohawk concept for "depressed" (e.g., *wake?nikù:rvs*, "I am depressed") is built on the verb root -*itv*- ("to be in poverty") and the noun root -*?nikuhr*- ("mind"). Thus a depressed person may be thought of as one who has had his or her ability to communicate squelched due to that person's spirit being lowered or "knocked down." The result of this is a poverty of mind or spirit. This impoverished state of mind would be achieved by being lonesome for a person, thing, event, or place. In some cases, as seen above, these feelings of lonesomeness may be maliciously imposed upon the bewitched person.

More specifically, there exists the Seneca verb root -*ataskanek*-, which means "to want a particular member of the opposite sex," while -*(e^n)?nikōhō*- means "to long to be somewhere else" (Chafe 1967:41, 51). There also exist specific Seneca concepts of -*kwe^{n}taœ?*-, meaning "to become lonesome and sad," and -*hate^n*-, meaning "to miss" or "to feel the absence of" (Chafe 1967:52, 68).

BORROWED NOTIONS

In certain instances European beliefs have entered into statements of disease causation. This is especially true in cases involving introduced diseases such as smallpox, tuberculosis, chickenpox, measles, malaria, yellow fever, scarlet fever, typhoid, and other historically introduced diseases. "Catching cold" is often cited as a cause of particular symptoms, and the use of such English disease names as "consumption," "TB," "dropsy," and "stricture" is quite common. In two cases (Herrick 1977:231) an English disease name (TB/consumption) is said to be caused by another English disease name (dyspepsia/biliousness). Besides "colds," the English terms "bile" or "gall" are frequently used in reference to malfunctionings of the liver. Such heavy reliance upon the idea that bile or gall is a major cause of a variety of symptoms is undoubtedly the result of European influences. This is clearly the case in statements attributing consumption to biliousness (Herrick 1977:231).

To understand the borrowing and application of such etiological notions, one must remember that it was not so long ago that many illnesses, such as tuberculosis (or consumption, as it was sometimes called), were understood in terms not dramatically different from those found in any traditional society. Consider, for example, statements concerning the causes of consumption mentioned in Dr. Pierce's *The People's Common Sense Medical Advisor* (1875), many of which were given to Waugh and Fenton by their Iroquois informants.

Causes of consumption: 1. *Predisposing Causes:* hereditary predisposition, scrofula, incompatibility of temperaments in parents, sexual exhaustion of parents, excessive sexual indulgence during pregnancy, climatic influences, sedentary habits, depressing emotions, any thing that impairs the vital forces and interferes with the perfect elaboration of nutritive material. 2. *Exciting Causes* (i.e., those capable of arousing the predisposing ones into activity, and which, in some instances, may themselves induce predisposition): spermatorrhea, dyspepsia, nasal catarrh, colds, suppressed menstruation, bronchitis, syphilis, retrocession of cutaneous affections, measles, scarletina, malaria, whooping-cough, small-pox, protracted fevers, pleurisy, pneumonia, long-continued discharges, masturbation, excessive venery, wastes from excessive mental activity, insufficient diet (both as regards quantity and quality), exposure to impure air, atmospheric vicissitudes, damp, dark dwellings, dampness with the absence of light, prolonged lactation, depressing mental emotion, insufficient clothing, improper treatment of other diseases, exhaustive discharges, tight lacing, fast life in fashionable society, impurity and impoverishment of the blood from any cause [Pierce 1875:490-491].

That some of the same explanations for tuberculosis are also found in both the Waugh and Fenton material should not be surprising in light of the fact that Dr. Pierce's World Dispensary was located in Buffalo, New York. Just how much influence Pierce and the Iroquois had on each other in western and central New York is not known. However, it would be safe to say that there was probably a great deal of contact between native Iroquois herbalists and those at the dispensary. Fenton (field notes 1939) noted that Dwight Jimmerson remarked that a particular root (294, *Collinsonia canadensis*) was specifically sought after by Dr. Pierce, it being "one of his main roots." I received similar information from two herbalists at Tonawanda in 1973. It was said that Pierce used to buy the roots of 258, *Angelica atropurpurea* (H1973B), for the making of a "blood purifier." The impact of Pierce upon the Iroquois cannot be denied.

Even though certain English disease names were used by many of the earlier informants, this in no way destroys the overall Iroquoian conception of health and medicine proposed in this study. Despite the use of English disease names, traditional explanations often accompanied Piercian ones. For example, one tells us that piles may be caused by (1) eating food that was prepared by a menstruating woman, (2) diarrhea, and (3) eating a lot of fruit (Herrick 1977:214). Pierce (1875:598) gave as two of the many causes of piles: "diarrhea" and "indigestible foods," but the notion regarding the harmful effects of menstruation is a native belief. Perhaps one explanation for the frequent use of gall or bile as a causal principle is that such a notion conforms with native ideas concerning object intrusion. The gall is something that must be expelled through vomiting. The notion also conforms to beliefs regarding vitality-containing fluids.

When the attitudes of the Iroquois regarding treatment of the patient are compared with those in Dr. Pierce's medical advisor, the similarities are so numerous that one is left with the impression that any differences between the two belief systems might be attributed merely to prejudice and an occasional failure of cross-cultural communication. Pierce's heavy reliance upon homeopathic medical practices and his use of herbal remedies conform well with Iroquoian conceptions of balance and good health. Even more striking parallels are found between European-based philosophies of medicine and those of the Iroquois when the treatment of what we now call psychosomatic, psychogenic, and chronic illnesses are considered.

Much of the Waugh material may be seen as a blending of Piercian and Iroquoian beliefs. For example, English names such as "consumption" and "gonorrhea" that were borrowed at the time of Pierce are still used by the Iroquois today, but with their original rather than their newer meanings. A more or less confused humoral theory involving such factors as masturbation or "a fast life in a fashionable society" was used by Pierce as an explanation for consumption. Today a physician would consider tuberculosis of the lungs as caused by *Mycobacterium tuberculosis*. Gonorrhea was thought by Pierce to involve an unknown virus, which, when it came into contact with an unbroken mucous membrane, caused gonorrhea. The same virus was thought by Pierce to cause syphilis or pox when it came into contact with an abraded mucous membrane.

Despite these seemingly erroneous explanations for tuberculosis, gonorrhea, and syphilis, they contain certain basic premises that conform to native Iroquoian conceptions regarding the causes of disease. In the case of consumption being caused by masturbation and a "fast life in a fashionable society," the native principle of taboo violation is quite applicable. In the case of gonorrhea and syphilis, both of which (according to Pierce) were thought to be caused by a single virus, the Iroquois similarly distinguished between types of venereal diseases on a symptomatic basis. The Seneca lumped all venereal diseases under the single concept *?oji:kweⁿs* (Chafe 1967:62). This parallels Pierce's lumping of gonorrhea and syphilis under the single viral cause principle. There were not, therefore, great differences between Iroquois medical beliefs and those of medical science at the time of Pierce.

INTERNAL MANIFESTATIONS OF IMBALANCES AS THEY RELATE TO NATIVE, ETIOLOGICAL CATEGORIES

In cases of taboo violation, unfulfilled desires, and coming into contact with things and events considered *ʔotkõ?*, the sick individual is, in most instances, thought to be directly responsible for the observed or experienced spiritual imbalances that result. In cases involving unfulfilled desires and things and events considered *ʔotkõ?*, the added dimension of infliction by others must be considered. This would not, however, exclude the possibility of a witch magically compelling an individual to violate a taboo. It must also be remembered that being unable to fulfill a desire is, in the broadest sense, an act that goes against the wishes of the Creator and, therefore, could be considered taboo violation.

Witchcraft, despite European conceptions of what this entails, was not exclusively inflicted by others. If the witch concept is extended to all things and events deemed evil, then acts of coming into contact with things or events considered *ʔotkõ?* would necessarily involve self-infliction. This would be true especially when the person does so knowingly. In other cases, certain animals or plants that are considered witches may inflict harm upon an individual without that individual being aware of it. For example, when a man urinates on an ant hill (presumably without knowledge that he is doing so), he may later experience a burning sensation while urinating due to the ants having "eaten his urine" (Fenton, field notes n.d.). At other times, an individual may be quite aware of having come into contact with an evil object, such as a corpse at a funeral. This dichotomy of awareness versus unawareness of the cause of imbalance would apply in all four etiological categories.

In addition to the self-inflicted/other-inflicted and the aware/unaware dichotomies of native etiological notions, the agent or mechanism involved in upsetting the internally contained vital force must also be considered. In cases of taboo violation, the spirit of a major or minor deity or the spirit force of any offended thing or event may be thought to be the agent responsible for displacing, upsetting, or in some way disturbing the inherent spirit force of the offending individual. Spiritual imbalance is thought to be brought about in cases of unfulfilled desires when the soul is strained as a result of going against the presages of the spirit or dream world. With things and events deemed *ʔotkõ?*, we find the disrupting effects of imbalanced spiritual forces being accompanied by the presence of physical objects. This should not, however, be inferred to mean that a natural/supernatural dichotomy is employed by the

Iroquois. Regardless of whether the sick individual's spirit becomes upset by an intervening spiritual force or by an intervening object, it is the spirit of the individual that is always affected. Traditionally the Iroquois did not distinguish between natural and supernatural types of illnesses. Spirituality was an essential ingredient in all illnesses.

In cases of witchcraft, the spirit of the victim may become upset in three general ways. The witch may, through contagious and imitative magic, (1) directly upset the spiritual balance of another. Plant 354, *Arctium* sp., provides a good example of such a practice. He or she may (2) magically implant an object or the spirit of an object. The end result of this second technique is an indirect disruption of spiritual balance via the obstruction of vitality-containing blood. Finally, the witch may (3) transform him- or herself into an object or animal deemed *ʔotkõ?* and directly invade the victim for the purpose of causing spiritual imbalance.

RELATIONSHIPS EXISTING BETWEEN ETIOLOGICAL CATEGORIES AND SYMPTOMS

As is true of Iroquois culture in general, there do not exist precise categories of disease causation for precise categories of symptoms. In the case of witchcraft, for example, it has already been observed that what we would call accidents are, in fact, thought to result from an other-inflicted, malefic exertion of power. An external, well-localized wound that we would consider the result of an accident would, in such a case, be considered the result of witchcraft, even though witchcraft is usually associated with either vague or localized internal pain.

In general, etiological statements that deal with what we would generally categorize as behavioral or mental illness are thought to be the result of witching. With rare exception, such as one that makes use of the borrowed concept of nervous breakdown, many of the other causes of behavioral illnesses either involve the direct or indirect mention of witchcraft. These usually suggest that there is an object, evil force, or poisonous plant (presumably ingested unknowingly) that must be expelled through vomiting. Excessive drinking and the craziness that results is also inferred as the result of witchcraft.

Unfulfilled desires are associated with what we would term chronic illnesses or, in some cases, psychosomatic or psychogenic illnesses. These are illnesses lacking well-defined causal agents that persist even after treatment, or illnesses that disappear suddenly or go into states of remission. Illustrating this would be grief or the depression that results from longing to be with someone or from wanting to be somewhere else. These feelings may augment or complicate any already existing malfunctionings or impairments afflicting an individual. By lifting the spirits of those depressed individuals, usually through some sort of communal show of support or compassion, the symptomatic discomfort accompanying chronic afflictions might be alleviated to a certain degree. In other cases, symptoms that were suggestively acquired might be removed altogether.

Modern medical science would consider this general class of illnesses psychosomatic or psychogenic in nature. The former concept is clearly related to the degree of experienced discomfort that accompanies chronic illnesses such as rheumatism, arthritis, certain respiratory and heart ailments, and peptic ulcers, among others (see Groen 1973; King 1955; Koch and Molnar 1974). However, this should not be interpreted to mean that the terms psychosomatic and psychogenic illnesses are not applicable or appropriate when considering other etiological categories. They may be involved in all other inferred causes as well. When an individual fears that a taboo has been broken, or that contamination by an item deemed *?otkō?* has occurred, this fear and any accompanying anxiety may lead to the individual suggestively acquiring (psychogenic illness) or suggestively intensifying the discomfort of already existing afflictions (psychosomatic illness). The same consequence can occur when one fears that another individual is practicing witchcraft against him or her. A lingering, mild illness thought to be the result of taboo violation could be suggestively intensified just as easily as one that is thought to be the result of an unfulfilled desire. There was an association of unfulfilled desires with diseases that we consider chronic illnesses such as rheumatism, arthritis, and respiratory illnesses. Perhaps this was a result of the fact that such illnesses were common among the Iroquois in aboriginal times, and that it was culturally more meaningful to credit such common illnesses to causes that avoided the more disruptive and antisocial inference of witchcraft and taboo violation.

An overlapping of etiological categories and symptoms is also seen in cases involving contact with things or events deemed *?otkō?* and witchcraft. Both may involve intense, well-localized, internal pain. Witchcraft would be suspected if the patient had reason to believe that someone else had implanted an evil thing inside of him or her. Otherwise, the intervention of an evil-minded or revengeful person would not necessarily be inferred, the evil object being thought to operate independently.

It is possible to make only general statements about overlapping etiological categories as they relate to various symptomatic characteristics. For the most part, the relationships existing between factors of etiology and symptoms are a function of the behavioral history of the person experiencing the spiritual imbalance. It is quite possible, provided the sick individual validates the diagnostician's theorized cause of imbalance, that taboo violation or unfulfilled desires could also be thought to cause illnesses characterized by a severe symptomatic pain and/or numerous symptoms. However, the likelihood of these diagnoses being made, as opposed to a diagnosis of witchcraft or things/events deemed *?otkō?*, would not be as great.

IROQUOIS "DISEASE" CATEGORIES

It can be inferred from a consideration of the uses of various plant medicines given in chapter 6, that plants are used primarily on a symptomatic basis. When specific disease names are used, these names are, in most cases, borrowed.

It is only after one goes beyond a search for specific disease names that traditional Iroquoian disease categories may be fully understood. Symptomatic descriptions do not constitute disease categories, and the application of English disease names to various symptomatic characteristics only serves to deceive the ethnographer by making it appear that there may also exist precise, native disease names. Precise disease names with precise sets of symptoms do not exist for the traditional Iroquois.

If one is to understand how the Iroquois traditionally conceive of types of illnesses, it is necessary to go beyond mere symptomatic descriptions and consider a whole sequence of behavioral events. Once this is done, the structured, yet flexible nature of the Iroquois cosmos is once again revealed.

Along with the four proposed, overlapping etiological categories discussed earlier, there exist certain other vague categories of illnesses. The following statements by Yankee Spring, a Tonawanda Seneca (Fenton, field notes 1934) illustrate these broad categories:

> Indians can tell by looking at the medicines whether or not the person will recover. The mixing of the various component roots depends on the judgement of the medicine man who has first seen the patient. There are many roots for fevers and other medicines are for wounds.

> There are two kinds of sickness. In the ten day's feast it is mentioned that the relatives noticed when the person was about to die. This is the only time that the fatal sickness is mentioned and the reference indicates the time when the illness changes from the first into the second or fatal sickness from which there is no recovery. The first, onōsodaiyō is a general term for illness; but, the second, sōdowe*hgo:wa:* or big sickness is fatal.

> This applies as well to the Little Water Medicine as to other medicines. When the doctor mixes the powder in water dipped down a running stream, he places one tiny dipperful on the surface of a cup of water. If it spreads out and sinks, this is a portent and the medicine will be effective. The patient then drinks it. But if the powder floats, the medicine is no good, and it will not work. They told me, when I took the guardianship of the medicine, that if it floats, the doctor should drink it. He should then try a second time, and if it still floats, he should drink it. If it continues obstinately to float a third time, he should dump it outside. Then he tries once more and gives up. The four attempts are because there are four days of restriction for the sick person. One does this for each day of confinement. During these four days of restriction, no one may see the patient.

> This all happens on the first night of the curing (wadíya?daniyat). This is when they sing and give the last dose to the patient during the singing. Before the administration of the medicine, one of the men makes a speech wherein he mentions everything connected with the case, including the time the doctor took over the case and how long he has been on it. Then he mentions the period of confinement that is to follow (wa?akiya?djawade*ʰ*?)

> The doctor usually administers the medicine morning, noon and night, and the fourth dose has to wait until the next evening when they come to sing. However, in most cases, the first dose is given in the afternoon or evening whenever the Keepers (hadane*ʰ*?t) and the singers and attendants (honōtcinō?ge*ʰ*?) arrive.

Chafe (1967:71) gives -nōhsota(æ)- for the noun root meaning "sickness," with -nōhsotaiy-/-nōhsotaiyō- being the verb root for "to make sick." Yankee Spring's description of the two basic types of illnesses are based on these concepts. The more

serious type of fatal sickness is simply denoted through the use of the augmentative. Barber Black (as recorded by Waugh, field notes n.d.) gave the Seneca term gono*ʰ*kdani as a general term for "sickness." This conforms to what Chafe (1967:72) has recorded as the verb root meaning "to be sore, hurt or feel sick," -nōōkte-/-nōōkt(a)-.

Beyond these general terms, it may be inferred that there exist certain diagnostic criteria that go into determining how a particular illness is classified. Included here would be such things as internal versus external illnesses, illnesses that affect children versus those that affect adults, and male versus female illnesses. In addition to these factors, I was told that some diseases are related or limited to certain weather, seasonal, or climatic conditions (cf. Kate Debeau's remedy for summer diarrhea given with 353, *Anthemis cotula*, W1912C). Others are thought of as being diseases that affect the upper body versus diseases that affect the lower body, the waist being the dividing point.

Generally speaking, it is possible to conclude that the treatment employed by the traditional Iroquois in attempting to restore balance to the spiritually imbalanced individual will be a function of symptomatic features (considered through time) in combination with the behavioral history of the patient. When the symptoms and behavioral history are considered, one may speak of the diagnostic criteria upon which the mode of treatment will depend. As is true for other traditional peoples in the world, for the Iroquois "diagnosis is a pivotal cognitive step in the solution of culturally appropriate responses to illness . . . [that will] . . . bear directly on the selection of ordinary, botanically-derived, medicinal remedies . . . the results of this selection, in turn, influencing efforts to reach prognostic and etiological decisions that govern the possible therapeutic need for a particular type of propitiatory offering" (Frake 1964:205). D'Andrade (1975:125) has made similar observations in studying folk disease concepts in Mexican populations, noting that "different belief clusters appear to be consequences and preconditions of the illness rather than the features that are used to define them." The behavioral history of the patient (preconditions of illness) and the features of symptoms through time (consequences of illness) are of great importance in non-Western or folk medical belief systems.

Once a particular case of imbalance or illness had been observed by members of traditional Iroquois culture, certain steps were taken in order to restore balance. An evaluation of symptoms in terms of (1) their number and severity, (2) their duration of existence, (3) whether they were internal or external, (4) whether they were localized or generalized, (5) whether they were experienced by a male or a female, (6) whether they were experienced by a child or an adult, (7) the climate or season in which they were experienced, and (8) their location on the body was undertaken. Also involved was an evaluation of the patient's behavioral history, including a consideration of previously tried medicines and techniques.

If the model proposed in this study is valid, then the past behavior of the patient is of importance in *all* cases of illness. The patient will be thought to have (1) violated a taboo; (2) gone against desires thought to be prophetic of future events while awake or while in the spirit or dream world; (3) come into contact with a thing or event thought to exude or radiate evil—including corpses, menstrual blood, and powerful charms; or (4) directly or indirectly provoked the wrath of an individual skilled in the manipulation of evil power. The degree to which these behavioral factors are verbalized, as opposed to being simply understood, would vary in relation to how symptomatic features are evaluated by both the patient and the health actor to whom the patient has appealed. Verbalizing the failure of previous treatments would be an important factor, especially if the therapist had no knowledge of them.

When the onset of symptoms is perceived as involving and/or observed to be of a relatively low degree of discomfort or pain, the role of the behavioral history of the patient would then be minimized. It would, however, still be implicitly recognized as being part of the illness's etiology. Only after the symptom or symptoms increase in degree of perceived discomfort or pain, or the symptoms persist for a culturally determined unreasonable length of time, would more specific etiological considerations come to the fore through explicit verbalization. The time factor, then, would be especially critical in such cases. With each passing day of symptomatic discomfort more and more specialized medicines and techniques would be used in the attempt to restore balance brought about by increasingly powerful, more particular, etiologies.

The symptoms might include no rain, no game from hunting, no heterosexual experiences, or no sales of baskets.

There is considerable ethnographic support for the importance of what Frake (1964:202) has called the "prodromal criteria." The time dimension of disease diagnosis and the ability of an experienced or perceived condition of imbalance to become reconceptualized through time is seen in several cases. In addition to the *onōsodaiyō* ("sickness") and *sōdowe^nhgo:wa:* ("big sickness") categories, there exist numerous statements by various informants to suggest that diseases are diagnosed through time primarily in relation to the symptomatic characteristics of intensity of discomfort and number of symptoms. Kate Debeau, for example, distinguished several kinds of sore throats (Fenton, field notes n.d.). They included (1) stinging when swallowing saliva, (2) diseased tonsils, (3) when the glands swell in the neck (also common mumps), (4) when lumps appear on the sides of the throat (diphtheria), and (5) the type of sore throat that, when neglected, becomes consumption. She also described three types of consumption. These distinctions were made on the basis of how symptoms changed through time.

Other incidences of symptomatic discrimination can be found in chapter 6. One example is A. Jonas's use of 348, *Achillea millefolium* (F1938A), for "when people fall and lie unconscious but limp." Another is Clara Red Eye's use of 247, *Rhus* ssp. (F1938C), for "swellings that increase and become red and spread." A third is her use of 189, *Rubus allegheniensis* (F1938B), for "coughing, thin, weak, fever every afternoon, sweats. When you catch cold, do not take care of yourself, and it turns to TB." David Jack used 27, *Polystichum acrostichoides* (W1912D), for " fever and cramps in children"; 42, *Asarum canadense* (W1912I), for "any kind of fever"; 75, *Ulmus rubra* (W1912C), for "when paralyzed—when you want to lie down and sleep, feel drowsy all the time and do not want to get up"; 40, *Lindera benzoin* (W1914A), for "when a person has cold sweats—not very sick at first"; 277, *Physalis heterophylla* (W1912A), "for *godjigwes* externally but not internally"; 356, *Aster novae-angliae* (W1914A), for "different kinds of fevers"; and 438, *Uvularia perfoliata* (W1914A), for sore eyes "of any degree."

It can be seen in chapter 6 that many responses center around symptoms or stages of symptoms. The traditional Iroquois did not have precise disease labels for precise symptoms or sets of symptoms. That is, they did not subscribe to the doctrine of specific etiology that characterizes Western medical science. Diagnosis of a particular case of imbalance depended on a wide variety of factors. Perhaps when precise disease labels were used by informants, they were used to satisfy Western ethnographers who insisted on knowing precise treatments for precise illnesses. In theory, it would not be possible for a native healer to tell what illness a particular herbal medicine was to be used for without first knowing directly or indirectly, through divination or dream, the behavioral history of the patient. Responses to questions regarding what illness certain plants were used for, therefore, may have concentrated on symptomatic features alone. For the traditional Iroquois, there would only exist disease phrases as opposed to specific disease names. These disease phrases would have taken into account the interrelationship of symptomatic, historical, etiological, and prognostic features.

Just as disease names involved complex statements that changed through time, so too did treatment change through time. One could, in fact, consider the type of health actor used to restore spiritual balance to be an integral part of traditional Iroquoian disease categories. Before these various health actors are discussed in detail, the observations of Stone (1935:535) regarding the processional nature of Iroquois disease treatment should be noted.

> If the Indian had exhausted his knowledge of herbs and the patient did not improve, it was felt that the disease was beyond the pale of natural maladies and must be due to some supernatural cause. The patient might have omitted some ceremony to an animal before he went on a hunt; or have failed to sprinkle the sacred tobacco over the carcass and the offended animal had cast the disease upon him. Or some ancestor's ghost was restless and was hunting the patient's body. Or, again, he may have failed to attend one of the six religious festivals, or, if present, might have made an error in a song or dance step and so brought down upon his head the ire of some slighted deity. So a clairvoyant would be sought, who, by careful scrutiny of the patient's past and interpretation of his dreams, would determine what spirit was causing the evil and prescribe the appropriate medical ceremony for the appeasement of the deity.

By considering the following etiological statements, as recorded by Waugh, Fenton, and myself, the reader may gain further insight into native

Iroquoian theory of disease causation. Plants are referenced to their positions in chapter 6.

David Jack as recorded by Waugh:

1. Birth marks are caused on a child by a pregnant woman wanting blackberries or raspberries to eat and not getting them.

Note added by Fenton (field notes n.d.): They claim that birth marks come from the woman when pregnant doing something (e.g., picking berries), and when child is born will have same birth mark. May be blackberries, etc. She is not supposed to go berrying. Case: woman recently stole and sold a heifer. When child was born, she had a calf birthmark on her leg. She got money and put it in her bloomers against her skin to hide it, hence baby had birthmark there on her skin.

2. Craziness, caused by drinking alcoholic beverages, is characterized by seeing snakes and using bad language.

3. When children are short of breath, this is the result of their mother's giving them the wrong food when they are little. It is like heaves.

4. When some woman does not like you, she may save some blood from her menses and give it to the man in some food. This will turn to consumption. He leaks inside after this (hemorrhages).

5. When a cow dries up, this may be caused by someone mixing butter with the cows food (this happening after the cow has had its first calf).

6. When a man can't make his water (stricture), this may be caused by connections (intercourse) with a woman when she has her changes. ?oji:kwe^{n}s ("venereal disease" cf. Chafe 1967:62) results.

7. If you are a good hunter, someone may use ?oye^n'kwa?ō:weh (276, *Nicotiana rustica*), fixing it so that you get no partridge.

8. A toothache may be caused by mice running over food, wetting on it, or even touching pies, cakes, bread, etc. May also be caused by cracking and eating nuts eatern by a squirrel. These are witched.

9. The body may swell up if strangers see a person making or taking medicines. Strangers should not see one with medicine on one's stomach. Neighbors are alright.

10. A man may sicken and die if another makes an effigy of him out of a certain kind of medicine (root or plant). Sharp instrument is stuck in where

you wish to injure the person—in the heart if you wish to kill. Boil the doll and then it will begin to work immediately. Save the liquid and also the doll left in it. Allow them to stand until putrid, then the man will sicken and die.

11. Abscesses in the body, legs are caused from being bewitched. The medicine used for this is also good for abscesses on the neck and for drawing out broken pieces of bone.

12. A little child may become spoiled and have bad blood if the mother has connections with another man.

13. The head goes wrong (dizziness or craziness?) when the heart "jumps" as the result of a love medicine being too strong.

14. Consumption may be caused by someone who is angry with you putting a cut-up bloodsucker in your soup. They will grow and cause consumption. Snakes and poison roots will do same.

15. When someone cannot make water freely, there is something wrong with the blood.

16. A milky juice sometimes comes out of a snake when killed. If the wind blows the smell of this toward one, a fever may be taken. This is much like typhoid. You generally lose mind.

17. A person may become sick if that person is homesick.

18. If a man is lonesome for his wife, the sickness that is caused by this may be corrected by cleaning out the stomach.

19. Suppressed menses in girls is caused by using mustard, pepper, or soda. Salt is alright.

20. A painful lump or swelling anywhere may be the result of a hurt or injury, etc.

21. Bad blood causes the hair to fall out.

22. Menses may be stopped if the woman goes with a man during menses, even if the menses have been suppressed for three years (i.e., for three years after menopause). Going with a woman during menses will ultimately prevent her from bearing children, it causes her to swell, have cramps, and get sore on one side of her abdomen.

23. When you catch cold in the intestines you become inflated and sore.

24. The same medicine is used for children when they have a fever, cramps, or when they are crying.

25. Sore or hollow teeth are the result of worms being in them.

26. When you have a rotten stomach and you do not know what the matter is, it is probably the result of someone having given you a poison.

27. There are four kinds of *?oji:kwe^{n}s* (stricture or venereal disease), three are bad, the other is not so bad. (a) Kind where lumps form and crack open in the middle, called "cut open with knife." (b) Lumps rotten at the top called "lumps rotten at top." (c) The old kind. It is inside and hidden but the person has a red nose. It attacks the end of the penis, as if burnt; present in both men and women. (d) The type when it has turned to worms.

28. When paralysis occurs, it is thought that that part of the body is dead. Paralysis is called "dead on the body."

29. A poison of menstrual blood may be mixed with fruits (blends with the color) or with red colored food and is used to bewitch another. This may be cured by taking an emetic.

30. If partridge feathers or intestines are thrown on the fire, this will result in bad luck for the hunter of partridges.

31. When butter is used from a young heifer (first calf) for frying potatoes, etc., and the butter is burned, the heifer will begin to dry up.

32. Restlessness and not feeling well are symptoms of being lonesome.

33. If young fellows go with women during their menses, the woman's breath will cause their faces to break out, which may turn to worms.

34. A girl may become sick, half crazy, or drown herself if a young fellow who has been refused by her parents takes a piece of her hair and places it in a hole in a tree and plugs up the hole. When the hair (tree) moves, she will become sick.

35. If someone has given you something bad to eat such as candy, tea, or some other food, your stomach will become sore and you will not know what is the matter. An emetic cures this.

36. If young girls work in the sun (too much?), they will get bad blood leading to suppressed menses.

37. Profuse menstruation may be caused by a flying squirrel eating corn soup or other food. Squirrels also give cramps and cause abortions.

38. If a man chews 402, *Acorus americanus*, and spits on place where woman's menses have fallen, it will cause profuse menstruation at once.

39. If a woman is unfaithful, it will cause her children to be loose and pass as fast as they eat.

40. If a menstruating woman washes clothing and then puts clothes to boil in water in which she has put her hands, it will "kill" or spoil her own blood so that she will have no children.

41. Piles are often caused by eating food prepared by a menstruating woman (worst kind), another by diarrhea, and another kind by eating a lot of fruit.

42. Suppressed menses results from a woman having a cold, the blood being affected.

43. A powdered medicine (see 277, *Physalis heterophylla*, W1914A) is good for *?oji:kwe*ⁿ*s* (venereal disease) sores externally, but not internally, and is also good for a bad stomach "when rotten stuff is vomitted."

44. A tea of 42, *Asarum canadense* (wild ginger), may be made to wash with and to sprinkle on one's clothing when visiting a person who is sick and who is using "good" medicine. The sick person will not be afraid if he is told that this medicine is being used.

45. When your sleep is disturbed it may be that people have died in the house and ghosts are pulling the quilt off of you, thus awaking you.

46. If a child is obstinate, it may be that it had a difficult childbirth.

47. When your woman goes off and won't come back, you may get lonesome and sick. An emetic will destroy this effect.

48. A yellow pond-lily (43, *Nuphar lutea*) may be hung in the house to ward off witches. It makes the witch blind.

49. Soot, mixed with snow, may be tied on a frost-bitten limb. It "sucks out the bad."

50. A sore mouth may be the result of kissing a girl with menses (this may also occur in children), or by smoking someone else's pipe.

51. A person's teeth or the teeth of his dog may fall out if the dog has bitten someone or if the person has bitten someone (in a fight) if the bitten person puts the crushed leaf of skunk cabbage (405, *Symplocarpus foetidus*) on the wound.

52. If a pregnant woman handles a man's rifle, he will not be able to hit anything afterwards.

53. Difficulty in urinating in women is caused by catching cold during menses.

54. Heaves (like those in animals) may be the result of a person catching cold. This condition is: "catches cold fits in the body."

55. Vomiting may be caused by a fever or by an unknown cause. This trouble may also be caused by eating gravy, butter, or other rich foods.

56. Lonesomeness is treated by giving the person a laxative, and this medicine is almost like love medicine for a person who is feeling that way.

57. Warts are caused by a secretion from glands on the back of a toad.

58. If you throw any of your hair out and a bird takes it and weaves it into her nest, when the wind blows the nest about, you will get dizzy. If a snake gets the hair and uses it in his nest (the snake knows as much as a person and is always trying some trick), you will get crazy.

Note added by Fenton (field notes n.d.): you will get crazy if you are not careful of your hair and throw it out where something will pick it up. It is the worst thing if they take your hair. There is no cure.

59. If you cut pieces off the finger or toenails and any person unfriendly to you gets them, he may bewitch you and make you sick by adding the nails to some kind of medicine.

Note added by Fenton (field notes n.d.): If throw away fingernails, dead people will pick them up and you will get sick.

60. If hair is pulled out of the middle of the head and this is wound around a certain root (name not given) and tobacco is used—the plant being told what to do—the person becomes crazy (something like a love medicine) and will follow after the person doing the trick.

61. A man may be angry at a trick played on his son. He will get this root (48, *Anemone virginiana*, W1912D) and tell the tobacco that he wants the man who played the trick to die. The man will go crazy or get run over or drowned.

62. Soreness all over with the skin or muscle twitching is an indication of being witched. A poultice (see 401, *Potamogeton* sp., W1912B) is placed on the twitching area (the "seat of the soreness") and it will draw out an object in the shape of a bug of some kind.

63. If a man is angry at anyone and wishes to punish him, he cuts a short length of boneset (see 369, *Eupatorium perfoliatum*, W1912G) and places it in a flask of liquor. The man will keep right on drinking until he kills himself.

64. A person may become injured or die when another person fashions a doll out of the root of the burdock (354, *Arctium* sp., W1912B).

65. If you are a woman and you are treating a man badly, you may break out like cancer if that man fashions a doll made from the roots of 102, *Amaranthus retroflexus* (amaranth pigweed), and sticks thorns in it where he wants the woman to be sick.

66. When you have consumption your stomach is quite rotten.

67. When the stomach is wrong, dry spit comes up, there is bad breath, a dry cough, and the contents of the bowels ferment.

68. Fever in children when the child sweats all along, may turn to cramps.

69. Sore eyes results from having something wrong with the stomach.

70. A cough or cold is caught from becoming overheated and then taking a chill. The "blood turns to water."

71. If blood from the menses drops on the ground and a snake or any kind of a wild animal eats it, it will make the woman go wrong and get sick.

72. A particular type of headache is caused by catching cold.

73. Girls or anybody when they throw up sour stuff, it is the result of having a bad stomach. It may lead to getting yellow and breaking out all over.

74. Cancer on the legs is thought to be due to the legs getting rotten.

75. A type of consumption is referred to as "quick consumption," which is characterized by "dying quickly." It is not clear whether this refers to the disease or the patient.

76. When you throw up yellow stuff, the stomach is bad and is sometimes accompanied by diarrhea.

77. When water hemlock (260, *Cicuta maculata*) is eaten it will make a man crazy and he will die shortly. When the root is chewed and swallowed, it becomes whole again within the stomach. You must vomit it out.

78. A typhoidlike fever may be caused by the odor from a dead snake.

79. If the leg of a certain spider ("death watch") is accidentally swallowed when it is in bread, you will be sure to get sick and die.

80. Blindness may be caused by touching the eyes after handling the milkweed (271, *Asclepias exaltata*).

81. Consumption may be caused by drinking a decoction of yellow pond-lily (43, *Nuphar lutea*).

82. When a person cannot make water freely, something is wrong with the blood.

83 *Ipomoea pandurata* (man-root) (see 282, W1914A) is not fit to be touched by children, and if you rub your fingers with it and strike anyone, you will kill them.

84. If a woman has menses (slightly) and a man goes with her then, she may put a pin or needle in her mouth. This will cause difficulty in urination and the penis becomes rotten inside.

John Jamieson, Jr. as recorded by Waugh:

85. Gathering medicines without using Indian tobacco will cause the joints to swell and stiffen (rheumatism).

86. The dead people can make you sick by staying around. They come in the afternoon and at night. If you are taking medicine, they may spoil it by putting one of their fingers on it. They want you to die and join them.

87. It will bring bad luck to kill a bat. Your cow will dry up or sickness will come upon some of the family.

Note added by Fenton (field notes n.d.): Killing bats brings bad luck. They are *?otkō?*. The rest will be against you.

88. A snake is always killed for good luck, but you must not look at anyone hanging up a dead snake or it will give you sore eyes. Toads should not be killed, but if anyone kills one and turns it over on its back, it will cause rain.

89. If a rattler bites a stick held in the hand, the poison will spread up the arm just the same.

90. A snake, particularly in June, is very dangerous. It may make one crazy or paralyzed. Snakes secrete a slimy or milky juice from the skin. A cat or dog which has killed one must not be allowed to be around, and especially where it could touch food.

91. Masturbation weakens a person and leads to insanity.

92. If the flower stalk of the dandelion is chewed, the teeth will rot because worms will be gotten off the stem.

93. When a person has had a sunstroke, he should not run any chances of exposure at the same date on the next or succeeding years. There is particular danger at such times.

94. If you defecate or urinate near an ant hill, the ants will eat it and you will either suffer from difficult

urination or you will become sore and rotten at the anus.

95. Poison ivy poisoning is sure to be contracted if you jump when you touch it.

96. When grass is dug as a medicine for a bad stomach, the hole should be filled as someone may come along and use tobacco and tell the grass to "rot his guts."

97. If a dog bites you, kill him right away and you do not "go crazy." He killed his dog by hanging him to a sapling for biting him at night.

98. Craziness may occur when the heart goes wrong. Laughing may cause a straining on the valves. John gave a fairly good description of the valves of the heart. He mentioned a case of a man who died of this a long while ago. A party of men had been betting about a game, betting for a pipe and tobacco according to custom then. The man laughed a great deal and was taken sick that night and died on the third night following. They did not know what was the matter. They used "medicine" to find out what kind of medicine to use, but it was of no avail. After the man died the Indian doctor bought the body for a gun, and an autopsy was performed (customary in former times when they did not know what was the matter). He took it to the place where it was to be buried and cut it open. He found all the bowels and other organs in good condition except the heart, which looked bloody around the upper part where it was pronounced rotten. This is supposed to have been where the trouble was. The man had had a bad pain in the heart. He had been crazy (delirious) with this and could do nothing but holler.

99. Pimples are the result of having worms in the face.

100. If a person gets pricked on the leg by one of the prickles of prickly ash (249, *Zanthoxylum americanum*) and says *agi* at the same time he feels it, it is sure to rot. June is the worst time. Say nothing at the time and it will be alright.

101. When a young woman catches cold in her kidneys, the blood and menses will get watery and strong. This is accompanied by paralysis sometimes and pains all over. It is caused by worms sucking the blood.

102. A decoction of yellow pond-lily (43, *Nuphar lutea*) will cause consumption.

103. Too much gall causes pains all over the body.

104. Stricture or difficult urination is caused by copulating with a menstruating woman. It is cured by a medicine made from the touch-me-not (252, *Impatiens capensis,* W1914B).

105. When a baby is lonesome or not satisfied with home, it lies still day and night as if thinking about something (they know as much as old people).

106. A little child may become sick if picked up by a pregnant woman. Baby becomes feverish and sick.

107. If a pregnant woman feels sore inside, legs or elsewhere, it is a sign that the baby is sick in the same place.

108. Swellings anywhere (on arms, legs, head, etc.) is caused by the blood catching cold.

109. A medicine made from wild buckwheat will cure or prevent the effects of adultry by the mother on the baby.

110. People should not use a species of *Viburnum* (346, *Viburnum* sp.) in preaching at the Longhouse or that person will never have a child again.

111. The bowels must be moved to cure rheumatism.

112. If a little boy urinates on an ant hill (belonging to the large red waisted ants), he will have a stoppage of urine and his limbs may become paralyzed. The ants are witches.

113. The leaves of vervain (291, *Verbena hastata,* W1914A) may be prepared in such a way that they will make an obnoxious son- or daughter-in-law leave.

114. Owl's oil is given for any kind of fits—for those caused by indigestion, etc. in children, as well as epileptic seizures in larger children.

115. If you don't wash your hands after handling 260, *Cicuta maculata* (water-hemlock), you will have fits.

Jim Davis as recorded by Waugh:

116. A woman may swell after copulation as a result of some bad medicine being administered to influence them, and may also result from a root having been made to represent the woman and pierced with thorns.

117. Suppressed menses is the result of catching cold.

118. When little children are peevish, there is something wrong with the back.

119. A cure for witchcraft which manifests itself in the stomach is made from horse gentian (340, *Triosteum aurantiacum,* W1914B).

120. A person is not fit to see a sick person who is using good medicine if that person is menstruating or has been to a funeral.

121. Cold blood leads to nonconception.

122. Diarrhea is caused by a menstruating woman preparing food.

123. A medicine for venereal disease (see 355, *Aster macrophyllus,* W1914A) will loosen the bowels.

124. Too much gall causes difficult urination.

125. Neglect of a proper feast will cause ghosts to come inside. At about sundown you will hear rattling. Sprinkle the medicine (see 258, *Angelica atropurpurea,* W1914C) inside and outside house where ghost is likely to go.

126. There is a medicine which cures a strong smell under the arms (see 405, *Symplocarpus foetidus,* W1914C).

127. Lonesomeness is characterized by being crazy and melancholy.

128. There is a certain medicine made from three plants which, when placed in a small coffin and pierced (the root) where desired, will kill a person in ten days.

129. A baby's navel will not heal if the baby has too much gall.

Robert Smoke as recorded by Waugh:

130. When a woman nursing a baby has anything to do with a man other than her husband, it will make the baby cross and sick. It will give the baby diarrhea and make it die.

131. A disease where the inside of the mouth and the throat get all raw is sometimes caused by fever inside.

132. A bad disorder where you pass blood is caused by going with a woman when she has her menses. A woman will sometimes do this on purpose. His urine turns to blood. Can make only a few drops of water at a time. Water also comes away involuntarily. A woman may get the same disease.

This same disease may be caused if a man is going with a young girl or woman and she sees him making water. She may go where he has urinated, pick up some of the dirt or soil which the urine has fallen on, and throw it on an ant hill. The ants will

eat this and will cause a prickling and hurting sensation in his bladder when he goes to urinate.

133. A bellyache may be caused by eating food prepared by a menstruating woman.

134. If you have been to a funeral, you are not fit to see the sick. If you drink and wash your hands and face with a decoction of wild ginger (see 42, *Asarum canadense,* W1914B) you can go in.

135. One cure for love medicine is to vomit. A fellow who took a particular medicine for this (see 447, *Corallorhiza maculata,* W1914B), threw up some white worms. Worms are caused by witchcraft.

136. If a pregnant woman whose child is to be born shortly, eats food prepared by a menstruating woman, this will give her pains in the sides (region of ovaries, perhaps), and may often cause an abortion.

137. Bad breath in older children and adults is a sign that there is something wrong inside.

138. When people have been away for some time and come back home again, there will be ghosts in the house (see 34, *Pinus strobus,* W1914C, and 43, *Nuphar lutea,* W1914B, for remedies).

139. Worms are caused by witchcraft or the occult means.

Sam Hill as recorded by Waugh:

140. Stricture is caused by going with a woman during her changes.

141. Kidney trouble is indicated by pains in the small of the back and dark circles and puffiness under the eyes in the morning.

142. The same remedy (unwilling to give) is used for inflammatory rheumatism and love medicine (witching).

143. A medicine made from 305, *Scutellaria lateriflora,* will prevent smallpox by keeping the throat clean. A throat becoming stopped up is fatal.

144. Pain after confinement (childbirth?) will go away if a certain medicine is taken (see 249, *Zanthoxylum americanum,* W1912D).

145. A child will become sick or spoiled if a woman (the mother) has anything to do with another man.

146. When a pregnant woman is drowsy or feverish, the woman is not sick—it is the baby who is sick.

147. If, when using a deer hunting medicine (see 383, *Prenanthes trifoliolata,* W1912B), one makes

noise (whistling or pounding) or has anything to do with a woman (or both, ha ha) for the twelve days preceding the hunt, the medicine will not work.

John Echo as recorded by Waugh:

148. If a menstruating woman eats salt it will suppress the menses. However, if the salt is put in a cloth and scorched a little (burning the cloth off), it may be used. The same restriction applies to a woman with a newly born child and to people of both sexes who are taking any kind of medicine.

149. "Legs" is a sign of bad luck and some folks will die. It will make you crazy to see one. Young people were told that if they went out at about sundown they would see it. Anyone who is plucky and catches it and holds it (by the hair) until morning will find he is holding only grass.

150. When you see a red bloodsucker in the water, the water is not fit for drinking at that place. Preachers at the Longhouse warn against these. They will spoil your blood and you will waste away and die with some disease no one can understand.

151. If a large thousand-legs, smooth on top, crawls up clothes and bites, it will cause the same trouble as spiders. If one of the legs is swallowed accidentally, it will cause consumption (no medicine to cure).

Note added by Fenton (field notes n.d.): belief exists at Allegany that moths cause consumption.

152. This kind of spider lives inside and it is hard to get medicine when they bite. If you scratch, will go all over body. There is another bad kind—has a big belly (dark or gray), lives in the grass. Will leave a mark on your flesh where they run (see 276, *Nicotiana rustica*, W1915A, for cure).

153. Warts are caused by secretions from toad's glands.

154. Hair combings should be thrown in the fire. If thrown out and a bird or a mouse makes a nest from them, it will make you dizzy. Rats also.

Elijah David as recorded by Waugh:

155. When the blood is bad, sores may break out on the neck.

156. Diarrhea may develop in children having no mother.

157. A gnag (nag?) may try to make you sick by witchcraft.

158. Constipation is the result of the guts being tied in lumps.

159. Witchcraft, which is manifested by a pain that moves about, may be cured by applying a poultice (see 341, *Viburnum acerifolium*, W1912A). This will take out ants, worms, charcoals, water beetles, crickets, broomsplints, hair, etc., which cause pains.

K. Gibson as recorded by Waugh:

160. If you have a sunstroke, you should stay in the same day the next year or you will "be done." It would make you blind.

161. Venereal disease turns to worms when too bad.

Levi John as recorded by Waugh:

162. Cramps are caused by worms (see 137, *Populus tremuloides*, W1912D).

163. A person who has attended a funeral will make a sick person worse. It will cause them to bloat up and die.

164. Dead people are the cause of bad dreams.

165. A person who sees a corpse or attends a funeral will spoil a medicine if he looks at it.

166. The same medicine (see 116, *Rumex crispus*, W1912A) is used for yellow jaundice (which is characterized by being bloated or being swollen across the body) and for cramps and pains in the abdomen (e.g., when appendicitis is suspected).

167. Worms in children cause convulsions.

168. Stricture is caused by going with a girl during the menstrual period.

Peter John as recorded by Waugh:

169. Peter knew of a saying to the effect that you must not point at a rainbow or else you will become hump-backed.

170. People are often made ill by dead people who wish the living to do something for them and keep bothering them until they do this. The dead are supposed to be able to return among us in this way (see 42, *Asarum canadense*, W1912J).

171. The root of 323, *Lobelia cardinalis* (W1913A), is made into a poultice and applied to a

spot where there is pain or trouble caused by witchcraft. This will draw out the bone, insect, or other article causing the sickness. A decoction of this taken inwardly will ward off witchcraft.

172. A medicine made from 227, *Euonymous europaea* (W1912B), and others will take the worms out of a man's body caused by venereal disease, which, in turn, is the result of breaking the solid food taboo. During treatment don't eat any meat—just soft foods like rice or corn. Don't go to the closet (out house), but dig a hole in the ground and use a box with a hole in it. You may see the worms jumping (if you put these in a bottle of water, they will live a year).

173. The blood may be purified by loosening the bowels, and this is used as a method of treatment to rid a man's system (i.e., get it out of his blood) of pox, clap, or buboes (see 212, *Dirca palustris*, W1912D).

174. There are two types of gonorrhea (?): one is when water comes slowly (difficult urination) and the other is when the water stops. To avoid return of the disease, avoid all relations with the opposite sex for one month. The first of these diseases is called "whites" by white doctors, and it is further added that this eventually goes in and rots the penis. It may, however, occur in both sexes.

John Silversmith as recorded by Waugh:

175. Sore mouth in children comes from their being kissed by their mother during her menses. Happens in adults too.

176. Young girls with the menses (called it "new moon") should not handle food that will be used by a woman near her term with a child. If they do, the woman will become very sore.

177. When small babies are cross and will not suckle, they twist around as if there were something wrong with their backs (see 332, *Mitchella repens*, W1912C).

Mrs. Sundown as recorded by Waugh:

178. Piles are caused by drinking after a girl who has menses.

179. An emetic (see 324, *Lobelia inflata*, W1912A) is used to cure sickness caused by witchcraft. A man who was crazy was cured by this medicine. The same remedy is good for colds or coughs as well.

Peter Sundown as recorded by Waugh:

180. A thousand-legs (the very large species of worm) is very poisonous in August. To bewitch and kill another, cut the worm in two and put its juice on tobacco or food and give it to the victim. The remedy for this involves burning a turkey feather (turkeys eat these worms), put a quantity of the black ashes in a glass of water and drink it if the inside of the throat is raw and swollen, or rub the water on the body sores.

181. Legs, cut off just above the hips, are sometimes seen along the road or elsewhere. It is a bad sign to the person seeing it, as all the other members of his family are sure to die within the year or a little more. The person seeing it, however, does not die.

David Key as recorded by Waugh:

182. Whoever sees "legs" will get bad luck. It consists of legs with no head or hair, but has dark markings on it like features. It has no eyes. When it runs you can hear a noise like a mouth slapping together. It does have a mouth. If you think you've caught it, you will have only grass, only rotten wood. If one is following you, you can get rid of it by stepping to one side and saying, "You can go by. You're in a hurry." Some of your folks will die whenever you see one. An old man won't see one. It is only young men who see them, as they wish to find a good runner (they are good runners themselves). Women are not followed.

183. To see even numbers of crows is good luck. Odd numbers are bad luck. If two crows fly across the path when hunting, this is good luck.

184. If a fox is heard barking near a hunter's shanty in the bush, as though he were telling something, it is bad luck. It means that someone in his family will die.

185. If a hunter sets a trap for muskrat (for example) and the animal scratches dirt over the trap (buries it), it is a sign that some of his folks will be buried.

186. If a muskrat puts a stick in a trap, it is a sign the hunter's wife is unfaithful (someone is putting a

stick in her). He had better quit trapping as he will have no luck.

John Jamieson, Sr. as recorded by Waugh:

187. "The two-legged one" is supposed to have two eyes and half the body. It will follow women too. It is bad luck to see one. The old people used to say to the young people inclined to run around nights that they would meet one. The fourth night they would meet him.

188. It is a bad sign for a snake to cross your path as you are going anywhere. You will not get what you are going after.

189. Two cows crossing your path are a good sign. Three are a bad sign.

190. If a fox comes close to your camp and barks, it will be bad luck. Someone will die.

191. Once he went out trapping muskrats in the spring. He found a place where there were lots of them. He set his traps here and went out in the morning. There was nothing in them. The muskrat had all left. His wife died shortly afterward.

192. A milk snake coming into the house and climbing somewhere is bad luck. Someone will die.

193. The Creator and the devil were quarreling. The devil took a stick and threw it down and it turned into a snake. He said he would put it there to scare people when they saw it. This is why people always kill them now.

Herbert Johnson as recorded by Waugh:

194. Soreness while urinating is brought about from going with a prostitute or a menstruating woman.

195. When a boy or girl is lazy or listless it is because the muscles lack rigidity (see 294, *Collinsonia canadensis,* W1912B, for cure).

Jake Schuyler as recorded by Waugh:

196. Consumption comes from the stomach getting rotten and spoiled from the use of too much vinegar, mustard, and salt. Worms form in stomach and consumption starts from this.

Jackson Jacobs as recorded by Waugh:

197. Palpitation or thumping of the heart, especially from fright, remorse (such as when you hear of a death), or anger will, when longstanding, weaken and kill you.

198. Worms in the lower part of the body may be caused by breaking the solid food taboo.

Peter Jacobs as recorded by Waugh:

199. White moths flying at night are said to be "witch." You may be bewitched if they brush against you leaving white powder (scales) on the clothing.

Note added by Fenton (field notes n.d.): Simeon Gibson (Ca.) has heard this.

Barber Black as recorded by Waugh:

200. A child may get diarrhea from its mother's adultery (see 303, *Nepeta cataria,* W1912A, for remedy).

S.A. Anderson (unknown place, time, and tribe) as recorded by Waugh:

201. Young people are told that if they break a cob of green corn into pieces instead of eating the corn from the whole cob, they will see or be chased by "legs."

Kate Debeau as recorded by Waugh:

202. When in the sugar bush a child frequently licks at an old wooden sugar spoon and gets a boil complaint (possibly, a "bowel complaint"), which causes a belt-like (welt-like?) swelling around the abdomen. The medicine that is used for this (see 400, *Sagittaria latifolia,* W1912A) is also good for general constipation in any aged person.

203. A runner's medicine (see 399, *Alisma plantago-aquatica,* W1912A) works because it "strengthens the blood veins."

204. Bad blood is characterized by chills, fatigue, and general lassitude because the blood gets cold. It starts with soreness and feverishness.

205. What some people call witchcraft among Indians really isn't. There are so many mixtures of blood today that during menstruation a young girl faints. Indians would say she was bewitched. No. The cause is the blood until such time as she gets married

and gives birth. Medicine converts her blood so she is alright.

206. If a witched person takes a tablespoon of a particular medicine (see 437, *Trillium* sp., W1912B) and spits it out and begins to use foul language, this is considered a sure sign of bewitchment. If a person recovers, he will never again be sick with this. This medicine, given with a physic, cured a woman suspected of being witched. Her excrement was like black dirt and contained broom splints.

207. Two young girls going through the bushes discovered these plants (437, *Trillium* sp.) and were surprised. One of them took one, and one night she felt ill. She thought Sam (?) had told her it was good medicine. She went to work and made a dose of it and drank it on several occasions the next morning. Her friend asked her, "Did you take it all?" She replied, "Sure I did." The other said, "Now you are going to get twins," and she did.

208. At menstrual periods a girl may lose her senses, the palms of her hands get sweaty and cold. Sweat also comes out around the mouth. It is thought to be caused by the blood being too strong.

209. When an adult breaks a baby's coccyx and it is neglected (i.e., no medicines are used), it turns into consumption.

210. The joints of bones may decay under fever. A poultice of white pine (see 34, *Pinus strobus*, W1912J) pulls this decayed matter out.

211. If a necklace of hazelnuts (see 97, *Corylus americana*, W1912A), which is placed around a baby's neck when its teeth are coming in, is thrown on the ground instead of being hung up in the house after the teeth have come in, the baby's teeth will decompose just as the necklace would.

212. Vomit is buried as it is considered dangerous in spreading disease.

213. A poultice of 266, *Sium suave* (W1912A), is used to remove pain in a broken limb. Two little boys once ate the roots of this plant and died. The plant then grew on their graves.

214. Certain girls have a disease of fainting suddenly. It is a blood disease which is brought about when the blood is either too strong or too weak. When there is too much blood the face gets dark in fainting; when too weak they become cold. This disease pulls all the blood from the hands, face, and all parts of the body.

215. There is another kind of fainting when a person foams at the mouth. Here, the medicine (see 57, *Ranunculus abortivus*, W1912B) is mixed with another ingredient. It occurs in any individual and is confined to families.

216. Grieving causes a sickness which "hurts the heart."

217. When the poultice that is used in treating cancer and fever sores (see 323, *Lobelia cardinalis*, W1912C) is removed, it is buried in the ground. This removes the disease from you.

Kate Debeau as recorded by Fenton:

218. One type of consumption is characterized by the lungs swelling and pains between the shoulders. It comes from neglected colds, rheumatism, tonsils—poison (toxin) is swallowed, then the lungs get white and then it is too late to cure.

219. Consumption may also be caused by billiousness.

220. TB may be caused by dyspepsia. Some people are affected by eating too much soup and belching.

221. Three kinds of consumption are as follows: (a) one with cough—a female's blood stops flowing by getting her feet wet, and that is strongest kind. She coughs and may last only three months; (b) one without cough deriving from tonsils. Coughs very little. Pains between shoulder blades. Coughs up slime. Lungs get progressively smaller. Long lingering; and (c) consumption of bone—female got it from her husband, now her lungs almost withered away. Consumption is an old disease.

222. Neglected colds produce "stitches" (pleurisy) on left-hand sides.

223. When the stomach starts to dry, it leads to consumption.

224. There is a medicine (see 262, *Heracleum lanatum*, F1933A) for when you get a bruise on the back of your stomach.

225. Young girls sometimes go crazy on account of menstrual periods.

Chauncey Johnny-John as recorded by Fenton:

226. Father said sickness travels on the wind. They say some sickness comes from flies.

227. The old people did not get so sick so often. We are not as tough as they used to be.

228. There are always lots of worms in any water.

229. Old Henry Phillips of Cattaraugus used to practice bloodletting. He sometimes cut on the forehead a little. He had an implement, a stick with a blade in it like a carpenter's gauge. He was an Indian doctor. Does not know sickness. Sometimes cut on bicep or thigh. The bad blood is darker, and the bleeding will stop on its own accord when the bad blood has passed out. No bandage. Heals up itself. No medicine used.

230. The worms come up when you are sick. You can hear them squealing inside when you eat something they do not like. The medicine taken kills them inside. One cannot tell when they are coming out. They usually go down. This medicine "white crooked root" (see 47, *Anemone canadensis,* F1938A).

231. Skin diseases are caused by bad blood and are treated by bleeding.

232. When there is weak blood, the person feels weak, the outer corner of the eyes will be red, there is general lassitude. Every time when the blood stops, you feel heavy (see 21, *Osmunda claytoniana,* F1933A).

233. When food you have eaten will not digest, you will experience pains in the back (see 380, *Polymnia uvedalia,* F1938B, for cure).

234. Dropsy, or watery blood, is characterized by swellings on the shins and claves of legs and the body. These are soft when you touch it and it leaves a dent when probed with the finger.

235. Consumption is characterized by a dry cough. People get hurt, then get the disease. Never saw it.

236. There are two types of consumption: (a) those who are hurt before and (b) those who get sick(?)—possibly referring to TB because this distinction is made in giving the medicine for TB (see 214, *Epilobium angustifolium,* F1933A).

237. A particular medicine (see 218, *Cornus amomum,* F1938A) is great medicine when man knows he is going to have possessions. Drink this medicine, an emetic, and it will stop.

238. Cramps are caused by worms coming up. You die if they come up in your neck (see 268, *Gentianella quinquefolia,* F1933A, for cure).

239. *Veronicastrum virginicum* (see 319, F1933B) is a medicine for fevers and too much gall. It is supposed to be poison if too strong and especially if someone sees you who has been to a funeral.

240. There is a medicine for women (see 306, *Plantago* sp., FndB) who cannot have children. It is good for the blood.

Sarah Snow as recorded by Fenton:

241. When using a particular emetic (see 389, *Solidago juncea,* FndB) one will get a headache afterwards unless one faces east while vomiting into a pail, throws the vomit in a stream in the direction of the flow of the stream, and ties a piece of cloth around the forehead.

242. A medicine made from the powdered root of the common teasel (see 347, *Dipsacus fullonum,* FndA) is used to kill another. Put a pinch of it in his tea or coffee cup. This may be used to take another's land. They get diarrhea and do not stop until they die. The same medicine was used by the old folks when they were jealous of a cute baby. They would dip their fingers in the medicine and put it on the baby's lips. The baby would die, then they would eat the ghost (?).

243. In changing of life, some may go deaf or blind (see 13, *Lycopodium obscurum,* FndA).

244. If someone dies and you cannot forget it, take an emetic (see 34, *Pinus strobus,* F1933A).

245. Low blood pressure, as your blood is boiling, or high blood pressure is a sign of impure blood (see 190, *Rubus idaeus,* ssp. *sachalinensis,* FndA).

246. The same medicine (see 267, *Gentiana andrewsii,* FndB) is used to cure oneself from witchcraft brought about by the jealousy of someone else, and for fever or headache.

247. When venereal disease gets in the blood, this medicine (see 370, *Eupatorium rugosum,* FndB) separates the disease from the blood.

Clara Redeye with Emma Turkey interpreting as recorded by Fenton:

248. A cold in the blood is characterized by no appetite, fever, and cramps in sides before menstruating (see 13, *Lycopodium obscurum,* F1938, for treatment).

249. Blood disease is characterized by (a) light, whitish color of blood during menstruation, or (b) dark, cloudy blood during menstruation, or (c) menstruation skips a month, poor flow.

250. TB is characterized by coughing, thinness, weakness, fever every afternoon, night sweats. When you catch cold, cough and don't take care of yourself, and repeat the process, it eventually becomes TB. It may also be caused by the indoor life and eating sweets.

251. A blood disorder is characterized by thinness, lack of appetite, headache, paleness, and menstrual cramps (see 117, *Rumex obtusifolius,* F1938B, for medicine).

252. A blood disorder is characterized by paleness, no appetite, and pimples on the face (see 261, *Daucus carota,* F1938C).

253. When worms are disturbed, they cause fever.

254. Piles may be caused by witchcraft.

Jonas and Josephine Snow as recorded by Fenton:

255. *Pteridium aquilinum* (see 25, FndB) is used as a medicine for TB in the blood—especially for women. The blood becomes red again after taking this medicine. (Fenton's note: she possibly refers to *leukoorrhea*).

256. When Sarah's children were born her mother stayed with her and Jonas went to work. Jonas heard of the idea that it's no use to go hunting when wife is about to have a baby (this was practiced by the Seneca). You can't catch anything, no use to go out and hunt.

257. Billiousness is when too much sugar makes sour in the stomach, yellow substance in the stomach begins to come up. If it comes up all the way, you die. The medicine for this (see 319, *Veronicastrum virginicum,* F1933A) is also used every spring to ward off spring fever (when you feel lazy and lethargic). It will break up gallstones, softening them so that they will pass off. This medicine is poison, and Sarah once took too much of it and got knots in her guts.

258. The same medicine (see 306, *Plantago* spp., F1933B) is used for cuts and wounds, for spider bites, and for the blood and pains anywhere in the body (Sarah's mother used it for the latter).

259. The blood gets bad sometimes and it used to make me sore all over my body. She took the medicine for this (see 185, *Prunus serotina,* F1933A) once when she had a sore hand and her face swelled.

Jemima Gibson as recorded by Fenton:

260. The same medicine (see 211, *Lythrum salicaria*) is used for fevers and sickness caused by the dead.

261. Female weakness is due to the blood getting out of order—the blood turns to water (see 258, *Angelica atropurpurea,* FndA).

262. A particular medicine (see 302, *Monarda* sp., F1939A) is used for fever and is also good for sickness caused by dead people.

Peter Hopps as recorded by Fenton:

263. There is a medicine (see 26, *Onoclea sensibilis,* F1939A) which is good for the early stages of consumption. It is not, however, effective in cases three or four years old.

264. *Thuja occidentalis* (see 38, FndB) is used for rheumatism. This medicine makes the rheumatism come out.

265. *Cicuta maculata* (see 260) is poison to drink. Sometimes it grows on graves.

Henry Redeye as recorded by Fenton:

266. Henry thinks Sherman, who is sick with piles, is being witched, so he gave "evil antidot" (a compound of powder made from several plants) to him.

267. *Smilax herbacea* (see 44, FndA) is good for the stomach and you can't tell what is the matter. This will kill anything in the stomach that is causing the trouble. Patient will get well in two to three days.

268. Bellyache is a sign of worms. If worms come up into the throat, the patient dies (see 268, *Gentianella quinquefolia,* FndB).

Albert Jones as recorded by Fenton:

269. Yellow bile in the body will cause pain (see 270, *Apocynum cannabinum,* F1938A).

270. The root of 306, *Plantago* spp. (F1938A), is good for what whites call "nervous breakdown," or when female overworks. When she has many children to look after, makes her nervous. She feels badly, dizzy in the head.

271. There is a particular medicine (see 72, *Hamamelis virginiana,* F1938A) that is good for anything that is evil. Supposing someone is sick, feels badly, do not know cause, take this medicine. We use

it when doctors can't find trouble with us. Indians would come home and take this.

Harvey Jacobs as recorded by Fenton:

272. Cancer is due to our diet getting a poison into the system, like greasy stuff, starchy stuff. Wherever there is a bruise or bump after this stuff gets in the system, cancer starts. It will start wherever it is tender (mouth, throat, womb). Wherever this is, it has to come out of the system. If sugar or acid in system, wherever this is, it won't heal. You must give patient mainly a blood purifier. Take a laxative first thing in the morning, when stomach is empty and before breakfast. Take blood tonic four times a day before meals and upon retiring. After eating, take a medicine that digests the food immediately. Keep from forming acid or sugar. What the patient eats will now be pure because these medicines have made it pure. In a few days it will stop the burning pain. It will draw that away because it is not forming poison from the food they eat. No diet necessary after taking this medicine—just don't eat too much pork or sugar or acid stuff. Can eat, but not enough to hurt. This blood purifying medicine (will not divulge its contents) is his main standby in cases of sugar diabetes, cancer, skin disease, and women's troubles.

Josephine Jimmerson as recorded by Fenton:

273. Dropsy involves the blood turning to water. The medicine for this (see 273, *Asclepias syriaca*, F1939A) turns the water back into blood.
274. Fever and chills are caused by going out in the winter and coming home chilly.

Simeon Gibson as recorded by Fenton:

275. Pitch Pine (see 34, *Pinus strobus*, FndD) is used in three main ways: (a) it is good for drawing out thorns and slivers. If eventually threatened by blood poisoning, also use a wash made from 68, *Sanguinaria candensis* (FndE). This will kill the blood poison; (b) when the needles and branches are picked up fresh after a wind, they may be taken home and hung over the bed of someone who has been wounded, and this will drive away the dead persons (note: this will not work if the branches have been cut off); and (c) when these windfallen

branches are put in a fire to smoke, a person who has been to a funeral and seen a dead person may fumigate and wash off his eyes (tear) and then will be fit to go home and attend a sick person. Then he will not spoil the medicine and make the sick person worse.

Esther Fatty as recorded by Fenton:

276. Esther told me that John Jacob's son sustained a broken spine and had lower limb paralysis, and subsequently he died last winter. At Hawthorne's first visit, John asked him if he had looked on any dead person that day or seen anyone else sick. John asked his wife first, then the boy was anxious to see Hawthorne. When the boy finally got to the hospital, it was too late to do anything.

Yankee Spring as recorded by Fenton:

277. Sickness (consumption) depends on overwork of certain parts of the body. Every time one catches cold, that is where it settles. Rough fellows who fought and drank are said to be subject to consumption—especially where they have been wounded. They merely wear down and it settles there.

Anonymous as recorded by Herrick:

278. Young girls get fits when they want a man, but cannot get one. These girls foam at the mouth.
279. Congenital clap (gonorrhea) is the result of hairlike worms being present in the urine.
280. A doctor once asked why Indians have no crippled babies. The answer was that one must marry outside one's clan.
281. A man once pulled up someone else's corn and potatoes. Later, he was dragged by a horse and suffered a year and a half.
282. Cancer is caused by thinking, worrying too much, and by loneliness.
283. Venereal disease is caused by having sex with strangers.
284. Blindness may be caused by collecting 403, *Arisaema triphyllum*, too early in the spring before it is ready to be picked.
285. Hives are caused by drinking wine.

286. If you are picking medicines and you find two plants growing next to each other, the plants are "married" and you must take both or else the one that is left will make you sick.

287. A certain plant (knotweed, smartweed), when near your bed, won't let you sleep.

4

Medical Treatments in Traditional Iroquois Culture

At least nine major types of health actors were used by the traditional Iroquois. These include: (1) self-treatment, (2) lay health treatment, (3) herbalists, (4) white doctors, (5) clairvoyants, (6) those involved in "doings," (7) members of medicine societies, (8) Indian doctors, and (9) witches (for the purpose of counteracting witchcraft). These relationships must be considered as they relate to the four overlapping etiological categories proposed earlier. If, for example, a mild sore throat is being experienced, the sick individual may not consider unfulfilled desires, coming into contact with a thing or event considered ?otkō?, or witchcraft as being possible causes of his or her discomfort. Such causes would be reserved for more serious symptoms. This would, of course, only be true if the individual in question did not have good reason to suspect a more powerful causal agent. Self-treatment or lay health treatment would suffice in such a case.

Should the sore throat increase in duration or in intensity of discomfort, and should the patient persist in excluding other possible etiological factors (witchcraft, etc.), he or she may then consider consulting what some have called a "professional" health actor (Polgar 1963). The services of an herbalist, who is often times a clairvoyant as well, might then be sought for the appropriate plant- or animal-derived medicine believed to have sufficient power to counteract the force thought to be responsible for the experienced symptom(s). If these measures were to fail, a clairvoyant (or, more likely, a highly knowledgeable or skilled herbalist who would play two or more health actor roles) might then be asked to contact the spirit world in order to discover the precise cause of imbalance. At this stage, one of two possible paths would be followed. First, the clairvoyant might sustain the patient's original theory that a taboo had been violated. The violation would now be thought to be a more serious one that was committed against a more powerful animal or other spirit force, such as abuse or neglect of a hunting charm or an angry dead relative. An appropriate individual or communal act of propitiation might then be prescribed. Or, he might discover that it was one of three other possible causes of imbalance: an unfulfilled desire, contamination by a thing or event that radiates or exudes evil, or witchcraft. Then, depending on the results of what Stone (1935:535) has described as the "careful scrutiny of the patient's past," an appropriate course of action would be recommended. Included among these would be: (1) individually performing or acting out suppressed desires, (2) communal participation in acting out suppressed desires (such as those carried out at the Midwinter Ceremonies), (3) individual acts of driving out malefic influences from the patient's body, (4) communal acts of driving out malefic influences from the patient's body, or (5) recommending a witch for the purpose of counterwitchcraft. Each of these prescriptions would, as discussed earlier, be a function of the patient's discovered or revealed behavioral history. As is the case with other elements of the Iroquois medical complex, these prescriptions may be seen to overlap in certain instances. This overlapping quality is seen as a consideration of each health actor category is undertaken (cf. Herrick 1983).

SELF-ADDRESSED AND LAY HEALTH ACTOR PHASES OF TREATMENT

There exist only vague references to the fact that self-treatment is followed when an individual initially perceives him- or herself to be in a state of mild imbalance. Like the theorized causes of very mild illnesses, self-treatment may not involve explicit verbalization. If it is understood that disease is the primary mechanism of social sanction in traditional societies, then any reluctance to discuss possible causes at the onset of illness should be expected.

Barring events in the behavioral history of the patient that might make him or her anticipate the involvement of more powerful spirit forces, the sick individual might well assume that the imbalance being experienced is simply a part of living in a continually shifting, delicately balanced universe. And since practically every adult member of Iroquois society knows the use of a few herbal medicines or other medical techniques, the self-application of these remedies and techniques might be undertaken in much the same way that someone today might take an over-the-counter medicine for an upset stomach, headache, or fever. It should be kept in mind, however, that even though these self-applied medicines may be utilized in a very offhanded manner, they should not be thought of as being a "pragmatic" or "naturalistic" aspect of Iroquois medicine. For the traditional Iroquois it was always the spirit or power of the medicine that was thought to operate in all types and at all levels of treatment.

The person believing his or her spirit to be in a mild state of imbalance might also seek the assistance of a friend or relative not considered an expert at manipulating various spirit forces. Fenton (1953:7) has observed that "when members of the family are taken sick, they try various herbal remedies suggested by the elders." Such behavior probably would be most common in cases involving young adults who had not yet become familiar with even the most widely known remedies. Nonexpert or nonprofessional elders would then serve as family reservoirs of medical knowledge.

PREVENTIVE CONCEPTS AND MEDICINES

It is appropriate to deal with Iroquois conceptions of preventive medicine at this point because such practices are carried out by both lay and professional health actors. The most general methods of preventing illnesses or imbalances in traditional Iroquois culture are seen to involve simply avoiding any or all of the theorized causes of illness. Spiritual balance could be maintained by (1) observing the standards of conduct established by the Creator, (2) fulfilling dreams or wishes of the soul, (3) avoiding things or events known to radiate or exude ?otkõ?, or (4) being careful not to incite revenge, envy, or hatred in others who might then directly or indirectly resort to acts of witchcraft.

Closely related to the first of these general methods of avoiding illness (i.e., avoiding taboo violation) is the practice of taking spring and fall medicines. The former are taken to ensure health during the summer months by protecting the individual from illnesses that are characteristic of warm temperatures. The latter ensure protection against cold temperature illnesses (cf. Rioux 1951:155). Generally speaking, it would be considered a violation of the established ways if an individual did not take these preventive medicines. If an individual were to get sick during the summer or winter months after having not taken these medicines, the cause might be attributed to this negligent act. If they were taken and illness resulted anyway, one possible explanation might be that not enough or, perhaps, too much (restraint being important in all things) of these fall or spring medicines were taken. The following analogy, shared with me by an individual at St. Regis, is useful in understanding the application of spring and fall medicines:

> Our medicine works the same way that a car radiator or motor oil works. In the spring we take medicines that keep us healthy all summer long. In the fall we take medicines that keep us healthy all winter long. The same is true for what we put in our car radiators and the type of motor oil we use—good for summer, good for winter [Herrick, field notes 1973].

Besides the spring and fall medicines (see 91, *Alnus incana*, Fnd A; 224, *Cornus sericea*, FndB; and 339, *Sambucus canadensis*, FndD, for examples of spring medicines), there also exist certain ball players' or runners' medicines that work on the same principle of "internal cleansing" through the use of cathartics, emetics, diuretics, and diaphoretics (see 224, *Cornus sericea*). In these cases, the imbalance prevented would be an unacceptable performance in a lacrosse game or running contest.

According to Stone (1935:532), "whenever one desired to communicate with the spirit world, it was universally considered necessary to be pure in mind and body. Therefore, prayer and ritual were entered on [sic] only after hours of meditation and physical cleansing, inside and out by the use of emetics and by taking sweat baths" (cf. Fenton 1942:513). Again, this act of ritual purification could be viewed as being a way of avoiding taboo violation, an unclean person being considered offensive to the Creator and other spirit forces.

Often, certain powerful animals were used in preventing illnesses. They were, of course, also used to counteract or cure illnesses. This latter topic will be treated in the discussion of medicine charms that follows. Regarding their use as preventatives, Allen Farmer reported to Waugh (field notes 1912) that both the blacksnake and screech owl would prevent rheumatism. The former was tied around the affected limb, while a necklace made from the humeri of the latter was dyed by boiling it in hemlock bark and worn around the neck. Anthony Day (Waugh, field notes n.d.) recommended scraping a turtle shell until a teaspoon of scrapings was obtained. This was added to a teacupful of water and drunk once a day. Repeating this process three or four times would prevent and cure smallpox. Similarly, John Jamieson, Jr. (Waugh, field notes 1912) said that the rattlesnake was used as a preventative for smallpox; while the grass snake, when held by the head and tail and chewed lightly all the way along the snake's body, would prevent the teeth from wearing out. Yet another technique for ensuring healthy teeth involved making a child who was throwing away a milk tooth say: "I will trade this for a weasel's [or other animal with strong, sharp teeth such as the mink or wolf] tooth" (Peter John, as recorded by Waugh, field notes 1914). The principle of sympathetic magic is clearly involved in this latter case. The same is true of the following bit of information, which, although not completely conforming to the concept of preventive medicine as we would use it today, does involve the manipulation of power for the purpose of influencing the outcome of some future event.

The navel string of children is taken and sometimes buried in a field, if the parents wish the child to be fond of working in the fields. If the string is buried in the bush, he will be a good hunter; if thrown in the river, he will be good at catching fish; if it is put away in a box, he or she will be always searching [Peter John, as recorded by Waugh, field notes 1914].

Besides the four general methods of prevention (1-4, above) and the use of certain powerful animals as preventive medicines, there existed three specific categories of illness that were frequently dealt with on a preventative basis. These included: (1) injuries, illnesses, or death thought to be brought about by the bites of certain animals deemed ?otkō?, most notably snakes; (2) very serious illnesses that were associated with whites, such as smallpox; and (3) illnesses thought to be brought about by neglected or offended ghosts of the dead.

In addition to snakes being used as preventatives, there also existed medicines that prevented being bitten by snakes. Both 177, *Geum rivale* (W1912A), and 176, *Geum canadense* (W1914A), were reported as being used for this purpose, while David Jack (Waugh, field notes 1912, verified by Fenton, field notes 1932) stated that "an eel-skin belt buckled around the legs will keep off rattlesnakes." A strong odor exuded from the belt and paralyzed the snakes. David Jack also told Waugh (field notes 1912) that "a circle used to be drawn with a black stick around a house to prevent children in playing from being bitten by snakes. Still done by some. Snakes are afraid because it looks like fire." John Jamieson, Jr. (Waugh, field notes 1912) reported a preventive medical technique used for avoiding "going crazy after being bitten by a mad dog." This involved "killing the dog right away by hanging him to a sapling at night."

At least six different plant medicines were used in the prevention of illnesses brought about by ghosts. Peter John, Jim Davis, and David Jack all held that 42, *Asarum canadense*, could be used for such a purpose. Jim Davis added that this medicine could also be used for anyone who was unfit to see the sick, regardless of the cause.

Nuphar lutea (43, W1914B), *Pinus strobus* (34, W1914C), and *Polymnia uvedalia* (380, W1914A, F1933A, F1940A) were also reported to be preventatives for illnesses thought to be brought about by ghosts. *Pinus strobus* (white pine), after having its leaves burned indoors so that the smoke permeated the whole house, "was considered a preventative of sicknesses of all kinds" (J. Schuyler, as recorded by

Waugh, field notes 1912). The same sort of general preventive application is seen in the use of 260, *Cicuta maculata* (W1914B).

The use of 258, *Angelica atropurpurea* (W1914C), as a preventative for the wrath of ghosts (resulting from neglecting a proper feast for the dead) illustrates the interrelatedness of individual plant medicines and the larger medical/religious ceremonial complex found in Iroquois culture. A more detailed discussion of the tenuous and culturally biased grounds for separating "medical" activities from "religious" activities will be undertaken later. It will be seen that most of what have been called "religious" ceremonies are best thought of as being preventive medicines that are carried out by groups, for groups (vis-à-vis, individuals for individuals, or groups for individuals). No better example of the role of individually applied plant medicines within a group or communally performed "religious" (equals "medical") ceremony is found than in Jim Davis's account of the use of 258, *Angelica atropurpurea* (Waugh, field notes 1914). This plant remedy was used as a medicine of last resort when the following ceremony had been neglected or improperly performed:

> Cakes, corn soup and corn bread are prepared. Pick out a speaker who begs dead people to stay away and tells them that a feast has been prepared for them. He then sprinkles tobacco in fire. A little of each food is placed on a plate for the dead and set on a table to one side. This must not be fed to dogs or chickens or they will become emaciated and die. The feast is held for only a couple of hours in the evening—no singing. Any of the food that is set aside for the dead which is left over in the morning is given to neighbors or eaten by person. At the feast all must be eaten or taken away. The same feast is used if a sick person has dreamed or imagined that he has seen white people come in hungry, with the difference being that foods are placed on the table like white folks, and English is spoken or attempted. It might be called *dohōdāsām?dōta?* ("use English language") or *gahsa?ih* ("take or eat all the food").

Certain other plants, such as 191, *Rubus occidentalis* (W1914A), and 336, *Lonicera dioica*, var. *glaucescens* (W1914A), were used to prevent infidelity on the part of a hunter's wife. A decoction of wild buckwheat (see 108, *Fagopyrum esculentum*, W1914A) was given to both the baby and its mother in order to prevent the baby from getting sick when the mother had committed adultery.

Diet also entered into Iroquois notions of preventive medicine. Lafitau (1724:369-370, as translated by Fenton 1942) reported that "diet is among them a great remedy as it is above all elsewhere; but it is not always far-fetched, general, and consists often only in the abstinence from certain meats, which they believe opposed to the disease with which they are attacked." Waugh (1916:131-133) has reported on a few of the animal meats that were to be avoided. It is not surprising to learn that "meat was absolutely prohibited when certain medicines were being administered, such as those possessed by the secret societies." One might suspect that the meat of the bear, birds, the otter, or any animal that was thought to make up the "real" Little Water Medicine would be avoided so as not to risk offending these animals. This action could be considered a type of preventive medicine.

In relation to diet as a preventive medicine, Harvey Jacobs (Fenton, field notes n.d.) reported the following information regarding food and cancer. I received much the same explanation at St. Regis in 1974.

> I believe our diet gets a poison into the system, like greasy stuff, starchy stuff, and wherever there is a bruise or lump after this stuff gets into the system, cancer starts. Wherever tender [mouth, throat, womb] it will start there. Wherever this is it has to come out of the system. If sugar or acid in system—wherever this is, it won't heal. In patient, you must give them mainly a blood purifier and a laxative first thing in the morning when stomach is empty, before breakfast. You take this tonic four times a day before meals and on retiring. After meals take a medicine that digests the food immediately —that keeps from forming acid or sugar. Drink a herb tea that is healing.

HERBALISTS/CLAIRVOYANTS

A distinction is drawn by some (e.g., Shimony 1961) between "herbal specialists" and "clairvoyants." This distinction is based on criteria involving the legitimacy and magical qualities of a medicine's power. We are told, for example, that "all medicine, no matter how common, is considered to possess 'power,'" while a distinction is made between *onōhgw?atha* (the spirit force of all herbal medicines) and another "particular" or "dangerous" type of medicine that "depends on magical or illegitimate agents" (Shimony 1961:263). Later, it is stated that "the boundary between the natural herbal preparations and the magical ones is often indistinct" (Shimony 1961:266). It is suggested here that Shimony may have ethnocentrically separated the "natural" from the "supernatural" or "magical" when, for the traditionally minded Iroquois (or for any people subscribing to animistic beliefs), such a distinction would not exist. This is an imposition of Western thought upon the Iroquois that stems from a misunderstanding of their beliefs in two regards: (1) a failure to realize that the person's behavioral history is believed to be involved in all cases of illness, even in cases of very mild illnesses; and (2) a failure to remember that for the Iroquois *everything* in the universe is thought to have power or life force.

Shimony's (1961) discussion of "particular" or "dangerous" medicines does, however, reflect traditional Iroquois beliefs regarding the "causes" and treatment of illnesses or imbalances. By imposing the natural/supernatural dichotomy upon these general types of medicines, however, the importance of this general to specific sequence of behavioral events appears to be only of secondary importance. This study holds that this distinction is the key to understanding both the causes and treatments of illnesses in Iroquois culture.

As discussed earlier, a delimitation of disease names occurs as one goes from "general illness" to "big sickness." Because it is a medicine's spirit, magical qualities, or power that is believed to be involved in *all* attempts at restoring (as well as disrupting or maintaining) spiritual balance, and since "big sickness" was thought to be brought about by the provocation of a "particular" very powerful spirit force, it follows that a more powerful, dangerous, or particular medicine was thought to be needed in order to restore balance. Particular medicines and particular illnesses necessarily required more particular or delimited names.

Probably the most significant factor contributing to whether or not an individual healer was considered an herbalist or a clairvoyant was the measure of success enjoyed by the healer as he or she treated increasingly more serious or particular cases of imbalance. It is not difficult to imagine how an herbalist who, having been *thought* responsible for restoring balance to someone believed affected by a very powerful spirit force, would become recognized as being gifted. This gift would be the ability to communicate effectively with the various spirit forces. As Shimony (1961:270) has also observed, some herbalists do earn reputations for being especially good at divining certain causal agents. The reputation of an individual herbalist would, then, be responsible for his or her being thought of as either a

clairvoyant, or simply a person with an above average knowledge of the powers of various plants. Therefore, any differences between herbalists and clairvoyants should be seen as being due to differences in degree, not kind.

The same factor of reputation through time might also qualify the clairvoyant/herbalist to be considered a witch (cf. Fenton 1940b:793). The knowledge to cure or reverse the effects of evil power would necessarily involve the potential for manipulating power for purposes of witchcraft. The herbalist/clairvoyant who was capable of counteracting witchcraft would not, however, be likely to admit such abilities. To do so would be to suffer socially.

By considering the collection techniques employed by various herbalists along with the plants used in divining certain illnesses, the distinctions that exist between herbalists and clairvoyants are further minimized. It must also be remembered that anyone who made use of the powers of plants, even at the levels of self-treatment and lay health treatment, was by implication also concerned with determining the causes of illnesses. Everyone, then, was thought to have the ability to diagnose and treat illnesses to some degree.

COLLECTION TECHNIQUES

At St. Regis I was told by my principal source of traditional medical knowledge that there are as many as 200 different ways to collect plant medicines. The same holds true for the techniques used in preparing various herbs for use. Personal techniques that proved to be effective were thought to be more effective than commonly known techniques. One collector, for example, might believe that plants should be collected in odd numbers, even numbers signifying a balance that would neutralize or inactivate the powers of medicines. Plants, therefore, are often collected in threes. Another might hold that even numbers are preferred (especially the numbers four, ten, and twenty; cf. Fenton 1949:234). All of this leads to a great deal of confusion when the idea of magical numbers is suggested in connection with the use of Iroquois medicines. The flexibility and innovativeness of the Iroquois herbalist would necessarily make a search for such numbers meaningless.

Techniques that seem less subject to individual whim include the taking of barks from particular sides of trees (most commonly, the east sides), and the respective upward or downward scraping of barks used as emetics and cathartics. The practice of boiling the most powerful plant of a particular formula first, or of giving it special recognition, may also be a fairly common practice.

Below are a few specific examples of collection techniques drawn from the Waugh and Fenton field data. The English is derived from interlinear texts. My most conservative source of cultural information made mention of such collecting invocations, but was reluctant to reveal them to a nonbeliever.

John Jamieson, Jr. offered the following information to Waugh (field notes 1912):

> Medicines should be picked in the morning. *Hawe^hní yu'* ("he whose word is good" or "God") commanded the people to use tobacco and to pick it in the morning. The picker asks or begs "power" from *Hawe^hní yu'* for the medicine, also for the water. Dead people come around in the afternoon and at night and may spoil the medicine by putting one of their fingers on it. They want you to die and join them.
>
> Plants and roots must be picked before lightning bugs appear. All kinds of insects appear at this time and spoil the roots and leaves.
>
> In gathering medicines you should use cotton. Place all the plants required of a single species on a strip of cotton cloth [sheet size]. The tobacco is sprinkled beside the first plant met and the invocation [see below] for help is repeated. This is done beside the first plant of each kind required, the roots or plants being bundled up in the cotton each time. The roots are not washed until the medicine man gets inside his own house. White cotton is used for anything wrong with the body and red for the blood, and the number of yards corresponds to the price to be paid the doctor. The invocation is as follows: "Well, I will want you—the one that will cure/help. You, medicine, are you willing to help? I want all that is mixed for this [medicine]. It is ready for the tobacco, this is my word [It is ready to pick.]."
>
> This invocation, however, is incomplete. The person's name and disease is omitted. The following invocation is just as good: "Well, this is for [sick person's name]. With some bark, you are the one to cure."
>
> The long invocation [above] is used for the strongest plant, which is sought first. A short form [below] is used for the less important plants: "You will help this principal medicine."
>
> An incantation over tobacco that is used as a cure for insanity caused by masturbation is as follows: "I beg for this . . . the way you made all people, our people. You didn't make all people·crazy. I expect that it will be as before. I give it all to you to cure this man."
>
> A medicine made from a transformed hummingbird is used both as a hunting medicine and a medicine for ill health. The invocation used for the hunting medicine is

as follows: "You will help me in this. To bring all the game to me. When I think, when I shoot, they will come to see what it is. I am ready to fire again, I will kill for everyone. For this I know you will give it the power. I know you are willing. I leave all to you, in your hands. For this is all for the good."

When this same medicine is used to cure illness, the following is said: "You are the one for this good luck medicine. You give me luck if you please, Christ. You are the one who can do it how I want. You give the power. The medicine is willing to help. I expect my desires are alright. I know it will help the people or anybody that is sick. I'll take it and give the medicine to him. They will take that. I know they will be sure to do it. No matter how sick, they will get around. They call to you to give him power. You are the one who is willing. You are willing if anyone begs for it. You are ready to give him the power that will help. The medicines are ready and willing. I leave it here now, what I am going to use."

Simeon Gibson shared these invocations with Fenton (field notes 1939):

> When a ballplayer's medicine is picked [see 224, *Cornus sericea*, F1939B], the following incantation is used: "This very real tobacco is as my message to you. They have chosen you to keep them. The young warrior men, these young and strong ones, they shall carry on the strength and they shall have a good mind when they play ball."
>
> A prayer said before a trapper's traps are set as follows: "Here this day I came with tobacco to sprinkle around the roots for you [i.e., the roots of 244, *Acer rubrum*, FndA] as I want you to help me catch the muskrats. So I will now take shoots home and boil it with the traps."

Simeon told Fenton (field notes 1939) that the Creator planted medicines so that when anyone goes for medicines he must take Indian tobacco:

> When they come to the first plant of whatever kind of medicine they are after, sprinkle that tobacco around the plants, making a speech and telling the medicine what it is to be used for. Name the person that is sick who is going to take it. When they have finished that speech then they can go on to the next plant. Pull that out [i.e., the next plant of a particular species] and take it home. Just drop the tobacco around the plant—needn't burn it. Can bury a little tobacco beside the root. Make offering at first, but skip taking it—take second one.

He was also recorded as saying that one should make separate medicines for each kind of disease from each plant.

David Jack as recorded by Waugh (field notes 1912):

> An invocation used for collecting a hunting [see 145, *Cardamine concatenata*, W1912B] medicine is as follows: "Thanks to the plants for medicine that are ready and filled. I've got to take to use them when I go hunting."

He also told Waugh (field notes 1912) that:

> all medicines have tobacco placed and covered over beside the first plant found. This is done so that the plant will know what you want it for. If a person were to come into the house and grab you and take you away by force you would not know what to do—the same way with a plant.

Jesse Cornplanter as recorded by Fenton (field notes 1933):

> The particular sickness is diagnosed by the physician before the medicine is sought. He goes out in the early hours before noon. He does not believe in storing up medicine for future use.
>
> Each medicine is sought individually and one travels until he finds it, concentrating on the medical problem he is about to solve. One passes up other medicines not of use to him. The symptoms were discovered and then medicines sought and used accordingly. We had no complicated diseases. The digestive organs were used to whole-corn foods which provided sufficient ruffage [sic].
>
> In offering tobacco, every detail is explained. So many words go with a pinch of tobacco as it is committed to the fire. The message goes into the smoke. More words follow and another pinch of tobacco punctuates the invocation. The message of the prayer rises to a particular agent. There are all kinds of spirits, but only the medicine spirits hear it through the medium of the smoke, because they alone are addressed.
>
> If I were to go for medicines I would first burn tobacco and tell the plants I was about to gather medicines. Then, all the plants would be ready for me to come. The length of the invocation depends on the power of the individual priest and the nature of the cause he is pleading. Immediately, those who are addressed commence listening. The tobacco is the medium of exchange which man has and with which he is able to procure the power of plants and animals; it is the vehicle of communication between men and all the spiritual powers. This is $oye^{n}?'gwa?õwe$—the real tobacco of the *ongwéõwe*, the real people.
>
> Wire grass for example, must be taken as an emetic only after throwing tobacco on the run and grabbing in passing a handful of the grass which is growing on the path. One must call upon the powers of the medicine to enter one's body in order that one may derive strength from the elasticity of the grass for a particular sport such as lacrosse. One should announce the name of the person who is to use the medicine. Then all the medicine people will attune their ears to the message that they may help a certain person at a certain time for a definite purpose.
>
> Supposing one were hunting boneset [see 438, *Uvularia perfoliata*] to mend a broken bone. The first plant you come to is where you place the tobacco offering on the ground but you leave that plant and go on to a second one of the same species, which you take. This tobacco offering is to the particular plant species and is secondary to burning tobacco to all the medicines before setting out to gather them.

Elijah David as recorded by Fenton (field notes 1934):

> Now, you tobacco that He released on earth to grow. He intended to create it lying on earth so that it would be helpful to us when He had finished making man. So it is, therefore, that I plead that you will help mankind. Whatever human may be sick, indeed, real tobacco that is used for pleading to the medicines is for helping those who are sick. Genuine tobacco, you all partake of it. So, then, as for you who rules, I beg of you that you will impart all your strength. I plead then that it will help them. So that is the sum of my words.

Chauncey Johnny-John as recorded by Fenton (field notes 1938):

> Chauncey Johnny John always takes along tobacco when going for medicine. It is the same thing as thanks for this world. Thanks for this medicine, we please it will help this person to get well again. We wish to God to help her.

He gave the following as being the invocation one uses when collecting plants that are to be used in the *nika:neká?a:h* (S.) or Little Water Medicine (Fenton, field notes 1938):

> Partake of the tobacco smoke here on earth. You, too, the medicine, keep us who are living on the earth right here [specifically]. Use all of your power as He, the maker, ordained you should help keep whoever is sick when we shall partake of the genuine tobacco. Be able to put all of you into action unreservedly as he intended you.

Herb Johnson as recorded by Waugh (field notes 1912):

> This kind of medicine which stands here, please help, right here, tobacco. All smoking, all water, make stout. That's all.

Peter John as recorded by Waugh (field notes 1912):

> When medicines are scarce he [Peter John] often takes 5 or 10 cents and leaves it in the hole where he takes a root out. When he comes back for more medicine he does not pay again. The money will do for this time. He says to the plant: "I have come to get you again. I hope you will do for me the same as you did before."

Baptist Thomas as recorded by Waugh (field notes 1914):

> In gathering medicines, when two went out together, they should dress up in Indian garb, and when they came to the medicines they were looking for, they should kindle a little fire.

Jackson Jacobs as recorded by Waugh (field notes 1912):

> You who lives in heaven, I ask of you the medicine that you have planted where we live. It grows all over the world. Whenever I get sick you have given me tobacco that I am to use. Now you will have a good smoke. You are to help all the people with all their diseases.

If the medicine is for a particular person, his name is substituted for *all the people*; and the same of the disease for *all diseases*. If there are several roots to be mixed, he will mention all the roots at every plant at which tobacco is placed.

K. Gibson as recorded by Waugh (field notes 1912):

When gathering medicines, should call in two or three men. Put some tobacco in the fire right at the house [this is for the "whole business"—meaning that it is for all the plants to be gathered]. After saying a prayer (below), take the gourd rattle and begin to sing [i.e., repeat the prayer]. "Help the sick, I hope it will be sure to be done. They will help, I will be sure to get better."

Levi Snow as recorded by Fenton (field notes 1933):

It tells in the *káiwi:yo:h* (the Good Message) that if one takes a medicine without telling it what it is to be used for, what it is to cure, it is no good as a medicine. It is up to the sick people to give you anything they want to for their work. Handsome Lake said you cannot ask them for remuneration.

Kate Debeau as recorded by Fenton (field notes 1933):

When collecting plants for curing an individual, tell the person's name and the disease first. If the medicine is for the curing of a private person, the prayer would be recited beside the first of each plant taken. When a lot of medicine is taken and the person taking it is not known to the one collecting it, one prayer is said for the whole thing. She does not pick the first plant of the first species sought. She comes to it, but puts tobacco of the garden variety and a piece of silver or copper beside it [e.g., Cardinal Lobelia 5¢, Mimulus 1¢]. The larger value is placed beside the best medicine. A little hole is dug beside the root of the plant. She just follows tradition or as she was taught in this. Mrs. Debeau employs the Lord's Prayer and the Hail Mary, ending with the words "imploring it" to be a good medicine, first having told the person's name to be cured and the name of the disease from which he is suffering.

John Silversmith as recorded by Waugh (field notes 1912):

When several things are gathered, you use only the tobacco for the first one.

Give us help for the sick, medicine. Help for [sick person's name].

Other noteworthy notions regarding the collection of medicines that are not concerned with acts of supplication as such, include the following:

Lightning Bugs. Fenton (field notes 1938) received from Chauncey Johnny-John the following information:

I collect medicines in the fall. I will have lots of medicines in the fall. In the spring the roots are too soft for collecting, and by fall they are harder, and it is alright to take them. Medicines should not be taken after the lightning bugs or *djistanohgwa?* [cf. *?ojistanohkwa?*,

Chafe 1967:62] appear, but it is alright to commence collecting again after their disappearance. I get my medicines early in the spring [contradicts earlier statement].

Levi John (as recorded by Waugh, field notes 1912) similarly reported that: "roots for medicines and also herbs must be gathered before lightning bugs come in the spring. After that they are not much good. This applies to medical plants appearing early or the first thing in the spring. The bugs spoil the medicine." Simeon Gibson (Fenton, field notes 1932) agreed with this remark, adding that "lightning bugs destroy the roots." He added that "you have to go early in the morning before noon. In the afternoon, the dead people are around and they would spoil the medicine." Jesse Hill (Fenton, field notes 1938) also made reference to this idea of not collecting medicines when lightning bugs were out, but no explanation was given. I received similar accounts of the effects of lightning bugs on medicines, the rationale being lightning bugs were "witch" (evil or *?otkō?*).

The Elusive Nature of Medicinal Plants. Quite often references were made by various informants to the mobility of medicinal plants. It has already been mentioned that some claim that plants can make themselves invisible when they are being sought for improper purposes. Martin Two-Axes (as recorded by Waugh, field notes 1912) made a similar statement: "Plants have the faculty of moving from place to place. You may see a plant in a certain place on one day, you may return in a short time and find no trace of them. They have gone elsewhere." Elijah David (Waugh, field notes 1912) and John Jamieson, Jr. (Waugh, field notes 1914) also reported that certain plants exhibit this ability to move about or disappear (see 228, *Euonymus obovata*, W1912A, for an example of a moving plant and 324, *Lobelia inflata*, W1914B, for an example of one that disappears).

Miscellaneous Techniques. It appears that there also exist certain idiosyncratic techniques for collecting certain plants. One such technique was that used by Jemima Gibson (recorded by Fenton, field notes n.d.) in collecting 328, *Lobelia* sp., or *egenadas:t*, "they hide" (FndA):

In the morning after the plant has stood in the water all night, take it out and go back to the place where it grew and return the plant to its place, laying it down. Then, get a fresh one. The second time take two plants. Stand them as before in cold water over night, and next morning return them to the place they grew. The third night, take three plants. That finishes the three day course.

Another plant that must be collected at night is 441, *Iris versicolor* (W1912A). This is according to Mrs. Sundown (Waugh, field notes 1912). Elijah David's method of collecting 147, *Cardamine douglassii* (W1912A), for the purpose of divining the perpetrator of witchcraft also involved collection at night.

A slight variation on the technique of placing tobacco near the first species of each plant that is to be collected is seen in Fenton's (field notes 1933, 1939) record of Sarah Snow's collection of the nineteen ingredients in her blood tonic (see 254, *Aralia nudicaulis*, F1933A):

> It takes about a pound and a half of tobacco to make this medicine. The tobacco is used at every plant that is taken for the formula. She gathers the plants during the last part of September, in the fall of the year. She always sprinkles tobacco at the side of the first plant of a variety, and this plant she does not take, but takes the next one. She makes a little hole with her hand parting the brush to clear a space on the ground next to the plant, and then covers it over afterward by placing the brush so that other people do not see it. Then she says a prayer.

The idea of there being "no one around" was also expressed by David Jack (Waugh, field notes 1914) in reference to collecting 234, *Rhamnus* sp. (W1914A). It is likely that a certain amount of secrecy would be exhibited by all collectors of herbs. The idea is that if a particular technique proved to be highly successful, it would become the exclusive property of that particular collector. Possession of especially effective techniques would, in turn, boost the prestige of individual healers.

Another technique that appears to be unique is David Jack's (Waugh, field notes 1912, 1914) practice of using red ribbons or silk in the collection of certain blood remedies. In one case, Indian tobacco was wrapped in a red ribbon and then buried between the roots of the tree. Another case involved getting on the east side of the tree ("where the sun rises"), sticking tobacco in the ground by the roots, and covering the hands with red cloth. Such techniques clearly involve the principle of imitative magic, a red ribbon being used symbolically to remind the plant of its intended use as a blood remedy. The practice of preferring barks or roots that point in certain directions may unknowingly combine

"magical" principles with more naturalistic principles. For example, certain sections or sides of plants may contain greater concentrations of certain chemical ingredients due to their receiving varying amounts of sunlight or water.

Of added interest is David Jack's method of collecting 406, *Juncus bufonius* (W1914A), for the purpose of aiding runners. The plants involved in this medicine had to be taken from a path running east and west, the direction of the old confederacy trail. They also had to be taken at precise intervals and in a definite order.

One collecting technique that I was able to record, but that was not already included in the Waugh and Fenton material, was that of not separating plants that are "married." Married plants are those of the same species that are growing close to one another. If the collector did not take both plants when collecting medicines, he or she would risk becoming ill. This would have been the result of the vengeful actions of the "mate" that remained.

It is likely that there were (and are) as many personalized techniques used in the collection of medicines as there were individual collectors. Jim Davis, for example, placed a single bead of wampum by a root of 178, *Malus coronaria* (W1914C), that pointed in the direction of the sun at noon. Both John Jamieson, Jr. (Waugh, field notes 1915) and Chauncey Johnny-John (Fenton, field notes n.d.) were recorded as using a special pointed digging stick that also served as a divining rod for locating medicinal herbs.

The important thing to remember is that the collection and preparation of herbal medicines will generally follow the familiar structured, yet flexible motif characteristic of most traditional Iroquoian cultural behaviors. Individual innovations (sometimes involving acculturative influences) in collecting and applying plant medicines will be seen to be based on the more general explanatory and organizing principles of imitative and contagious magic.

DIVINATION TECHNIQUES

Certain plants were thought to be useful in determining the causes of particular imbalances. In keeping with the processional nature of Iroquois medical practices, these divination techniques were used primarily after previous therapeutic techniques had failed. When this failure of initially applied techniques occurred, help was required in determining or pinpointing the behavioral act responsible for the experienced or observed imbalance. Everyone was thought capable of reading or interpreting certain signs of future events, such as weather signs, signs of good or bad luck, or signs of impending illness. However, because of their past successes, some individuals were thought to be especially good at reading signs, including signs revealed in dreams.

The role of the fortuneteller may be better understood if Chafe's (1967:76, 86) analysis of the Seneca word for this type of health actor is considered. It is seen that ta:ya?tówetha? is given as meaning "fortuneteller." This word is based on the verb root -oweht- meaning "distress" or "disturb," and the noun root -yá?ta(a)- (with the duplicative) meaning "picture" or "the form it is in." The fortuneteller, then, could be seen as being a person who was thought to have special abilities to read or interpret signs (including dream events) indicating future events and/or have special abilities to determine past causes for present disturbances or imbalances.

The examples of divination given below clearly indicate that *any* individual is believed capable of looking into the past or the future. Again, the degree of expertise would be the critical factor separating the professional from the lay clairvoyant.

Jemima Gibson as reported by Fenton (field notes n.d.):

> A person who is suffering sleeps on 328. *Lobelia* [see FndA], to learn the cause of his suffering. Put it under your pillow at night and dream about your sickness during the night to discover what is wrong with you. Only the sick person does this and the next morning he will know what medicine to use.

Nathaniel Hawthorne as reported by Fenton (field notes n.d.):

> Sickness and death may be determined by setting a chair for bait. Whoever sits on the chair caused the death.

John Jamieson, Jr., as recorded by Waugh (field notes 1912), gave the following divination techniques, both of which involve animal medicines:

> There is a small kind of skunk [the long-tail variety] which is "witch." This is good medicine for luck. Save the heart of this and dry it. Carry it when hunting. Put tobacco beside it and say: "I am going to save this for medicine for good luck for all kinds of game." Put it

under pillow at night and you will dream where to go for skunk.

> A particular hummingbird (*djie^n(n)ho^nwádjie^n*) [cf. *jithõwe^ntõh*, Chafe 1967:83] which is blue and brown is considered to be a witch. It will sometimes fly into a hole in a log and when you turn the log over you would find a snake, mouse or some other animal. The bird has the power of transforming itself instantly into these. The person should kill the creature found, covering it up with his handkerchief or hat. Upon removing the cover, he will find the bird once more. This bird should be cooked and dried, and then saved where no one else can see it. It should be wrapped in silk. Cut a little piece off and place it in a small vessel of water. Tobacco is employed. The person performing the divination can tell what medicine to use, if he is ill. In addition, he can tell fortunes.

Waugh (field notes 1915) received from John Jamieson, Jr. the following rather unique method of foretelling a future event:

> The old people say that some, when they killed anyone, dried and smoked the brains (*dji?seweda?*) [cf. *?oji?syõwõhta?*, Chafe 1967:63]. They noted which way the smoke blew and this was considered the right way to go to kill some more people. This was called "doubling one's brains."

David Jack (as recorded by Waugh, field notes 1912) seems to have developed divination skills in the area of love relationships. *Lilium philadelphicum* (430, W1912A), for example, was used to find out whether or not one's wife had been unfaithful. The same plant was used in deciding a case of love between a young man and woman (see also 369, *Eupatorium perfoliatum*, W1914A). Another technique for discovering faithfulness involved the suspicious man placing a root of 48, *Anemone virginiana*, under his pillow for the purpose of dreaming the truth.

For divining illness, David Jack (Waugh, field notes 1914) used 161, *Ribes americanum*, and 257, *Panax trifolius*. A drink was made from the former plant. Tobacco, along with a personal article of the sick individual, was placed under the pillow before retiring. This drink allowed the diviner to "dream of the correct medicine." The latter plant was put in water, with the sick person drinking and washing with the liquid. It was added that "if the roots sink, you will die. If they float you will get better. If they spin around quickly, you will recover quickly. If two go across the vessel and then back [like goals in lacrosse], that means that a lacrosse game must be played." A similar technique was used by Robert Smoke (Waugh, field notes 1914). *Cardamine concatenata* (see 145, W1912D) was floated in water and tobacco was burned, the "root...told what it is to be used for." This last comment reaffirms the idea that certain plants could be used to divine the causes of imbalances by *anyone,* and that plants need only be asked for this to be accomplished.

Jim Davis (Waugh, field notes 1914), after disclosing the use of 282, *Ipomoea pandurata* (W1914B), as a witching medicine, reported a technique used for divining the perpetrators of witchcraft. Divining the prognosis of a person who had been witched was also discussed. Below is Waugh's account of these specialized divination techniques.

> Had a small piece of material—square, black, thin—looked like a little piece of blackened straw or stem of some kind. This was wrapped in a lot of colored rags. It is floated in water and the reflection will tell fortunes and will show likeness of person who has bewitched you.
>
> A medicine called *u?na:yunt*—said to be taken from an abnormally small deer, where the muscle twitches after killing. Box containing it must be tied with buckskin string. A medicine is made from this after it has been cooked. A minute particle was also floated in water, into which were first placed small particles of pyrites called "gold." The medicine was said to feed on these. The water of this is drunk for almost any kind of disease of which cause is not known [this being a typically witchcraft-induced sort of illness]. Take the water in which the small particle of [deer] medicine has been floated. Can also tell by reflection who has practiced witchcraft. If medicine sinks, you will die; if it floats, you have a chance to get better. If the particle floats over one stone, then sinks, will die in a short time; if two stones, not so soon; if it goes over all the stones and keeps on floating, chances of recovery are good.

Elijah David (Waugh, field notes 1912) detailed the use of 147, *Cardamine douglassii,* as a counterwitchcraft medicine (see W1912A). He also included a technique for divining the perpetrators of witchcraft.

> Get it at night and at the last quarter of the moon. If the collector is a man, he takes a girl with him. They go, say about 4:00 p.m. and remain until dark. You will see lights where the plant grows. These will all go to one spot. Stick a stick into the ground here and depart. Return the following day at anytime and gather the roots of the plant where the stick stands. Mash these and put them into a small wooden cup and cover with a black cloth. Let it stand for a half hour and you will see the face of the perpetrator of the witchcraft. This is fixed so that it is looked at by lamp light. When you see the agent in the glass, take a needle and stick it into the eye. This will make his eye sore.

Another technique used in discovering the perpetrator of witchcraft involved "putting a teaspoon of [an unspecified medicine] into the mouth of the corpse of the person who has died of bewitchment. You will then hear of somebody else's death in a short time—namely, the person who has done the bewitching. A little liquid is also put on the clothing of the corpse" (source unknown).

David Jack (Waugh, field notes 1912, verified by Fenton, field notes 1932) reported the following method of divining one's life span. It does not directly involve plants, but is worth noting simply because it supports the idea that everyone was thought capable of reading signs of future events. "A corn husking pin made of a bear's bone is used when a young man or woman is 17 years old. They try to bend it in their fingers. If it breaks easily, they will die soon. If it bends, they will live long."

More traditional methods of scrying were also shared with Fenton (field notes 1938). Albert Jonas (reporting on the techniques used by Levi Baptiste) gave the following three divination techniques:

> (1) Indian tobacco. Puts in tea saucer, fills ½ full of water, grinds up tea in palms of hands. Puts it in water a few minutes. Leaves on table and covers it for a few minutes. Goes back, uncovers, and is careful not to disturb water. Looks and tells the pictures on it. He says different things about it. If you are not feeling well, he might tell you you have to have *yèi?to:s*.
>
> (2) *jiske:h* (ghost) he uses. He burns the tobacco and talks to the ghost if there is anything the person wishes to find out. They used to sit in the dark at night. There is a person on each side of him. Position the patient in front of him, holds hands, keep quiet for 5 minutes. After 5 minutes, fortune teller starts to tell the subject everything. Joins his hands before the lights go out. Does not get rigid. Burn tobacco first, talks to ghost as to what he wants to find out about the patient. "You tell me what is to be done if a man or woman is not feeling well. Tell me what must be done to get well again." Then sit down and hold hands and put light out. Hold with one hand. Quiet for 5 minutes, then commences to tell: "I see a man or a woman" and the ghost tells them who are inside not to be frightened.
>
> (3) Tea method. Pours 5 leaves into cup. Drains water out slowly leaving tea leaves in cup. When empty, put teacup upside down a few minutes. Turn it up and look at leaves. Start telling what to be done. Predicts sometimes in this way. They say he can tell it pretty close and sometimes he is right.

Although my principal source of information regarding the use of medicines was usually reluctant to discuss traditional ways in any great detail, he did offer the following information dealing with the diagnosis of imbalances. The English word that he used to describe himself when he was engaged in such practices was "pathfinder."

> A patient puts medicine on a plate. Then I pour hot water on it and read it. I can also diagnose sicknesses from hair samples sent to me in envelopes.

There is also Beauchamp's (1922:33-34) classic account of the use of "The Great Medicine" or *nika:neká?a:h* for purposes of prognostication.

> A wooden cup is taken to a running stream, and filled by dipping down the stream. When brought back to the house it is placed near the fire, with some native tobacco. There are prayers while the tobacco is gradually thrown on the fire. The smoke is grateful to the Great Spirit, and with this American incense their prayers rise heavenward. The medicine man then places a piece of skin near the cup, and on this the medicine is laid. He takes up a little of the pulverized compound with a wooden spoon and dusts it on the water in three spots in the form of a triangle. This is closely watched. If it spreads over the water and whirls about the surface, the sick person will recover. If it sinks at once he will die and nothing can be done. In the one case the medicine is given; in the other all the water is thrown away.

It is important to remember that the power possessed by the fortuneteller was and is simply an extension of powers thought to be possessed by everyone, with the clairvoyant being distinguished by "knowing a lot of things" (Rioux 1951:155). To Shimony (1961:270), "it is not at all clear how the practicing fortunetellers become aware of their own powers." To Parker (1928), it is apparent that it is *others*, vis-à-vis the fortuneteller him- or herself, that is the critical factor in determining this status position. If a particular individual exhibited even a modicum of success, it is likely that he or she would be consulted in the future. This would be especially true in cases involving highly individualized techniques of divination that refined or went beyond those known to nonspecialists. It was, as Parker (1928:9) pointed out, "the *formula* that counted for success," not any one specific herb or ritual.

Those in our culture who might attempt to explain the success of traditional or folk health actors in terms of a scientific model might also rely upon the behavioral concept of "erroneous cause-effect pairings." That is, a particular formula employed by a health actor in attempting to restore balance to a sick person might become accidentally paired or associated with the recovery of the patient. Although this is true in terms of a behavioral model, the overall effectiveness of folk medicine derives from the anxiety-reducing effects of beliefs and expectations. Medical beliefs and expectations in

general would consequently be a function of these accidental associations as they become integrated into the overall cosmological scheme of the people under investigation.

With regard to the effectiveness of clairvoyants specifically, Shimony (1961:270) has remarked that "a convincing fortuneteller may be very effective in helping to cure an ailment that is psychogenic, and a record of past successes will enhance his reputation, thereby increasing his effectiveness in dealing with a new psychosomatic case." It should be added that this would not only be true in cases of psychogenic and psychosomatic illnesses. It would apply in all cases of illness because all illnesses involve a psychological component. Similarly, all methods of treatment are, to some degree, accompanied by the placebo effect (cf. Frank 1961; Herrick 1976).

It was this heavy reliance upon novel combinations of techniques by the fortuneteller, as he or she went about discovering the various causes of distress or disruption, that might account for the failure of ethnographers to isolate any single systematic procedure used by all clairvoyants. The wise and successful diagnostician was one who was able to see any or all possible relationships existing between the condition of the patient and the myriad of things and events that might somehow be used to restore balance. Thus, the varied use of dream content, tea leaves, cards, psychoactive drugs, interrogation, scrying, or any other individualized techniques should be expected (cf. Rioux 1951; Fenton 1953; Shimony 1961). Furthermore, one would expect to find the Iroquois readily accepting and adopting new techniques of neighboring communities. The same would hold true for techniques employed by Europeans. Timing would be seen to play a large role in the acceptance or rejection of introduced techniques. A record of recent failures on the part of the most renowned fortunetellers of a particular community might, for example, make that community particularly receptive to new curing methods.

WHITE DOCTORS

A good example of the processional nature of Iroquois medical treatment is found in the roles that white doctors have played in health care. The use of such healers may also be seen as reflecting the overall receptivity of the Iroquois to innovative medical techniques.

Rioux (1951) investigated the circumstances under which the Iroquois consulted white doctors or clairvoyants. He also studied the conditions under which the services of both were sought. He noted that "people usually resort to the fortune-teller for illnesses that appear mysterious to them, mostly internal diseases with many symptoms" (Rioux 1951:156). If, however, the disease was external and well diagnosed, "the patients consulted the doctor first before going to the fortune-teller or to the witch" (1951:156). Of course the availability of nearby doctors or clinics would also enter into the selection of a traditional versus a nontraditional healer, as would the religious beliefs (e.g., traditional vs. Christian) of the patient in question. The most telling feature of this white doctor/clairvoyant selection process from a cultural point of view, however, is the fact that in certain instances some individuals used (and still use) both. This "double-security system," as Rioux (1951) has called it, would be typical of the traditional Iroquois mentality that stressed innovation and flexibility within loosely structured patterns of behavior (cf. Weaver 1967).

In the opinion of some of the more conservative members of Iroquois society, however, this reliance upon both native and the white man's medicines was seen as having deleterious effects upon the sick individual. This was especially true in cases involving a patient who had first gone to a white doctor, received no relief, and then called upon the for-tuneteller as a last resort. A good example of this practice of turning to more traditional medicines (not necessarily involving a clairvoyant) after white medicines had failed is found in the account of Charles Gordon (Fenton, field notes 1939):

> I had been sick a great deal. When I first took sick I walked around, gradually getting weaker, then bed ridden, then could not eat. Dr. Flat used to come daily from Coryda. I took tablets every ½ hour . . . did me no good. Encourage me (?) I knew old people had used *ojidjowa:* [see 319, *Veronicastrum virginicum*, F1939B]. I told my wife that I will not live longer. Told her to fix that. . . . That's what I took three mornings she give me that. That will make you physic [laughed]. Next day I got up and walked around again.

If, after reverting to more traditional medicines, the patient was still not cured, "he was apt to think

that he had not enough confidence in the traditional dances and feasts and blames it on himself for not being cured" (Rioux 1951:157). Snyderman (1949:219) documented a similar attitude:

> Ed Curry, the present keeper of the "Little Water Medicine," told me of a man who had been given their medicine but who would probably not get well because he had first tried all of white man's medicine and called for "real medicine" only as a last resort. He decried the fact that many Indians "did not have faith" in their own medicines and used them only when it was too late. "People," he said, "had faith in the medicines, so the medicines worked."

Snyderman (1949:218-219) received from Lena Snow the same sentiments: "It's a shame that the Indians no longer utilize their own medicine." She added that the present ineffectiveness of Indian medicine was due to the "lack of faith in their power."

Before moving on to the other general categories of health actors or activities, it is useful to look at a few examples of how the elements of "faith" and "confidence" enter into Iroquois medical practices. The words of Jesse Cornplanter (as recorded by Fenton, field notes 1933) may be seen as representing the general attitude held toward medicine in Iroquois culture:

> The patient must be in a certain frame of mind—it must be flexible. He must work in conjunction with the medicine. He must have faith in its power in order to help it. In the Indian philosophy of sickness, it is thought that one's mind must be freed of worry and distrust in order that the patient may get well.

FAITH, CONFIDENCE, AND PEACE OF MIND MEDICINES

As discussed in chapter 2, there is good reason to believe that there are psychotherapeutic benefits associated with treatments involving both "physical" illnesses and behaviorally induced "mental" illnesses. It is significant to note that traditional Iroquois theories regarding illness and treatment included a recognition of the importance of anxiety reduction. Cleansing the mind or spirit of fear, stress, and anxiety would necessarily enhance built-in, biological mechanisms of recovery and defense.

Related to this process of anxiety reduction is the often expressed, generalized belief that "everything is good for something." Fenton (field notes 1938) has recorded the following relevant statements made by Floyd Johnson of the Allegany Reserve:

> I cut myself with a saw several years ago going up the hill west of the house. I reached over and stripped off a handful of leaves [anything]. I chewed to a pulp, applied, tied it with a handkerchief. It burnt. I waited. Finally, it stopped bleeding. I adjusted the bandage. I went up and got the tools and went home. When I took off the poultice, it had adhered to the wound. I rebound it with hot water and in three days it was alright. I never discovered what leaves I grabbed.

If one believes that everything is good for something, and if one expects all plant substances to have curative powers after they are told what they are to be used for, these beliefs and expectations will, theoretically, play as important a role in the process of recovery as any of the so-called "practical" benefits of treatment. This is irrespective of the empirical, physiological, or biochemical effects of the plants. This principle of "faith" and its anxiety-reducing effects could also account for the effectiveness of panaceas prepared from several (or perhaps *all*) plants or animals. The Little Water Medicine, originally made up of all or several animal-derived medicines, is one such powerful panacea.

An example of a plant-derived cure-all is found in David Jack's (Waugh, field notes 1912) account of a medicine made from the finely shredded or powdered barks of "nearly all" trees. It was to be used in open spaces, a big tablespoon being steeped in two quarts of water. A dose consisted of "drinking it all," followed by rest. It was said to "remove soreness" and, because it was ordered by *He-wen-i-yu* ("God"), it "is the oldest of the lot." This notion of a powerful cure-all having the ability to relieve vague or generalized discomfort may be seen as conforming to the more general relationship existing between nonlocalized, internal illnesses and powerful causal agents. Again, a powerful cause necessitates a powerful restorative medicine.

Also involved in many of the medicines used by the Iroquois is what we might call the "power of positive thinking." These "confidence" medicines were numerous and just as real and effective as any other kinds of medicines. It should be remembered that for the traditional Iroquois "medicine" simply involved the exertion of "power" for the purpose of influencing future events or things in either a beneficial or harmful way. Examples of such medicines follow.

For added strength, Jim Davis recommended a medicine made from 255, *Aralia racemosa* (W1914B). Similarly, 282, *Ipomoea pandurata* (W1914A), was said to be so powerful that "if you touch it and strike someone, you will kill them." This latter plant gave a hunter so much strength that he could "carry two deer with ease." Because of this strength, however, it is believed unfit to be touched by children.

If the root of 109, *Polygonum arenastrum* (W1914B), had a circle of stems attached to it, the root could be dried and powdered and kept on one's person for the purpose of making one seem like the best looking in a crowd. Jim Davis reported that the same properties were characteristic of 323, *Lobelia cardinalis* (W1914B). These plants are considered "love medicines" that could be applied to oneself. The former, however, also affected others when the powder was surreptitiously put in someone's tea. The suggestive influences of these medicines could be seen to parallel the "sex appeal" oriented cosmetic industry in contemporary American society. They instill in the individual a degree of confidence that is often necessary for successful interpersonal relationships.

In some instances, a hunter would not only be given physical strength and well being through the use of a particular plant medicine, but would be given "mental" strength or "peace of mind" as well. *Rubus occidentalis* (see 191, W1914A), for example, was used "for when a man is hunting and his wife is fooling around." Once the hunter and his wife had taken this medicine, "she will then stay home no matter if he stays away a year." For obvious reasons, it is unlikely that a wife would ever make the claim that this medicine was ineffectual.

David Jack similarly told of a medicine that could be used in the event that a man's wife did "run off and wouldn't come back." This medicine (see 86, *Quercus bicolor*, W1914A) was "made by the person's folks who are left." It "takes away any lonesomeness caused and destroys the effect for the one who has run away."

Yet another type of "confidence medicine" involving "positive thinking" is seen in John Jamieson, Jr.'s use of 72, *Hamamelis virginiana* (W1914B). This medicine was used for "when a pregnant woman has fallen or got hurt in any way and fears child will be born prematurely." In this case, the suggestive influence of the medicine (viz., take this medicine and you will not lose your baby) may be in operation. This would serve to decrease anxiety and allow normal physiological functioning in the expectant mother. The "pragmatic" effects of the pharmacological properties of this plant (*Hamamelis* or witch hazel, which is an astringent that acts upon the blood vessels of the uterus) would also have to be considered.

In the application of what have been called "confidence" or "peace of mind" medicines in this section, and in the application of all "medicines" in Iroquois culture, the element of individual and group *faith* is in operation.

COMMUNAL MEDICAL PRACTICES

The medical beliefs and practices of the Iroquois were remarkably similar in many ways to those of nineteenth- and early twentieth-century medical science. This was especially true in regard to the use of drug materials. The following writings of Pierce (1875:383-384) further reveal similarities between traditional Iroquois theories of therapy involving a "freeing the mind of worry and distrust" and those of physicians of his time. Such attitudes do not presently prevail within medical science, but there appears to be an increasing appreciation of the therapeutic benefits deriving from the anxiety-reducing effects of any sort of therapy (cf. Frank 1961; Kanfer and Goldstein 1975).

> Mental impressions made upon the sick exert a powerful influence upon the termination of disease. The chances of recovery are in proportion to the elevation or depression of spirits. Pleasant, cheerful associations animate the patient, inspire hope, arouse the vital energies and aid in his recovery. While disagreeable and melancholy associations beget sadness and despondency, discourage the patient, depress the vital powers, enfeeble the body and retard recovery.
>
> Unless persons, who visit the sick, can carry with them joy, hope, mirth and animation, they had better stay away. This applies equally in acute and chronic diseases. It does not matter what a visitor may *think* with regard to the patient's recovery, an *unfavorable opinion should never find expression in the sick room*. Life hangs upon a brittle thread, and often that frail support is *hope*. Cheer the sick by words of encouragement, and the hold on life will be strengthened; discourage, by uttering such expressions as, "How bad you look!" . . . etc., and the tie which binds them to earth is snapped asunder. The visitor becomes a *murderer*!

Regardless of what type of communal medical ceremony was being performed, it was the uniting of minds or spirits in a show of support for the patient that served as a basis for such activities. "Medical" acts in this case mean those concerned with the manipulation of power for the purpose of restoring,

maintaining, or upsetting spiritual balance to a spiritually balanced/imbalanced person, place, thing or event. The principle of "faith in medicine" was simply fortified or maximized by group participation (cf. Shimony 1961:275). As Pierce (1875) might say, the spirits of the patient would be lifted so that the tie that bound him or her to earth would not be snapped asunder (cf. thematic statement number 10).

We would expect to find communal ceremonies centering around chronic or wasting diseases or symptoms such as rheumatism, fevers, headaches, paralysis, and so forth. These diseases or symptoms involve a generalized pain or discomfort that is subject to alleviation or removal resulting from the countersuggestive or placebo effects of various ritualized therapeutic techniques. Periods of remission, or recoveries that are enhanced through the reduction of distress or anxiety, are then logically paired with a particular ceremony or ritual, thereby strengthening the cultural belief in their effectiveness.

Relating these communal medical practices to the value orientations and themes discussed in chapter 1 (see thematic statements 6, 9, 14, 17, and 19) is Shimony's (1970:246) remark that "the ideal culture stresses the theme of being an Indian with the expectation that the individual will feel toward his fellow Iroquois a strong sense of in-group loyalty ('we must be of one mind, one heart, one body'). Thus, one should display the characteristics of accommodation, concession, helpfulness, and reciprocity, and finally there should be a consensus or unanimity of minds." The same notions extend into the realm of medicine. Appropriately, the Seneca concepts for "help" and "cure" are represented by a single concept comprised of the combination of the verb root -*keha*- and the noun root -*yá?t(a)*- (Chafe 1967:65).

It is likely that *orenda* was the concept used to represent this process of conjuring up power through the uniting of minds or voices. The Huron form of the Seneca concept for "song," for example, is *?oeⁿnõ?*, "its song," (Chafe 1967:50) The technique for doing this clearly involved chanting or singing in some cases. Such activities possibly served the purpose of providing the patient with a monotonous stimulus with which to interact, thus heightening relaxation and suggestibility to the communally expressed wishes for recovery and well-being.

If one looks at the Seneca concept meaning "to perform the Dance for the Dead" (*-ahkiwe-*) and the noun derived from this verb root meaning "Dance for the Dead," "Feast for the Dead," or "Chanters for the Dead" (*?ohki:we:h*), a very subtle example of the link between communal singing and dissociational states is found. Dissociational states feature behavior characterized by intense attention to a monotonous stimulus and heightened suggestibility. With the progressive morpheme, this verb root concerns itself with communal singing. The same concept, however, is used in reference to someone who is intoxicated (Chafe 1967:37). For further information regarding this connection between altered states of consciousness and the medicine societies of the Iroquois, see Lex (1973). The existence of a specific Seneca concept meaning "to benefit (someone) through a medicine ceremony,"*-at(C)e$^{n'}$h-/-at(C)e^{n}-*, further attests to the importance of such activities to the Iroquois (Chafe 1967:42, 54).

THE RELATIONSHIP BETWEEN INDIVIDUAL AND COMMUNAL MEDICAL PRACTICES IN TRADITIONAL IROQUOIS CULTURE

Three basic types of interactional situations existed in traditional Iroquois culture. First, there were cases that involved an individual seeking the restorative help of another individual, the latter individual being a lay or professional health actor. When the individual health actor was of professional status, the patient assumed a much more passive role in treatment. He or she then relied on the special abilities or powers of one thought to possess varying amounts of skill at manipulating or communicating with spirit forces. It was this specialist who performed the rituals necessary for restoring balance to the patient. How he or she went about performing these rituals has already been discussed in this chapter under the heading "Herbalists/Clairvoyants."

It was the clairvoyant who was responsible for diagnosing the cause of an imbalance and prescribing the required treatment for the observed imbalance. This was done *after* a particular stage or kind of illness was determined on the bases of symptomatic criteria and the behavioral history of the patient, including previously tried treatments. The clairvoyant might then sustain the patient's theory of disease causation, such as taboo violation or unfulfilled

desires. Alternatively, he or she might divine another etiology, such as contact with a thing or event deemed *?otkō?*, or witchcraft. By the time the patient had felt it necessary to consult a clairvoyant, the symptomatic discomfort being experienced would be great enough to warrant theorizing that (1) the taboo broken was very serious, (2) the unfulfilled desire was such that it must be professionally discovered and fulfilled, (3) the thing or event deemed *?otkō?* must be neutralized or removed through a counteractive measure, or (4) counterwitchcraft must be employed. If cases 1, 2, or 3 were divined, the clairvoyant might then (1) prescribe an appropriate propitiatory act to be performed by the patient (in cases of taboo violation); (2) prescribe an appropriate propitiatory act to be performed communally, such as a Feast for the Dead, the Skin Dance, Dew Eagle Dance, or other rites of any ceremony; (3) recommend that the patient perform certain dream fulfillment rites him- or herself; (4) recommend that the patient participate in a communally performed dream guessing/fulfillment ceremony, such as those carried out at Midwinter Ceremonies; (5) prescribe a medicine designed to counteract the effects of *?otkō?* upon the patient; or (6) prescribe a communal ceremony designed to counteract the effects of *?otkō?*. It is when the clairvoyant prescribes communally performed therapies that we may begin to consider the second and third types of interactional situations: an individual seeking aid from a group, and a group seeking aid from a group. From a cultural perspective in which medical matters are distinguished from religious matters, the third type of situation would constitute religious behaviors.

Again, caution must be taken in thinking that these therapies are mutually exclusive. As in the case of notions of disease causation, the treatments used in restoring balance must be seen as being interrelated. They may also vary with the patient's history and the imagination and skill of the clairvoyant. The violation of a serious taboo could, for example, be equated with "coming into contact with a thing or event deemed *?otkō?*" if the thing or event encountered involved the improperly supplicated spirit of the otter or a neglected False Face. It is even possible that through the malefic exertion of power an individual (i.e., a witch) could deceive the spirit of the otter and make it believe that an innocent fisherman had in some way offended it. The otter is very powerful and "particular," and is therefore a potential source of evil. At this point, taboo violation, a thing or event deemed *?otkō?*, and witchcraft would become possible theorized etiologies. The skilled clairvoyant would be aware of these possibilities. In the event that the Otter Society or other communal therapy was of no help in restoring balance to the patient, the clairvoyant might resort to techniques of counterwitchcraft (case 4, above) and either perform these rites him- or herself or recommend someone who could.

Despite the fact that there existed a myriad of circumstances or conditions that went into determining the prescription of various therapies for the range of possible, theorized etiologies, the same basic pattern of behaviors was involved in all therapeutic options. In cases involving an individual seeking treatment from a group, and in cases involving a group seeking treatment from another group, the sequence of behaviors carried out roughly approximated those carried out by an individual herbalist/clairvoyant in treating an individual patient. Confession on the part of the patient, acts of supplication and the thanking of various spirit forces, naming the illness or discussing the theorized cause of the illness (when known), naming the patient (when known), preparation and/or application of the medicine, or the performance of a medicine ritual or ceremony, and the repetition of thanks all made up the generalized sequence of events followed by both individual herbalists/clairvoyants and the various communally performed treatments. Finally, just as there existed variations in individual techniques employed by individual clairvoyants/herbalists, both between and within local communities (cf. Fenton 1940b:793), there also existed variations between and among local communities as they performed various communal medical practices (cf. Fenton 1936:4, 1951). A discussion of these various communally performed medicines is presented in the paragraphs that follow.

DOINGS

Included in the category of communal medical practices referred to as doings are such social events as games, dances, feasts, adoption ceremonies, songs, and other rituals such as the tobacco invocation.

According to Wallace (1972:72) even social dances may have curative functions. All of these activities might or might not necessarily be performed at regularly scheduled Longhouse ceremonies (Shimony 1961:274-281). Speck (1949:60), for example, reports that the War Dance, although more likely to be performed in the summer during the Thunder Ceremony, was also used "to make ceremonial friendship for a child who is sick when the Medicine Man prescribes 'need of a friend' to give relief." The same dance, when performed at the Midwinter Ceremony, "is a cure for a variety of ailments, including vague disposition" (Kurath 1968:50). The Dew Eagle Dance, a rite of the Eagle Medicine Society, might be performed to create a ceremonial friend for someone in need of friendship. This friendship bond, in turn, was believed to cure and prevent serious illness (Wallace 1972:78). In conforming to the general Iroquois way of doing things, flexibility within structure dominates. The actual performance of these doings depended on such factors as the degree of imbalance being experienced by the patient and his or her economic situation. The patient who was diagnosed as needing a communal event of this sort could postpone the event to a time when a regularly scheduled performance of it was due. This was possible when the patient's condition was not deemed critical. On the other hand, if the patient lacked the necessary resources or was critically ill, a rush ceremony might be performed on demand (Shimony 1961:275). The patient receiving benefits from such hastily performed, often incomplete, medical ceremonies was required to vow that it would be conducted properly as soon as possible. Failure to do so was considered a violation of the established ways and a possible cause of future imbalances.

Even though communal medical ceremonies in the doings category involved an individual receiving help from a group, both the individual *and* the participants in the social event were thought to benefit from its performance. Communal guessing at Midwinter, for example, was not directed at merely appeasing the frustrated souls of individuals; they were also concerned indirectly with the well-being of the community or nation as a whole (cf. Blau 1963). Thus, such medicines could be considered preventive in nature for some participants. Furthermore, such medicines could be requested by an individual with or without consultation by an herbalist/clairvoyant. They could also be arranged by others for an individual or for an entire group.

To appreciate the interrelationship of all medical activities, it is useful to look at some of the plant-derived medicines used specifically in the doings that involved games. Many game medicines were designed to give strength or magical abilities to individual participants. Overall, however, the preventive or restorative aspects of these types of medicines were gained from the communal act, or from participation in the communal act. We find, then, an overlapping of individual-for-individual medicines, group-for-individual medicines, and group-for-group medicines.

MEDICINE SOCIETIES

Also straddling the group-for-individual and the group-for-group categories of medical activities is that class of communal medical practices called "medicine societies." Traditionally, these societies include the Otter, Bear, Husk Face, False Face, Little Water, Pygmy, Mystic Animals (Shake Pumpkins or Medicine Men's), Buffalo, and Eagle societies. Parker (1909) also included the Chanters for the Dead or ?ohki:we:h (Chafe 1967:37), the Women's Society, and the Our Sustenance or tyōhéhkōh (Chafe 1967:77). Many students of Iroquois culture would not include these last three ceremonies under the heading "medicine societies," although Parker (1909) felt justified in doing so. The basis for the confusion surrounding what is truly a religious as opposed to a medical matter (cf. Shimony 1961:277) is found in the already discussed, ethnocentric conceptualization of medicine. If one subscribes to the thesis that, for the Iroquois, "medicine" referred to the manipulation of a universal power on a group or individual basis for the purpose of maintaining, restoring, or upsetting the spiritual balance of any person, place, thing, or event, then questions dealing with whether or not a particular ceremony is religious or medical in nature becomes a problem only for those who already make the cognitive distinction between these two types of activities (cf. Homans 1941). The same would hold true when considering the imposition of magical/religious distinctions upon most other cultures of the world by members of Western culture. If the distinction is dropped, inclusion of the Chanters for the Dead, the Women's Society, and the Our Sustenance Ceremony under the heading of "medicine societies" or "medicine" becomes understandable. They are all clearly involved in maintaining and restoring the general welfare of individuals or groups of individuals. Therefore, they are good medicines. The spirits of the dead, women, and the entire League are the beneficiaries of such activities. Similarly, the Bush, Maple, Thunder, Strawberry, Raspberry, and Green Corn ceremonies may be considered both preventive and restorative communally performed medicines.

When what are usually considered religious activities are performed at regularly scheduled times they, like the medicine societies and all other calendrical ceremonies, take on the characteristics of preventive medicine. Certain linguistic evidence further supports the idea that religious ceremonies were equated with medical activities in the minds of the traditional Iroquois. The Seneca concept for "Longhouse officials" (teyōkhiya?towéhtanih, meaning "they deliberate for us") is based upon the same concepts used in reference to the "clairvoyant" (Chafe 1967:76). Broadly speaking, the act of reciting and remembering the various spirit forces in the kanō':nyōk may be considered a preventive medicine.

If one accepts the position that the Iroquois also conceived of people in general as medicines (especially when brought together for the purpose of uniting their minds), then it would be fair to speculate that women were thought of as being more powerful sources of medicines than were men. Such a statement may be made because of the important economic, social, and political roles played by women in traditional Iroquois society, along with their reproductive function. The existence of a specific Women's Society (called deswadenyationdottu' by Parker 1909) supports this view. This is the society that "preserves the ritual by which good fortune and health are obtained for women" (Parker 1909, cited in Fenton 1968:126). Communal performance of these rites for the benefit of women as a whole would make the Women's Society more of a religious ceremony in the usual anthropological sense of the word. It is likely that such rites were performed on demand for certain individual women at certain critical points in their lives. Critical points would be at times of birth, at times of infertility, and so forth. Such uses would certainly make them more medical in the usual sense of the word. Lacking ethnographic evidence, one might predict that group-for-group, group-for-individual, and individual-for-individual use of the rites of the Women's Society existed in much the same way that it did for the other medicine societies and/or doings.

I said earlier that acts of hospitality and reciprocity were greatly valued in traditional Iroquois culture. Both could be subsumed under the broader sharing ethic. In no other aspect of Iroquois life was this sharing ethic more prevalent than in the realm of communal feasting. The act of giving food was in itself an act that symbolized the spirit of communal sharing. Linguistic support is found for this in the Seneca verb root meaning "to give food to" or "to feed," -nō(:)te^n-/-nō(:)t-. With the reflexive morpheme, this verb root comes to mean "to be generous" (Chafe 1967:72). In the broadest sense, sharing and generosity are, like food itself, medicines that maintain and renew the gift of life. This sharing was to take place not only among the earth-dwelling Iroquois people themselves, but was also expected to occur between human beings and the various spirit forces in and beyond the sky. Sacrifices, the ceremonial unification of minds, the feeding of various ritual objects such as charms or False Faces, the feeding of the dead, and so forth, serve as means to these ends. Related to this is the Seneca concept of ?oiwa:níyehkōh that literally means "it is used for solidifying the ceremony." This concept is based on the verb root -(h/:)niy-/-(h/:)niye- meaning "to be solid" and the noun root -(C)í:w(a)- meaning "ceremony, message, or word" and is used in reference to the food provided at a ceremony (Chafe 1967:54). Similarly, kahsá?ō:?, which is built on -hsá?-/-'s?a-, "to eat" and means "all eaten up," is the name given to a particular ceremony for the dead (Chafe 1967:56).

In the individual-for-individual mode of medical treatment, we find an individual lay or professional health actor treating an individual patient. Herbal medicines were and are primarily used in this mode. The group-for-individual mode involves individuals seeking help from the performance, on demand, of a particular rite that is but one segment of a larger ceremony, which is in turn normally performed at regularly scheduled intervals. In the case of what

have been termed "doings" by Shimony (1961), the sick individual extracts, on demand, certain rites of the regularly scheduled Longhouse events for the purpose of mustering social support and/or pleasing certain spirit forces. Although requested by the patient him- or herself, or by the friends or family of the patient (with or without clairvoyant participation), all participants are thought to benefit from these events. Plant medicines may also be involved here.

It is also possible that on the advice of a clairvoyant, an individual could request, or have requested for him- or herself, the performance of a particular rite by the Bear, Buffalo, Otter, Pygmy, Eagle, False Face, or Husk Face societies. The same herbal medicines could be used in conjunction with any of these rites. The Women's Society, the Feast for the Dead, Our Sustenance, and any of the other so-called religious ceremonies would be included here provided that an individual was permitted to extract particular rites and apply them on an individual basis. These ceremonies, like Longhouse-derived doings, were also performed at regular times and were open to the public. On ceremonial occasions their functions went beyond the curing of individuals. They also provided thanks for and renewal of cures already effected by them.

An individual cured by a medicine society was required periodically to renew the ritual so as to prevent recurrence of the original illness (Shimony 1970:241). The same principle held true for the performance of rites usually referred to as religious ceremonials, thanks and renewal being critical to their future effectiveness.

It is in considering the Little Water Medicine Society and the Society of Mystic Animals that we begin to find group-for-group concerns coming to dominate over group-for-individual ones. The former was and is performed privately by members only when treating individuals, while being open to the public when performed at regularly scheduled renewal ceremonies (Fenton 1936:17).

The Society of Mystic Animals (yi:?to:s) is also performed by members only, membership being comprised of all previously cured members of the various medicine societies. It is best understood as a society having as its main goal the propitiation and supplication of the animals (the bear, buffalo, otter, and eagle) who first revealed the uses of plant- and animal-derived medicines to human beings. The medicine animals were also associated with the Little Water Society, the society that honored and renewed the powerful *nika:neká?a:h* (cf. Fenton 1979). This medicine consisted of a secret admixture of animal-derived powders and water. It is extremely rare and perhaps nonexistent in modern times, but there does exist a substitute for the "real" *nika:neká?a:h* made from plant materials. The reader is referred to *Mentha spicata* (301, F1938C) for the ingredients in and preparation of this medicine. The primary goals of both the Little Water Medicine Society and the Society of Mystic Animals are thanks, renewal, and prevention, as opposed to the restoration of balance of any one individual or group of individuals. However, the latter function might sometimes operate. The occasional use of the Little Water Medicine or *nika:neká?a:h* on an individual basis indicates that certain rites or medicines might be extracted on demand from these ceremonies as well.

As one moves from the two powerful medicine societies to a consideration of the False Face, Eagle, Bear, and other societies, curing powers are seen to decline. The same is true as one moves from regularly scheduled ceremonies, involving group participation, to the extracted, communally performed segments of those ceremonies called "doings." Finally, there are the least powerful medicines and rituals of individual herbalists or clairvoyants who rely heavily upon the use of plants.

The Otter Society may be thought of as being a more accessible branch of the highly secretive Society of Mystic Animals. Although principally concerned with propitiating the tutelary of small water animals and fish, it has the additional function of influencing the health, fortunes, and destinies of male human beings (Parker 1909, in Fenton 1968:121). The otter was thought to be a member of both the Mystic Animal and the Little Water societies. The unique position of the otter in nature may account for its membership in both of these powerful societies. The otter is both a land and a water animal, as well as its being both a hunter and a fisher. This is especially significant when it is considered that water was thought to be, in itself, a medicine or source of power (cf. Mooney and Olbrechts 1932). Mooney's (1890:335) statement that "Bathing in the running stream or 'going to the water', as it is called, is one of their [the Cherokee]

most frequent medico-religious ceremonies, and is performed on a great variety of occasions, such as at each new moon, before eating the new food at the green corn dances, before the medicine dance and other ceremonial dances before and after the ball play, in connection with the prayers for a long life, to counteract the effects of bad dreams or the evil spells of an enemy, and as part of the regular treatment in various diseases" supports this idea. The restorative or revivifying qualities of water to living things clearly made it a source of power to the Iroquois. Mythologically, it was the dew from the hollow of the eagle's back that revived the scalp of the Good Hunter. The Dew Eagle Ceremony of the Eagle Society symbolically commemorates this event. Furthermore, springs, rivers, and lakes were thought to be medicines for the living, ever-expanding world (Foster 1974:296). Contradicting this conception of water in traditional Iroquois culture, however, is Foster's (1974:103) statement that "water was neither a medicine nor a food." However, I infer that water was both a medicine and a food. Likewise, solid foods were considered medicines for the earth's growing things as well as for various spirit forces. The practice of feeding masks, spirits of the dead, and medicine charms illustrates this idea of food as medicine or food as a restorer of balance and, as discussed above, it was food that was used for solidifying Iroquois ceremonials.

Linguistic evidence for the idea that water is a fundamental type of medicine for the Iroquois is found in the verb root $-at(C)e^{n'}h-/$ $-at(C)e^{n}-$ (Chafe 1967:42). This means "to benefit (someone) through a medicine ceremony." When combined with the nominalizer, the verb root becomes a noun that refers to a "medicine ceremony" $(?ote^{n}:sh\alpha?)$. If, however, the verb root is combined with the nominalizer and the verb root $-o-$, meaning "to be in the water," the noun comes to mean "the preparation of a (medical) ceremony." For example, $hote^{n}:syo?$, with translocative and purposive, means "he has the ceremony ready." It is likely, then, that water was and is a medicine to the Iroquois people. It will be seen later in this study that medicinal plants growing in or near the water were, in general, thought to possess a great deal of power. Water is power, and power is medicine.

Like the Otter Society, the Pygmy Society was a mechanism that permitted access to the powerful medicines of the Magic Animals, which included White Beaver, Blue Panther, Great Horned Serpent, Pygmies, Exploding Wren, Great Naked Bear, and others, and the charms made from them. The "little folk," from whom this society derived its name, were the custodians of the extremely powerful medicine charms of the magic animals.

The False Face and Husk Face societies were concerned both with individual curing on demand and the communal celebration of past cures by individuals who had been cured by their powers. Masks of the former societies represent a wide range of potentially harmful spirits (e.g., Flint, Forest Spirits, Stone Giants, Whirlwind), while those of the latter societies represent powerful agricultural and fertility spirits and act as heralds for the False Faces. The feature that best characterized their existence in Iroquois culture, however, was their symbolic unification of the following themes: (1) people may be protected from harm (disease and natural disaster) through the manipulation of certain powerful elements of the universe, (2) there exists a link between the power of animal medicines and plant medicines, and (3) water is the key ingredient to unlocking the secrets of medicines as revealed to man by the animals.

Iroquois folklore has the False Faces agreeing to "drive away sickness from among the new people, and to protect them from storms by causing the winds to go high into the sky" (Beauchamp 1922:13). These masks also had preventive and disease causative powers and thus conform to the broad conception of medicine as used in this study. The distinction between medicine and religion is seen to be blurred when we consider Blau's (1966) analysis of these spirits. He reports that of the eight rites associated with the False Face Medicine Society (viz., the new mask rite, feeding the masks tobacco, guessing desires of the masks, feeding corn pudding to the masks, curing an individual of symptoms associated with the masks, the entertainment rite, the traveling to cleanse homes rite, and the preventive blowing of ashes rite), four are designed to benefit the spirits of the masks and four are designed to benefit people. In some cases an individual mask may cure an individual person. In others, a group of masks may cure a person or group. When False Face Society members gather to renew the power of the masks or to guess the desires of the masks, we find a group of persons exerting their power to influence a spirit force or class of spirit forces which, in turn, reciprocate to benefit human beings. The power or medicine of the False Faces is further evidenced by their ability to cause, if neglected, various illnesses and misfortunes.

The Husk Faces, according to Parker (1909, in Fenton 1968:129), "endeavored to cure certain diseases by spraying and sprinkling water on the patients." They also provided medicine for "good crops and many births" (Fenton 1936:12). They were woven from corn husks and were worn by men disguised as women as they heralded in the False Faces at the Midwinter Ceremony. All of this creates a conceptual and symbolic linkage between the power of water, food, and women. Water was a medicine for corn, corn was a medicine for women, and women, as sources of life, were medicines for human beings.

INDIAN DOCTORS

Discussion of the Indian doctor in Iroquois culture is often neglected. This neglect may be due in part to the dubious reputation of certain of these so-called "show business" healers, as well as to a lack of a clearly defined role associated with them. Rioux (1951:155), for example, has stated that the Indian doctor "often dresses in ceremonial garments, takes part in exhibitions, county fairs, and sells medicine outside the Reserve." He added that "the folk are inclined to talk about them with a little disdain and to remark that this kind of a doctor is primarily a moneymaker."

In contrast to this moneymaking image of the Indian doctor that had developed at the time of Rioux's (1951) observations are the following remarks received by Waugh and Fenton. An entirely different image of the Indian doctor is found in these comments.

Chauncey Johnny-John, as recorded by Fenton (field notes 1938), says:

Old Henry Phillips was an Indian Doctor. Does not know sickness. Sometimes cut on bicep, upper arm, thigh or forehead. When let out bad blood, then stop. Bad blood darker. Will stop of its own accord when bad blood has passed out. No bandage. Heals up itself. No medicine used. Skin diseases from bad blood treated by bleeding. He had an implement—a stick with a blade in it like a carpenter gauge.

Waugh (field notes 1912) related the following information received from John Jamieson, Jr.:

> A theory of one kind of craziness is that the heart goes wrong. Laughing may cause a straining of the valves, as in valvular heart disease. John gave a fairly good description of the valves of the heart. John mentioned a case of a man who died of this a long while ago. His grandfather told him of it. A party of men had been betting about a game . . . betting for a pipe and tobacco according to custom then. The man laughed a great deal and was taken sick that night and died on the third night following. They did not know what was the matter. They used medicine to find out what kind of medicine to use, but it was of no avail. After the man died the Indian Doctor bought the body for a gun. This was customary in former times when they did not know what was the matter. He took it to the place where it was to be buried and cut it open. He found all the bowels and other organs in good condition except the heart which looked bloody around the upper part, where it was pronounced rotten. This is supposed to have been where the trouble was. The man had had a bad pain in the heart. He had been crazy [delirious] with this and could do nothing but holler.

In reference to the highly poisonous plant 260, *Cicuta maculata*, David Jack (Waugh, field notes 1912) made the following report concerning the practice of performing autopsies. Indian doctors were not mentioned.

> When eaten, it will make a man crazy—he will die shortly. When the root is chewed and swallowed, it becomes whole again within the stomach. A preacher at the Longhouse will say that it will grow in your head. By this it is meant that it is willing to kill you. People used to often make post-mortem examinations when a person died suddenly without the cause being known, but not now, against the law.

Dwight Jimmerson (Fenton, field notes 1939) made a more generalized statement about Indian doctors as he talked about the use of 294, *Collinsonia canadensis*.

> Heard that it was not used, but has also heard that Indian root doctors use it. I got it from my father. Alfred, my brother, was also an Indian Doctor. They either cured or were killed by the people.

From these characterizations of the Indian doctor, such a health actor seems to have been more than a charlatan or moneymaker in the eyes of the Iroquois. The one specialized feature that separated this health actor from others was that of surgery. Acts of cutting the body after death for the purpose of performing an autopsy and acts of cutting the body for the purpose of letting bad blood were apparently the criteria upon which the Indian doctor was distinguished.

That the Indian doctor either cured or was killed by the people is an indication of the fact that this health actor dealt primarily with illnesses of the witchcraft variety. He or she was probably called upon when the cause of an illness was unknown. It should be remembered that in such cases witchcraft was usually the presumed cause of illness. Since all other previously tried treatments and presumed etiologies had failed, it became the job of the Indian doctor to remove or at least relieve the effects of internally implanted objects. If the Indian doctor succeeded, a great deal of power was attributed to him. If he failed, the door would be opened for accusations of witchcraft. Such accusations would, of course, accompany the failure of any health actor thought to have great powers. The Indian doctor would be especially suspect because of his or her experience and knowledge of where and how objects could be implanted so as to bring about death to a victim. This healer's knowledge would be more particular and hence more dangerous.

As to why and how disdainful attitudes developed toward the Indian doctor in modern times, one explanation may be found in the increased acceptance or adoption of scientific surgical techniques by nonconservative Iroquois. For the nonconservative, these nonscientifically trained Indian doctor surgeons were probably seen as frauds or dangerous charlatans. To the conservative element, the Indian doctor in his traditional role was probably still appealed to when needed.

At the same time, however, there existed the image of the Indian doctor that had traveled with medicine shows to frontier regions. This kind of Indian doctor was often a white who claimed some Indian ancestry. Their knowledge of secret herbal medicines was, in many cases, known both to whites and Indians (see Vogel 1970:130-143). This moneymaking, "county fair" type of Indian doctor may have become confused with the traditional type in the minds of certain ethnographers and Indians. The uncritical lumping of two very distinct types of health actors may account for the differences that exist between the Indian doctor referred to by Rioux (1951) and the Indian doctor discussed by Waugh's and Fenton's sources of cultural information.

The Indian doctor health actor category used in this study is of the type that was described in the Waugh and Fenton field notes. It is likely that this health actor was appealed to under one of four circumstances. The first occurred when witchcraft was not suspected while the patient was alive, but *was* suspected after death and verification through autopsy was needed. The second occurred when witchcraft was suspected while the patient was still alive and magical means of object extraction either did not work or were not available. Such circumstances necessitated the surgical removal of an implanted object. The third circumstance occurred as a therapeutic measure to accompany any or all theorized disease etiologies that involved seriously spoiled blood, that is when bloodletting techniques were required (cf. Stone 1935:532). The fourth circumstance occurred whenever other physical mutilation was required, such as in the case of amputation or scarification.

COUNTERWITCHCRAFT

Shimony (1961, 1970) has approached witchcraft and counterwitchcraft in Iroquois culture from the point of view that such topics are only indirectly related to the entirety of the medical complex. She has theorized that the close association of witchcraft with the medical complex has made witchcraft *the* mechanism largely responsible for the preservation of traditional ways among today's Iroquois people. If, however, the line of thought presented in this study regarding the concept of medicine in Iroquois culture is followed, witchcraft (bad medicine) and counterwitchcraft (good medicine) become more than closely associated with medicine. They are and were medicine. Witchcraft as a source of imbalance was only one of four interrelated etiologies that could, in turn, be treated in a wide variety of ways. One of these ways involved counterwitchcraft.

It was the entire medical complex, including those events such as the religious ceremonies concerned with sustenance, which are not traditionally thought of as being medical in nature, that was responsible for the conservation of traditional beliefs among the Iroquois. These beliefs, in combination with the Iroquois philosophical penchant for elaborating upon existing structures, have accounted for the high degree of cultural persistence of the Iroquois into the present.

Related to a discussion of counterwitchcraft is the use of charms. It should be remembered that charms

are powerful medicines derived from various parts of very powerful animals. These charms, when in the hands of an evil person, could be used to bring about great harm to someone else. These same charms were also thought to be of great help to the sick when properly utilized by their holders or caretakers. There appear to be only two circumstances under which these powerful charms may exert malefic influences: (1) when they have been knowingly or unknowingly neglected, and (2) when they are in the possession of an individual who wishes to make use of their power for evil purposes.

Not unlike other things and events deemed *?otkō?* because of their great power and potential for causing harm when not used properly, the medicine charms of the Iroquois may be viewed as being both good and bad medicines (cf. Parker 1909, in Fenton 1968:120). They are, in simplest terms, reservoirs of magical power that may be tapped by groups (medicine societies or their subsets) or individuals for good or evil purposes. They were and are the tools of medicine societies and evil-minded witches.

There existed several other ways to inflict harm on others that involved less powerful items. These items were discussed in the section on witchcraft in chapter 3 of this study. Included among these items were certain very powerful plants. The most important characteristics of these plants with regard to a discussion of counterwitchcraft is the fact that they, like charms and other things or events that exuded or radiated *?otkō?*, were thought capable of both causing and curing spiritual imbalances.

The primary feature separating plants used in witchcraft from charms and other more powerful things and events used in witchcraft was the element of human manipulation. Because of the great amounts of power exuded or radiated by charms and certain other animal-derived medicines, it was thought that they could bring about spiritual imbalances in humans simply through contagion. This does not mean that they were never manipulated by human beings. In the case of charms, menstrual blood, the legs of spiders, and so forth, the potential for human involvement was certainly present. However, in general, human manipulation did not have to be involved. The reverse was generally true of plants used for purposes of witchcraft. There did, however, exist some plants that were thought to be so powerful that they were thought capable of upsetting ones spiritual balance simply by being close to them. Examples of these would be 43, *Nuphar lutea,* and 282, *Ipomea pandurata.* The former was thought to be a cause of consumption, while the latter was not to be touched by children. The jack-in-the-pulpit (403, *Arisaema triphyllum*) seems to have been another of these powerful plants, with its upsetting influence being limited to spoiling other good medicines. I was told that this plant was not to be kept indoors with other medicinal plants. It was also thought capable of causing a collector to become blind if it was picked too early in the spring.

As in the case of the things and events deemed *?otkō?* discussed previously, plants believed to be useful as mechanisms of witchcraft were often thought to be useful in counteracting witchcraft as well. David Jack (Waugh, field notes 1912, 1914) reported that the following plants could be used in such a way: 47, *Anemone canadensis;* 55, *Hepatica nobilis;* 354, *Arctium* sp.; 10, *Marchantia polymorpha;* and 272, *Asclepias incarnata.* In the case of *Anemone virginiana,* counterwitchcraft for a love medicine was involved. It should be remembered that making someone lonely for someone or lonesome for some place may constitute witchcraft. Antilove medicines, then, *are* antiwitchcraft medicines. Other examples of antilove (antiwitchcraft) medicines include: 325, *Lobelia kalmii;* 327, *Lobelia spicata;* 50, *Caltha palustris;* and 217-219, *Geum* sp.

There exist specific antiwitching remedies in cases involving the roots of certain plants that are fashioned into dolls intended to represent the potential victim of witchcraft (see 354, *Arctium* sp.; 102, *Amaranthus retroflexus;* and 180, *Physocarpus opulifolius*). For counteracting the *Amaranthus* doll, *Crataegus punctata* (171) was made into a tea and wash. Dolls made from *Arctium* were counteracted by a tea and poultice from the leaves of this same plant (see 354, *Arctium* sp., W1914A). For dolls made from 180, *Physocarpus,* a poultice was made from the barks of 241, *Staphylea trifolia* (W1914A) and 34, *Pinus strobus.*

There are specific antiwitchcraft herbal medicines. *Veronica officinalis* (318, W1914A) is used for counteracting bad luck in hunting. *Ranunculus abortivus* (57, W1914A) is used for counteracting "poison another has given you." *Gentiana andrewsii* (267) was sprayed on the patient for headache, pain, and sore eyes, the types of illnesses usually associated with witchcraft. In addition, there also existed herbal remedies for witchcraft in general. Clara Red Eye (Fenton, field notes 1934) reported that "there is a bundle of medicine in the house called *waitke^n? teyotoke^n'tdo* which is a compound of several medicines . . . there are five or six in it." Henry Red Eye similarly reported that 250, *Oxalis stricta* (FndA), was one of ten ingredients in an antiwitch medicine.

5

Powerful Medicinal Plants in Traditional Iroquois Culture

Having tried to understand the interrelatedness of the medical-religious complex in traditional Iroquois culture, we now turn to the task of determining the relative amounts of power possessed by the various medicinal plants. Plants used on an individual basis for the purposes of restoring, maintaining, or upsetting the spiritual balance of an individual human being are medicinal only in a limited sense of the word. It is seen from the discussion above that there also existed Iroquois medicines at the group-for-individual and group-for-group levels. With these distinctions in mind, it is now possible to point to certain structural and functional characteristics that make some plants more powerful than others (cf. Herrick 1978).

One class of medicinal plants are those that were used primarily on a symptomatic basis in cases of mild discomfort. That is, they were used to intensify or counteract certain symptoms such as fever, diarrhea, weakness, chills, diuresis, amenorrhea, and vomiting. These plants were associated with a wide variety of possible taboo violations, or they were thought to have preventive value. Many were blood purifiers, diaphoretics, spring and fall emetics, or laxatives. It was presumed that most mild illnesses could be attributed to any number of possible taboo violations, and that such offenses were largely understood with no verbalization of specific causes being made. Specific causes and a concomitant verbalization of these causes occured only as symptomatic features increased in duration or intensity. As symptomatic intensity and duration increased, other more powerful treatments were employed, such as the use of charms and various communal treatments. These might or might not have involved the use of herbal medicines. Of this general class of plants,

those possessing the greatest amount of relative power were those that were considered panaceas or strengtheners of other medicines.

A second class of plant medicines consists of those thought to be of use in divination. The point was made earlier that many of these plants could be used by both lay and professional (clairvoyant) health actors. As symptomatic intensity or duration increased, the professional, having expertise at looking into the past and future, was called upon to make a more precise diagnosis and prognosis. In more serious cases involving a diagnosis of unfulfilled desires, communal acts of fulfillment or guessing were relied upon, with or without the use of plants. In general, plants used in divination were considered to be more powerful than those in the class discussed above because they were thought to possess special qualities that enabled them to be used in determining past or future events in the spirit world. All plants, however, were thought to possess power or spirit.

The idea of a plant being particular or dangerous serves as the primary criterion upon which the power of medicines at the individual-for-individual level is based. Even though a certain plant might have nutritive and/or utilitarian value in addition to medicinal value, these more generalized powers would qualify them for positions intermediate between highly nutritious, utilitarian, and symbolic plants and highly particular plants. Nutritious, utilitarian, and symbolic plants include the maple, the pine, the elm, corn, beans, and squash. Particular plants are those constituting witching medicines.

I noted in chapter 3 that certain types of illnesses were associated with witchcraft in traditional Iroquois culture. A consideration of plants used in the treatment of these types of illnesses would, therefore, be an indirect way of determining which plants were used in practicing witchcraft as well. These inferences may be made because things and events that were thought to be powerful enough to be used in countering the effects of witchcraft were also thought to be powerful enough to be used in practicing witchcraft, and vice versa. Plants used in either bringing about or curing the various types of illnesses would, therefore, be considered witchcraft/counterwitchcraft medicines. The types of illnesses covered include bad luck in general, bad luck at love, loneliness, infertility, infidelity, bad luck at hunting or fishing, insanity, alcoholism, paralysis, dizziness, convulsions, epilepsy, vague internal pains or swellings, blindness or sore eyes, and intense internal pains such as those associated with piles, cysts, or boils.

Plants that were in themselves "witch" or *?otkō?* were also thought to be very particular and dangerous. These plants, although capable of being manipulated by evil-minded or revengeful individuals for the purpose of practicing witchcraft, were not thought to require human involvement in bringing about harm to a person. This class of plants was and is thought to be useful in counteracting the evil or ill effects of other things (including other plants) and events deemed *?otkō?*. Examples are spiders, snakes, and worms. As discussed in chapter 4, this same rule applied to all powerful, particular, or dangerous things and events of the universe. This bivalent property of witching medicines is best exemplified in the case of love medicines, in particular, and medicines that compel a thing or event to do something, in general The last include basket medicines, hunting medicines or charms, and certain game medicines.

Certain other structural/functional features correlate with the more powerful witchcraft and counterwitchcraft plant medicines. These features generally involve the principles of imitative and contagious magic. These principles, in turn, serve as unifying themes upon which relationships among various things and events of the Iroquois cosmos were explained. Included among these structural and functional features were such things as whether or not a plant was eaten by, resembled, killed, or counteracted certain plants or animals that were deemed *?otkō?* or evil. A second consideration was whether it grew in or near the water, water in itself

being an important source of medicine. A third consideration was whether it grew on or near graves, the dead being considered a source of *?otkō?*. A fourth consideration was whether the plant was poisonous, acrid tasting, or generally considered to be malodorous or obnoxious, these characteristics, signifying an item of the universe that was evil. A fifth consideration was whether the plant possessed thorns, barbs, or burs. A sixth consideration was whether it possessed hooklike structures or other catching, ensnaring, or collecting qualities, such characteristics being associated with love and/or basket (witching) medicines. A seventh consideration was whether the plant possessed structural features that resembled all or part of the human anatomy, these plants being useful in practicing witchcraft and counterwitchcraft. Finally, an eighth consideration was whether the plant stayed green the year round, these plants being considered immune from the power of seasonal forces.

A further criterion for establishing the relative degree of power possessed by certain medicinal plants was the effectiveness of the plant in curing highly acute, contagious, or extremely debilitating diseases referred to by their English names. Examples of these diseases included smallpox, syphilis, cholera, malaria, typhoid, yellow fever, cancer, and tuberculosis (consumption). Whether or not these diseases were correctly diagnosed in terms of modern medical science is immaterial. Also immaterial to the present discussion is whether or not any or all of these diseases were introduced (cf. Parker 1928; Hrdlicka 1932) or native (cf. Jarcho 1964). The same may be said regarding native versus introduced plants. What is important, from a cultural standpoint, is the belief that certain plants were and are thought to be effective in treating certain very serious, English-named diseases.

It must be remembered that any of the diseases mentioned above may have been considered extensions of less serious, native disease or illness categories. For example, boils, cysts, pimples, and pustules were generally thought to involve worms (or other objects) in the blood. Syphilis and smallpox may have been thought to be extreme cases of worm-infected blood, the blood further being considered the result of internally contained or implanted *?otkō?*. Fever, on the other hand, was a symptom frequently associated with death or ghost sickness.

The fever (and other accompanying symptoms) of typhoid may have been interpreted as being an extreme case of ghost sickness. In the case of a typhoid epidemic, the widespread occurrence of fevers may have been considered a communal punishment for offending a particularly powerful ghost or ghosts. The same probably held true for any of the more serious communicable diseases, with epidemics being interpreted as communal punishments for violating tradition in one way or another. In such cases, it is likely that communal treatments were used along with various plant medicines that were known to counteract or intensify certain symptoms. These plants would necessarily have to be very potent. Their use in treating serious, English-named diseases would set them apart from other plants used in treating spiritual imbalances with similar symptoms.

By considering all of the above-mentioned structural and functional criteria, it is possible to construct a checklist of the more powerful medicinal plants in traditional Iroquois culture. This checklist is presented below. A quantitative indication of which plants were thought to be the most particular or powerful could be arrived at by simply noting the presence of various witchcraft/counterwitchcraft structures and functions. From the standpoint of a people who have traditionally exhibited a great deal of flexibility in thought, however, a more meaningful approach to this question might involve an analysis of each plant's potential for either bringing about or curing severe cases of spiritual imbalance. For example, a plant having thorns or burs that grew near the water and possessed a root structure that was human-shaped would have a much wider range of potential uses (both good and evil) than one that was lacking any one of these structural features. The successful herbalist or clairvoyant was (and is) one who could recognize and appreciate a plant's potential uses. This skill was then combined with the faith, beliefs, and expectations of the patient in order to bring about harm or a cure. Much depends upon the imagination, creativity, and sensitivity of the skilled herbalist/clairvoyant, along with the degree of faith possessed by the patient/victim. These factors of human involvement necessarily build obsolescence into any particular list of the most powerful medicinal plants in traditional Iroquois culture. Like the earth upon which they dwell, the traditionally

minded Iroquois will continue to grow and expand. Their evolving conceptions and uses of medicinal plants will reflect this growth and expansion. In light of this, we should not be surprised to find that some of the more powerful plants listed below are species that were introduced into the New World.

The list of species in this chapter are checked against the criteria given below. The number appearing before each species refers to its position in chapter 6.

A. Eaten by, resembles, kills, or counteracts animals or plants that are considered *?otkǫ?* or evil
B. Grows in or near water
C. Grows on or near graves
D. Is poisonous, thought to be poisonous, generally acrid or obnoxious
E. Has thorns, barbs, or burs
F. Has hooklike structures or other ensnaring/capturing qualities
G. Is totally or partially human-shaped
H. Is evergreen
I. Is thought to be a panacea or general strengthener of other medicines
J. Is used in divination
K. Cures bad luck in general, bad luck at love, loneliness, infertility, infidelity; or is thought to be useful in selling various items (e.g., baskets)
L. Cures bad luck at hunting or fishing
M. Cures or prevents snake and spider bites
N. Cures insanity
O. Cures paralysis, dizziness, convulsions, epilepsy, or alcoholism
P. Cures death or ghost sicknesses, or is an antighost medicine
Q. Cures vague internal pains, swellings, sore eyes, blindness, cysts, boils, piles
R. Thought to be effective in curing or preventing certain severe, English-named diseases (e.g., smallpox, syphilis, tuberculosis, malaria, typhoid, yellow fever, cancer, cholera)
S. Is an ingredient in the substitute Little Water Medicine
T. Used to bring about bad luck, accidents, death
U. Used to bring about vague internal pain, blindness, sore eyes

V. Used to bring about infidelity, infertility, loneliness, depression

W. Used to bring about dizziness, convulsions, paralysis

X. Used to bring about insanity or alcoholism

Y. Reference is made to the fact that the plant is in itself "witch" or *?otkõ?*

348. *Achillea millefolium* A, O, Q
402. *Acorus americanus* B, J, O
24. *Adiantum pedatum* K, M
168. *Agrimonia gryposepala* F, K
91. *Alnus incana* L, Q
102. *Amaranthus retroflexus* G, Q, T
47. *Anemone canadensis* A, F, Q
48. *Anemone virginiana* F, J, K, R, T, V, X
258. *Angelica atropurpurea* A, B, D, P, T
259. *Angelica venenosa* A, D
354. *Arctium* spp. E, Q, T, Y
403. *Arisaema triphyllum* A, D, Q, V, Y
271-274. *Asclepias* sp. A, T, U, Y
355-359. *Aster* sp. L, R, V
356. *Aster novae-angliae* V
404. *Calla palustris* B, M
50. *Caltha palustris* B, K
144. *Cardamine bulbosa* B, D, T
145. *Cardamine concatenata* B, D, J, L, V
146. *Cardamine diphylla* B, D, K
147. *Cardamine douglassii* B, D, J, Q, U
411. *Carex platyphylla* F, T
412. *Carex prasina* C, F, L
310. *Chelone glabra* A, Q
157. *Chimaphila umbellata* A, H, I, Q, R
260. *Cicuta maculata* B, C, D, W, X
213. *Circaea lutetiana* S
363-364. *Cirsium* spp. E, Q, R
103. *Claytonia virginica* A, K, O
426. *Clintonia borealis* B, K
447. *Corallorhiza maculata* A, G, J, K, L, P
224. *Cornus sericea* A, M, N
98. *Corylus cornuta* K
171. *Crataegus punctata* E, Q
2. *Cronartium ribicola* A, M, R
448. *Cypripedium acaule* B, Q
449. *Cypripedium arietinum* A, B, Q
450. *Cypripedium calceolus* B, W
275. *Datura stramonium* D, E, X
200. *Desmodium glutinosum* F, K
347. *Dipsacus fullonum* E, T, Y

212. *Dirca palustris* R, Q
214. *Epilobium angustifolium* I, K, Q, R
227. *Euonymus europaea* A
228. *Euonymus obovata* A, Q
367-368. *Eupatorium* sp. B, S, R
369. *Eupatorium perfoliatum* B, J, N, R, T, Y
307. *Fraxinus americana* L, M, Q
295. *Galeopsis tetrahit* A, K
329, 331. *Galium* sp. F, K, Q, V
267. *Gentiana andrewsii* B, K, N, O, Q
251. *Geranium maculatum* F, K, Q
175. *Geum aleppicum* F, K, O
176. *Geum canadense* A, F, S, V
177. *Geum rivale* A, B, F
372. *Helianthus strumosus* A
54-55. *Hepatica nobilis* G, J, V, X
262. *Heracleum lanatum* A, L, Q, R
56. *Hydrastis canadensis* K, Q, R
284. *Hydrophyllum virginianum* A, B, Q, R
230. *Ilex verticillata* B, N
375. *Inula helenium* O, Q, R
282. *Ipomoea pandurata* G, I, Q, R, S, T, Y
441. *Iris versicolor* B, D, K, O
77. *Laportea canadensis* D, K, O
430. *Lilium philadelphicum* B, D, J, K
323. *Lobelia cardinalis* B, I, K, Q, R, S, V
324. *Lobelia inflata* B, K, N, Q
326. *Lobelia siphilitica* B, Q
327. *Lobelia spicata* B, K, Q
336. *Lonicera dioica* K, V
337. *Lonicera oblongifolia* K
297. *Lycopus americanus* B, D, T, Y
298. *Lycopus virginicus* D, Y
211. *Lythrum salicaria* P
122. *Malva neglecta* Q, V
10. *Marchantia polymorpha* F, K, V
432. *Medeola virginiana* L, O, S
300. *Mentha canadensis* A, B, Q
299, 301. *Mentha* spp. A, B, Q, S
312. *Mimulus ringens* A, O
332. *Mitchella repens* A, H, N, Q, R, V
165. *Mitella diphylla* A, L, Q
302. *Monarda* sp. B, N, P, S
276. *Nicotiana rustica* A, J, N, R, T
43. *Nuphar lutea* B, N, O, P, Q, R, U, Y
215-216. *Oenothera* spp. O, Q, S
378. *Onopordum acanthium* E, Q
454. Orchid sp. B, F, Q
263. *Osmorhiza claytonii* L, K

20. *Osmunda cinnamomea* B, P, T
99. *Ostrya virginiana* K, Q, R
256. *Panax quinquefolius* G, I, L, Q, R
236. *Parthenocissus quinquefolia* A, D, Y
5. *Peltigera polydactyla* F, K, T
314. *Penstemon hirsutus* K, V
1. *Peziza succosa* G, S
100. *Phytolacca americana* Q, V
33-34. *Pinus* sp. H, I, P, Q, R
306. *Plantago* sp. G, K, M, N, Q
451. *Platanthera grandiflora* B, P
453. *Platanthera psycodes* A, B, S
67. *Podophyllum peltatum* A, D, Q
239. *Polygala paucifolia* H, Q, R
109. *Polygonum arenastrum* V
114. *Polygonum punctatum* B, D, N
380. *Polymnia uvedalia* G, O, P
132, 133, 137. *Populus* spp. A, K, O, R
104. *Portulaca oleracea* A, Q
401. *Potamogeton* sp. B, L, O, Q
381-383. *Prenanthes* sp. A, L, O, Q, V
304. *Prunella vulgaris* A, E, I, P, Q, R, Y
158. *Pyrola elliptica* H, O, P
57. *Ranunculus abortivus* D, M, O, Q, R
247. *Rhus* spp. A, O, Q, R
9. *Rhytidiadelphus triquetrus* F, O, Q
161. *Ribes americanum* A, D, J
203. *Robinia pseudo-acacia* E, T, Y
187. *Rosa acicularis* B, E, U, Y
191. *Rubus occidentalis* E, K
116. *Rumex crispus* I, Q, R, V
400. *Sagittaria latifolia* A, B, Q

68. *Sanguinaria canadensis* B, I, Q, R
123. *Sarracenia purpurea* B, F, K, O, V
166. *Saxifraga pensylvanica* B, S
414. *Scirpus tabernaemontanii* A, B, E, M, Q
164. *Sedum telephium* H, Q
266. *Sium suave* A, B, C, O
434. *Smilacina racemosa* B, M, Q, T
445. *Smilax hispida* B, E, Q, T
421. *Sparganium eurycarpum* B, E, O, Q
193-194. *Spiraea* sp. K, Q, R
241. *Staphylea trifolia* A, Q
396. *Taraxacum officinale* A, Q, V
63. *Thalictrum dioicum* Q, X
167. *Tiarella cordifolia* L, N
120. *Tilia americana* I
248. *Toxicodendron radicans* D
207. *Trifolium repens* Q, V, X
437. *Trillium* sp. D, J, K, L, Q
340. *Triosteum aurantiacum* Q, T
35. *Tsuga canadensis* A, I, O, P, R
422. *Typha latifolia* B, Q, R
75. *Ulmus rubra* A, R, W
79. *Urtica dioica* D, T
291. *Verbena hastata* A, B, T
318. *Veronica officinalis* A, L
319. *Veronicastrum virginicum* D, I, Q
341. *Viburnum acerifolium* A, Q
208-209. *Vicia* sp. F, Q, V
125-129. *Viola* sp. A, K, Q, X
195. *Waldsteinia fragarioides* A, M
398. *Xanthium strumarium* E, V

6

Native Names, Uses, and Preparations of Plants

The plant list that follows is arranged according to the system found in Mitchell's (1986) *Checklist of New York State Plants*. Species are numbered consecutively. The plant's authority is given after each Latin binomial. Common English names follow that. The field notes of Waugh (W), Fenton (F), and Herrick (H) are recorded chronologically and verbatim for each species. If the field notes contained no date, "nd" is inserted where a date would normally appear. The initials of the Iroquois authorities appear next; a key to these authorities follows this introduction. The principal language of each authority is indicated.

Readers wishing to know plant names in the Iroquoian languages of their authorities should consult the dissertation from which this book is derived. Native plant names were reproduced in precisely the same ways as recorded in the original field notes. The orthographies used by the ethnographers differed in many specifics. These differences are also explained in the dissertation.

They were eliminated from this publication for linguistic and technical reasons.

The uses of each plant are provided, often as recipes along with other ingredients. The additional plants are identified by number if they appear elsewhere in the list. Formulas for preparing medicines are given, as are modes of application and dosages. This information is often abbreviated, for here too the entries remain faithful to the original notes.

IROQUOIS AUTHORITIES

The initials identifying Iroquois authorities in the plant list are keyed to full names below. All names are as they appear in the original notes. One authority, a Mohawk speaker, asked not to be identified. That individual is listed as anonymous. In each case, the principal language of the authority is indicated, along with his or her location and ethnographic worker.

AD	Amanda Doxdater	Seneca	Tonawanda	Fenton
ADy	Anthony Day	Oneida	Oneidatown	Waugh
AJ	Albert Jonas	Seneca	Allegany	Fenton
AJJ	Amos Johnny-John	Seneca	Allegany	Fenton
AJm	Arthur Jimmerson	Seneca	Allegany	Fenton
Anyms.	Anonymous	Mohawk	St. Regis	Herrick
BB	Barber Black	Seneca	Tonawanda	Waugh & Fenton
BT	Baptist Thomas	Onondaga	Six Nations	Waugh
CG	Charles Gordon	Seneca	Cornplanter	Fenton
CHJJ	Chauncy Johnny-John	Seneca/Cayuga	Allegany	Fenton
CRE/ET	Clara Redeye with Emma Turkey interpreting	Seneca	Allegany	Fenton
DB	Deforest Billy	Seneca	Allegany	Fenton

DC	Delia Carpenter	Onondaga	Onondaga	Herrick
DJ	David Jack	Cayuga	Six Nations	Waugh
DJn	Dwight Jimmerson	Seneca	Allegany	Fenton
DK	David Key	Seneca	Six Nations	Waugh
DW	David Williams	Oneida	Oneidatown	Waugh
ED	Elijah David	Seneca	Tonawanda	Waugh & Fenton
EH	Ellen Hill	Cayuga/Mohawk?	Six Nations	Waugh
EP & BC	Edna Parker &			
	Betsy Carpenter	Seneca	Tonawanda	Herrick
FJ	Floyd Johnson	Seneca	Allegany	Fenton
FK	Frances Kittle	Seneca	Cattaraugus	Herrick
FS	Foster Sundown	Seneca	Tonawanda	Herrick
FSt	Fannie Stevens	Seneca	?	?
HH	Hilton Hill	Seneca	?	?
HJ	Harvey Jacobs	Seneca	Cornplanter	Fenton
HJim	Howard Jimmerson	Seneca	Allegany	Fenton
HJk	Howard Jackson	Cayuga	Allegany	Fenton
HJm	Howard Jamieson	Cayuga	Six Nations?	Fenton
HJn	Herb Johnson	Seneca	Tonawanda	Waugh
HRE & SRE	Henry &			
	Sherman Redeye	Seneca	Allegany	Fenton
HS	Hanover Spring	Seneca	Tonawanda	Fenton
HW	Hiram Watts			
IL	Ike Lyon	Onondaga	Six Nations	Waugh
JA	James Alfred	Mohawk	Caughnawaga	Waugh
JC	Jesse Cornplanter	Seneca	?	?
JD	Jim Davis	Cayuga	?	Waugh
JE	John Echo	Onondaga	Six Nations	Waugh
JG	Jemima Gibson	Cayuga	Six Nations	Fenton
JH	Jesse Hill	Seneca	Tonawanda	Fenton
JJ	Jackson Jacobs	Cayuga	Six Nations	Waugh
JJJr	John Jamieson Jr.	Cayuga	Six Nations	Waugh
J & JS	Jonas &			
	Josephine Snow	Seneca	Allegany	Fenton
JJm	Josephine Jimmerson	Seneca	Allegany	Fenton
JSch	Jacob Schuyler	Oneida	Oneidatown	Waugh
JSl	J. Silversmith	Onondaga	Six Nations	Waugh
JT	Jimmy Thompson	Mohawk	St. Regis	Herrick
JW	J. Walker	Cayuga	Six Nations	Waugh
KD	Kate Debeau	Mohawk	Caughnawaga	Waugh & Fenton
KG	K. Gibson	Cayuga	Six Nations	Waugh
L & AP	Lyle Kinivy &			
	Alice Pierce	Seneca	Allegany	Fenton?
LJ	Levi John	Mohawk	Six Nations	Waugh
MJW	Mrs. J. Williams	Mohawk	Caughnawaga	Waugh
MR	Mrs. Reuben	Seneca	Tonawanda	Waugh
MS	Mrs. Sundown	Seneca	Tonawanda	Waugh
NH	Noah Homer	Oneida	Oneidatown	Waugh
NL	Noah LaFrance	Mohawk	St. Regis	Fenton

ON	Oscar Nephew	Seneca	Allegany	Herrick
PH	Peter Hopps	Mohawk	St. Regis	Fenton
PJ	Peter John	Onondaga	Six Nations	Waugh
PJb	Peter Jacobs	Mohawk	Caughnawaga	Fenton
PS	Peter Sundown	Seneca	Tonawanda	Waugh
PT	Peter Taylor	Mohawk	Caughnawaga	Waugh
RB	Roy Buck	Cayuga	Cattaraugus	Herrick
RS	Robert Smoke	Cayuga	Six Nations	Waugh
RSn	Rhonda Snyder	Seneca	?	?
SG	Simeon Gibson	Cayuga	Six Nations	Fenton
SH	Sam Hill	Onondaga	Six Nations	Waugh
SJ	Solan Jones	Onondaga	Onondaga	Fenton?
SS	Sarah Snow	Seneca	Allegany	Fenton
TL	Tom Lewis	Onondaga	Onondaga	Herrick
WP	Windsor Pierce	Seneca	Cornplanter	Fenton
WS	Windsor Snow	Seneca	Allegany	Fenton?
YS	Yankee Spring	Seneca	Tonawanda	Fenton

FUNGI
PEZIZACEAE (CUP FUNGI)

1. *Peziza succosa* Berk.

F1938A CHJJ See 301, *Mentha spicata*, F1938C.

PUCINIACEAE (RUSTS)

2. *Cronartium ribicola* J.C. Fisch   WHITE PINE
BLISTER RUST

H1973A Anyms. For cancer. Also kills spiders.
Scrape off rust from pine tree, sprinkle on sore,
put cloth over it lightly.

POLPORACEAE (PORE FUNGI)

3. *Fomitopsis officinalis* (Fr.) Faull   FELTED
HEART-ROT CONK

W1912A ?? Blood purifier. Cut very fine and boil a
little, 2 tablespoons, in — qt. warm water. Take a
tablespoon 4 times per day.

W1912B SH A physic. Take from the pine or pine
knot. Powder or steep. If powder, take dose the
size of a nickel.

W1914A JJJr This fungus makes a loud noise before rain or snow. Therefore, it is a weather
medicine. It is also used as an ingredient in
medicines. It was formerly shot with a bow and
arrow or a bullet before removing from tree.

LYCOPERDACEAE (PUFFBALLS)

4. *Calvatia gigantea* (Pers.) Lloyd   GIANT PUFF-
BALL

W1914A DJ For sore navel in children and pregnant women. Use when white, cut in ½ and tie on.

PELTIGERACEAE (LICHENS)

5. *Peltigera polydactyla* (Neck.) Hoffm.   DOG-
LICHEN

W1912A DJ For when a man is lonely, when his
wife has gone away. Acts as an emetic to clean out
the stomach. Put a piece about 3-4 in. square in a
cup of warm water, cover well. Drink ½ of it right
away, followed by 2 qt. of clear water. Wet the
head all over and drink remaining ½, followed by
another 2 qt. water. Vomit both times. Woman
will surely come back.

W1914A DJ Antilove medicine. Put inside cloth
and boil to make a tea. Use as an emetic. Give
cupful and vomit, following with warm water. The
person who bewitched you will become as poor as
a dishcloth.

FndA DJn See 185, *Prunus serotina*, FndB.

6. Mold, unspecified

W1914A ?? An unspecified, unidentified "white,
moldlike growth on damp soil" that glows at night.
This general category could include *Peziza ostra-
coderma* and many other fungus species.
u^ntgahiyúsda^{h}kwa? 'you can see better' Lets you
see at night. Dry it, powder, mix with a little water
and rub on outside of eyes. Also rinse mouth with
it. This makes fire appear to come out. Will light
you at night and also scare away animals.

PARMELIACEAE (LICHENS)

7. *Parmelia conspersa* (Ehrh.) Ach.

W1912A RS Trench mouth, raw throat, caused by
fever inside some times. With: (1) 53, *Coptis trifolia*, W1912E; (2) 308, *Fraxinus nigra*, W1912G.
Take lichen off bark of (2) and about a handful of
the bark of (2) as well. Add 1 plant of (1). Put ingredients in 1 cup cold water. Take a teaspoon of
this and leave in mouth until water gets warm,
then drink and take another teaspoon. Repeat until you drink all of it.

POLYTRICHACEAE (MOSSES)

8. *Polytrichum commune* Hedw.   PIGEON
WHEAT

W1912A ED See 12, *Lycopodium lucidulum,* W1912A.

HYLOCOMIACEAE (MOSSES)

9. *Rhytidiadelphus triquetrus* (Hedw.) Warnst. SHAGGY MOSS

W1912A DJ See 401, *Potamogeton* sp., W1912B.

MARCHANTIACEAE (LIVERWORTS)

10. *Marchantia polymorpha* L.   MARCHANTIA

W1914A DJ Cure for caddis fly love medicine. Make a small cup of tea from it. Drink it, then drink warm water and vomit.
Also, a love medicine itself. Chew or make tea and give it to another as a love medicine. If sprinkled on articles, they will be bought.

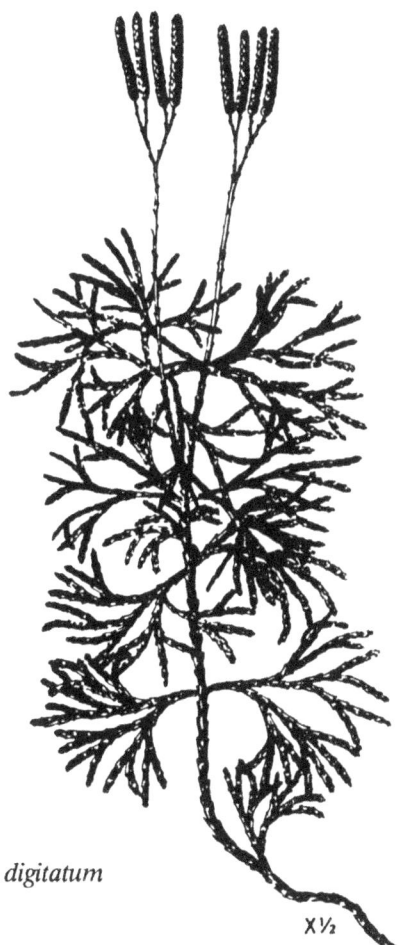

L. digitatum

X ½

LYCOPODIACEAE (CLUBMOSS FAMILY)

11. *Lycopodium digitatum* A. Br.   RUNNING-PINE, CHRISTMAS GREEN, GROUND CEDAR (RARE)

W1912A ED To induce pregnancy. With: (1) 82, *Juglans cinerea,* W1912A; (2) 212, *Dirca palustris,* W1912F; (3) 116, *Rumex crispus,* W1912F; (4) 354, *Arctium* sp., W1912F; (5) 307, *Fraxinus americana,* W1912A; (6) 308, *Fraxinus nigra,* W1912D; (7) 157, *Chimaphila umbellata,* W1912B. Use a handful of each. Boil in 8 qt. water until 4 qt. remain. Take a tablespoon before meals. Use a handful of bark from the east side of (1), the same for the roots facing east of (2)-(4), a handful of bark of (5) and (6), and a whole plant of (7).

12. *Lycopodium lucidulum* Michx.   SHINING FIR-MOSS OR CLUBMOSS

W1912A ED When the blood is bad and sores break out on the neck. With: (1) 212, *Dirca palustris,* W1912I; (2) 158, *Pyrola elliptica,* W1912E; (3)

L. lucidulum   X ½

307, *Fraxinus americana,* W1912E; (4) 190, *Rubus idaeus,* W1912C; (5) 157, *Chimaphila umbellata,* W1912D; (6) 8, *Polytrichum commune,* W1912A. No preparation given.

W1914A DJ For the blood. When woman catches cold due to suppressed menses. Put small handful in 6 qt. water. Boil. Drink whenever you can.

13. *Lycopodium obscurum* L.   GROUND PINE, EASTERN TREE CLUBMOSS

F1933A CHJJ See 21, *Osmunda claytoniana,* F1933A.

F1938A CRE/ET Cold in blood. With: (1) 26, *Onoclea sensibilis,* F1938A. Put ½ cup of plant with 2 roots of (1). Boil in 1 gal. water until water turns red. Drink abundantly during day.

FndA SS Medicine for change of life. Some go blind or deaf in change of life, and this prevents these. Put small bunch of the root in 3 qt. water, bring to boil, drink like water.

L. obscurum   X ½

14. *Lycopodium sabinifolium* Willd.   CYPRESS CLUBMOSS, GROUND FIR, HEATH CYPRESS

W1912A PT See 21, *Osmunda claytoniana,* W1912A.

15. *Lycopodium* sp.   CLUBMOSS

W1912A ED See 37, *Juniperus virginiana,* W1912A.

W1912B KD See 57, *Ranunculus abortivus,* W1912C.

W1913A PJ Nosebleed. Smoke from it (i.e., spores) will stop bleeding when sprinkled on.

EQUISETACEAE (HORSETAIL FAMILY)

16. *Equisetum arvense* L.   FIELD HORSETAIL, COMMON HORSETAIL

H1973A DC Heard it was good for diabetes.

H1973B Ep & BC For teething babies. Baby chews on stem.

H1973C FS Rheumatism, joint aches, headaches, minor aches and pains.

17. *Equisetum fluviatile* L.   WATER HORSETAIL, PIPES

W1912A DJ Snowsnake medicine. With: (1) 411, *Carex platyphylla,* W1912A; (2) "fishweed," which is probably 210, *Myriophyllum verticillatum,* W1912A. Steep 1 plant of each in 2 qt. water. Pick out best snowsnake, wet and wash it with this liquid. Let dry.

18. *Equisetum hyemale* L. var. *affine* (Engelm.) A. Eat.   SCOURING RUSH, STONE SCOUR-ING RUSH

W1912A PT See 21, *Osmunda claytoniana,* W1912A.

W1912B DJ For white spot on the eye. Mash stem of whole plant and place in ½ cup warm water. Let stand ½ hr. Take cloth, soak, and drop in eye.

W1912C BB The devil made this plant for wampum for the Indian people, but it is a poor substitute.

F1939A JG For kidney trouble (backache or else urinating too much or too infrequently), for summer complaint, and for scrubbing pots with. Put 6 plants in 2 qt. water. Boil 3 min. Drink lukewarm as often as you can take. Make stronger for summer complaint.

H1973A DC Venereal disease.

H1974A Anyms. Excessive urination in women who are ruptured. Boil up like tea.

OPHIOGLOSSACEAE
(ADDER'S-TONGUE FAMILY)

19. *Botrychium virginianum* (L.) Sw.   VIRGINIA GRAPE FERN OR RATTLESNAKE FERN

W1912A SH See 397, *Tussilago farfara,* W1912A.

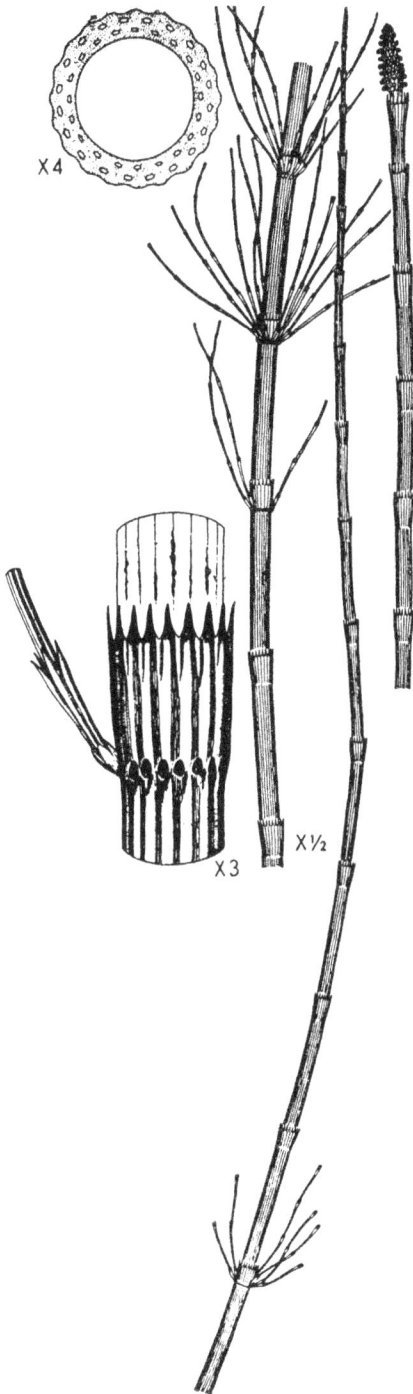

E. fluviatile

B. virginianum

OSMUNDACEAE (FLOWERING-FERN FAMILY)

20. *Osmunda cinnamomea* L.   CINNAMON FERN

W1912A KD See 26, *Onoclea sensibilis,* W1912F.

W1914A ?? For difficult birth of a calf. With: (1) 421, *Sparganium eurycarpum,* W1914A; (2) 131, *Sicyos angulatus,* W1914A. Chop up 1 plant of each and mix it with cow's food.

W1914B RS For cold, headache, and malaise from *gutuwiniōsi* 'a cold'. Put 1 root in 3 qt. water, boil down ½. Drink a lot, cover up in bed and keep warm. Make sweat and keep covered.
Also for rheumatism or pain in joints. Used as above, but put 2 roots in 6 qt. water.

FndA KD Use not given.

H1973A Anyms. See 28, *Thelypteris palustris,* H1973A.

21. *Osmunda claytoniana* L.   INTERRUPTED FERN

W1912A PT Gonorrhea. With: (1) 190, *Rubus idaeus,* W1912D; (2) 12, *Lycopodium lucidulum,* W1912A; (3) 233, *Rhamnus alnifolia,* W1912C; (4) 18, *Equisetum hyemale,* W1912A; (5) 191, *Rubus occidentalis,* W1912B; (6) 336, *Lonicera dioica,* W1912A; (7) 333, *Diervilla lonicera,* W1912B; (8) 31, *Larix laricina,* W1912B. Take root, add root of (1), a handful of (2), (3), (4), (5), (6), (7), and (8). Put all in 1 qt. water. Boil down to 1 finger. The more you drink, the quicker the recovery. Taboo on liquor and sweets.

F1933A CHJJ Weak blood. With (1) 254, *Aralia nudicaulis,* F1933C; (2) 25, *Pteridium aquilinum,* F1933A; (3) 27, *Polystichum acrostichoides,* F1933A; (4) 13, *Lycopodium obscurum,* F1933A. Take 4 roots, add 2-3 roots of (1), roots of 1 plant of (2), (3), and (4). Pound all together. Boil in 2 qt. water, strain. Drink liquid cold, as much as you can.

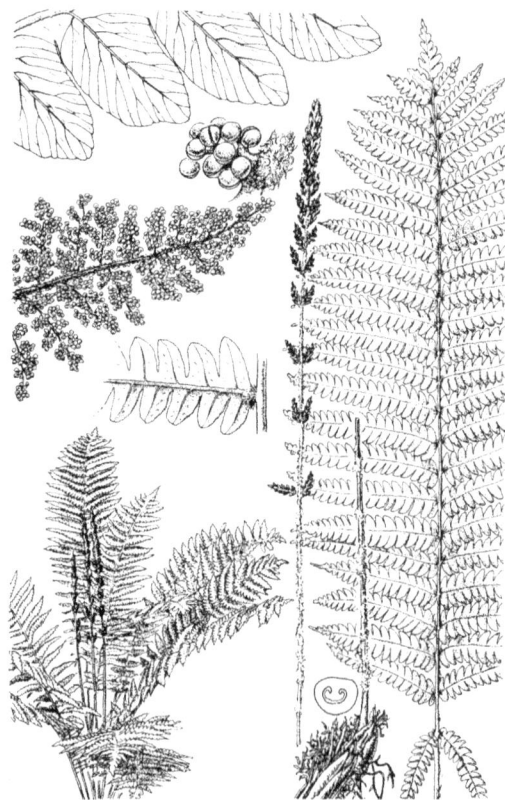

O. cinnamonmea

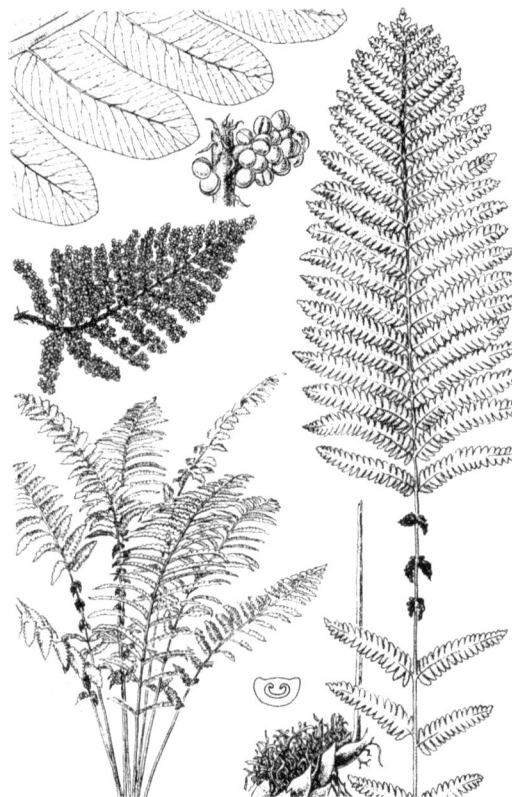

O. claytoniana

22. *Osmunda regalis* L. var. *spectabilis* (Willd.) Gray
ROYAL FERN, FLOWERING FERN

W1912A JJJr For the blood, when a woman catches cold in her kidneys and her blood gets like water. Menses watery and strong. Paralysis, some times. Take small handful of the fronds, bundle them and boil in 1½ qt. water, down to ½. Take ½ cup before breakfast, ½ cup after breakfast. Do the same at lunch and dinner, 6 times in all. Use all in 1 day.

W1914A DJ See 375, *Inula helenium*, W1914A.

POLYPODIACEAE (POLYPODY FAMILY)

23. *Polypodium virginianum* L.   ROCK POLY-PODY, COMMON POLYPODY

F1938A CHJJ Cholera. With: (1) 86, *Quercus bicolor*, F1938A; (2) 35, *Tsuga canadensis*,

F1938A; (3) 186, *Prunus virginiana*, F1938C. Take whole plant, add bark scraped downward of (1), (2), and (3). Boil in 2 qt. water down to 1 qt. Take small portions.

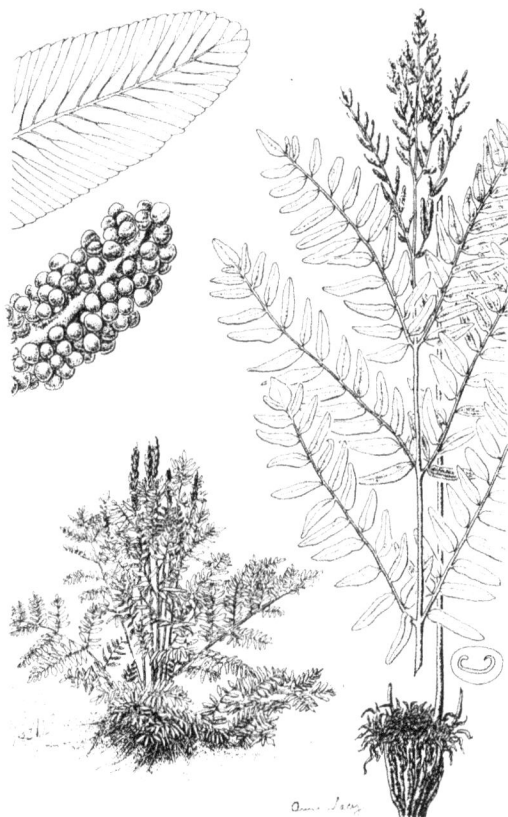

P. virginianum

ADIANTACEAE (MAIDENHAIR FAMILY)

24. *Adiantum pedatum* L.   MAIDENHAIR FERN

W1912A ED Sore back in babies. Smash plant and apply as poultice (see 27, *Polystichum acrostichoides*, for an alternate medicine).

W1912B MS Excessive menstruation. Steep a handful of roots until water gets dark. Use 2 qt. water. Take any amount, anytime.

W1912C KG Cramps in children and gonorrhea. Put 2 small bunches in 2 qt. water and boil down

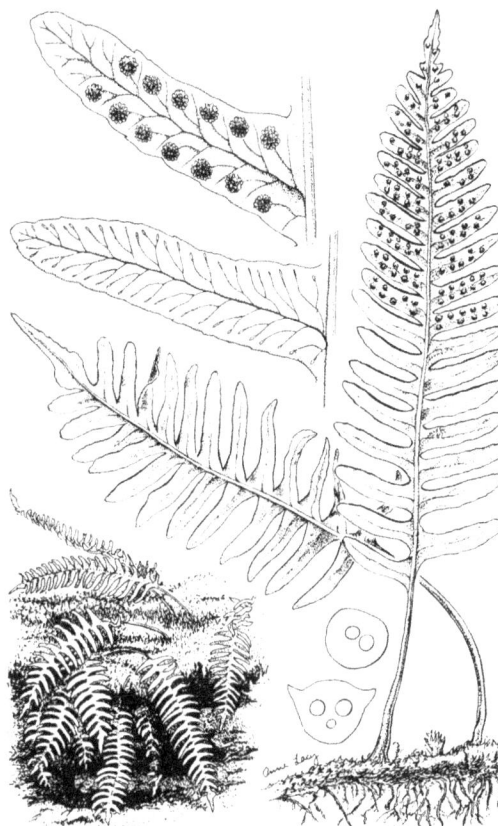

O. regalis

A. pedatum

½. For 1-month-old infant, give ½ teaspoon 3 times per day. For gonorrhea, use as a wash.

W1912D KG Cessation of urine due to gall. Take roots of 3 plants, put in 3 qt. water and boil to about 3/4. Take 1 cup twice after meals and again before going to bed.

W1912E DJ Remedy for love medicine. Steep 3 plants in 5 qt. water until 4½ remain. Drink until you feel like vomiting.

W1912F JJJr No use given.

W1912G SH Use not specified, possibly a venereal disease. With: (1) 308, *Fraxinus nigra*, W1912F. Take handful of smashed roots of 2 plants, add bark of (1).

W1914A JD Snake bites. Smash a couple of fronds, wet with warm water, tie on.

W1914B JJJr Venereal disease. Decoction of root. Drink some and wash sores. Dries up disease temporarily without really curing.

F1933A CHJJ See 294, *Collinsonia canadensis*, F1933A.

FndA KD Excessive menstruation. With: (1) 171, *Crataegus punctata*, FndA, as an alternate medicine. Take a handful of roots in 1 qt. of water. Boil down to ½.

FndB SS For abortal pains; for pain when about to deliver. Drink whole plant. No specifics given.

FndC CG Did not know.

H1973A DC Gonorrhea. Use as a douche.

H1973B Anyms. For ladies to get period. Cleans out. Also for abortions up to 2 months. Boil root.

CYATHEACEAE (TREE FERN FAMILY) (DENNSTAEDTIACEAE)

25. *Pteridium aquilinum* (L.) Kuhn   BRACKEN, BRAKE-FERN, BRAKE

W1912A KD Prolapsus of uterus, also good for an old man who cannot retain urine. With: (1) 255, *Aralia racemosa*, W1912E; (2) 247, *Rhus* sp., W1912K; and, for the second use, (3) 91, *Alnus incana*, W1912D; (4) 333, *Diervilla lonicera*, W1912C. Take a pinch of bracken, a pinch of (1), and a pinch of the bark scraped upward of (2). Add a pinch of ingredients (3) and (4) for the second use. Boil all in 1 or 1½ qt. water until ½ remains. Drink ad lib.

W1914A JD See 282, *Ipomoea pandurata*, W1914B.

F1933A CHJJ See 21, *Osmunda claytoniana*, F1933A.

F1939A PH See 26, *Onoclea sensibilis*, F1939A.

FndA CHJJ Rheumatism. Other ingredients and preparation not given.

FndB J&JS Makes good blood after menses. Also used after baby is born by the mother. For TB in women, too. Wash, pound 1 handful, boil in water, and drink.

FndC PH Not given.

FndD NL For bedding when camping.

FndE CHJJ See 198, *Baptisia tinctoria,* FndC.

H1973A Anyms. Used by people who make thunder and lightning.

H1973B DC Used with 2 other unspecified plants (roots) to clear out infection (probably venereal disease).

H1973C Anyms. For when suffering too much after birth. Boil, drink, and poultice. Also for diarrhea. Rub in hands, put in cold water, drink.

ASPLENIACEAE (SPLEENWORT FAMILY)

26. Onoclea sensibilis L.   SENSITIVE FERN

W1912A DJ For cracking of the breasts when there is plenty of milk but it will not flow. Steep the whole plant or the roots alone in 6 qt. water. Drink.

W1912B SH Venereal disease. With: (1) 254, *Aralia nudicaulis,* W1912C; (2) 255, *Aralia racemosa,* W1912D; (3) 81, *Juglans cinerea,* W1912D; (4) 134, *Populus xjackii,* W1912. Break the roots of 1 plant of each. Put in 5 qt. water, boil down to 4 qt. Drink a cupful 4-5 times per day. Do not eat fish, pickles, pork, fat, or drink liquors. Use ulcer remedy made of (3) and (4) as a wash for sores.

W1912C MS To remove pain after childbirth. Take a handful of root, steep in 4 qt. water. Drink warm. Drink 1 qt. every 4 hr.

W1912D KD Makes blood. With: (1) 225, *Celastrus scandens,* W1912C; (2) 238, *Vitis vulpina,* W1912C. Boil ingredients to a teacupful in ½ gal. water. Make wine from grapes (2). Mix. Take 2 tablespoons before meals and bed.

W1912E PT Use not given. One handful or 2-3 roots in 1 gal. water. Boil well, do not drain. Drink as much as you can.

W1912F KD Gonorrhea. With: (1) 20, *Osmunda cinnamomea,* W1912A; (2) 91, *Alnus incana,* W1912C. Preparation for external use: Take 1 handful of bark of (2) scraped upward, put in cup of cold water. Let stand 6 hr., squeezing bark occasionally. Wash affected parts. Preparation for internal use: Take handful of *Onoclea sensibilis* roots and a double handful of roots of (1). Add 12 gal. soft, cold water, cover, boil down to ½. Take a wineglass about 3 hr. apart.

W1914A DJ For the blood that causes the hair to fall out. Boil the root in 5 qt. water until ½ qt. remains. Drink and wash the head with it pretty often.

W1914B DJ To promote menses stopped by a woman going with a man during her menses. Also for swellings, cramps, and sore abdomen. Boil root a little in 7-8 qt. water. Same root may be used 3 different times. Second time boil down 1 qt, third time boil down 2 qt. Take a good drink whenever dry.

W1914C DJ For trouble with the intestines, when you catch cold and get inflated and sore. Preparation not given.

W1914D JD To give strength after childbirth. Put the roots of 3 plants in 3 qt. water. Boil quite a bit. Drink as much as you like.

F1938A CRE/ET See 13, *Lycopodium obscurum,* F1938A.

F1939A PH Consumption. With: (1) 185, *Prunus serotina,* F1939A; (2) 72, *Hamamelis virginiana,* F1939B; (3) 35, *Tsuga canadensis,* F1939A; (4) 25, *Pteridium aquilinum,* F1939A; (5) 30, *Abies balsamea,* F1939A; (6) 29, *Taxus canadensis,* F1939A. Take roots, add young shoots of (1), bark cut upwards of (2), outer bark of (3), root of (4), young shoot tops of (5), and young growth of (6). Put in

8 qt. water, boil down to 4 qt. Take out water and save. Take only during early stages of consumption.

FndA CHJJ Fertility drug for woman, also good for the blood. With: 342, *Viburnum lantanoides*, FndB. Take root, peel root of *V. lantanoides*, mix. Boil, drain, cool. Drink as much as you can as often as you want.

H1973A JT For troubles after childbirth.

H1973B DC Arthritis, gets rid of infection.

H1973C Anyms. For deep cuts. Top part for poultice.

O. sensibilis

27. *Polystichum acrostichoides* (Michx.) Schott
CHRISTMAS FERN, SHIELD FERN

W1912A ED Sore back in babies. With: 167, *Tiarella cordifolia*, W1912D. Take a handful of each, smash them, add a little hot water, apply as a poultice (see 24, *Adiantum pedatum*, for an alternate medicine).

W1912B MS Child's spinal trouble. Smash roots, apply to back and feet as poultice. Also an emetic for dyspepsia. Steep roots to any strength. Drink first thing in morning, repeat for 3 days.

W1912C ED Rheumatism in back and legs. With 294, *Collinsonia canadensis*. Smash roots, steep in 12 qt. water. Put feet in as hot as can stand. Keep in ½ hr., covering up with a blanket. Do this 3 times a day for 3 days. Stop 3 days and repeat. Go to bed after treatment and cover.

W1912D DJ Fever and cramps in children. With: (1) *nayunra^hduzu^h* 'small leaves'. Boil a small plant of each in 1 qt. water until down ½. The fern is for cramps, the vine (1) is for fever. Take a teaspoon very often.

W1912E JJJr Diarrhea. With: (1) 181, *Porteranthus trifoliatus*, W1912A. Take a small handful of leaves and 2 switches of (1), which are broken into pieces about 4 in. long. Boil in 1 qt. water down to ½. Take ½ cup before meals. Finish the quantity.

W1914A DJ Listlessness in children, some times mother takes too. Put a small plant in 2 qt. water and boil quite a bit. Give or take a teaspoon whenever child will take it.

W1914B ?? Young fronds were formerly eaten for greens.

W1914C JD Convulsions and red spots on bodies of children. Smash a double handful of roots, wet, put in a narrow bag, and place along back and head. Leave on all night if possible.

F1933A CHJJ See 21, *Osmunda claytoniana*, F1933A.

FndA KD No use given.

FndB JG See 263, *Osmorhiza claytonii,* FndB.

FndC CG Emetic for consumption. Take handful of roots, steep. Make 3-4 qt. Drink and vomit.

H1973A JT Take before and after baby to clean womb.

H1973B DC Lady's medicine, for inside. Use roots.

H1973C Anyms. When a man can not talk. Powder, inhale, cough up.

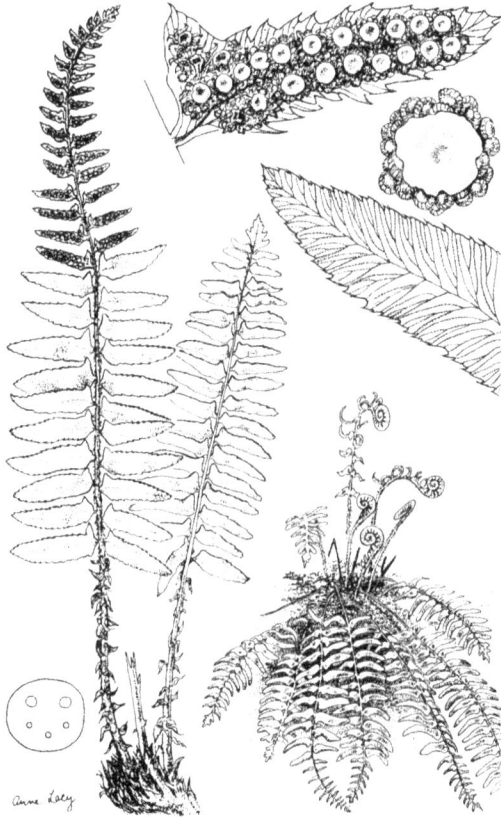

P. acrostichoides

28. *Thelypteris palustris* Schott   MARSH FERN

H1973A Anyms. Young men deliver 7 rusty stems (may be referring to 20, *Osmunda cinnamomea*) to where any kind of trouble is. Tobacco is burned and in 24 hr. lightning, earthquakes, etc., will punish people making trouble. Roots for woman's troubles.

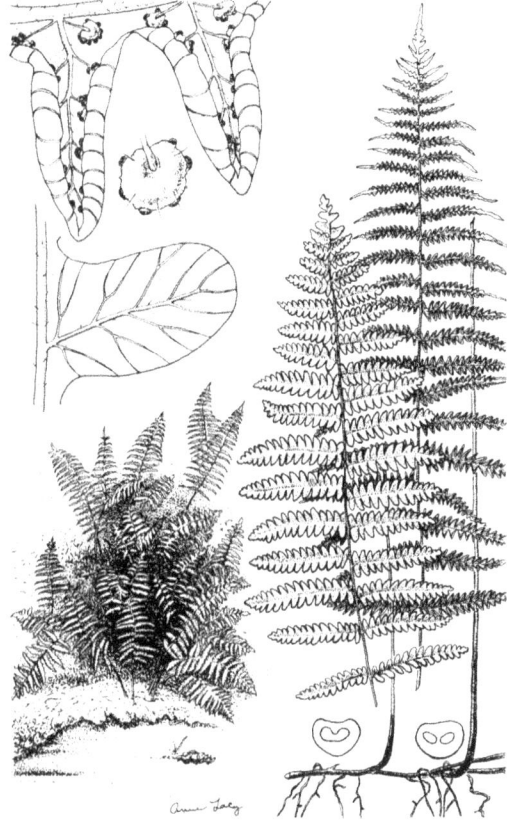

T. palustris

GYMNOSPERMS
TAXACEAE (YEW FAMILY)

29. *Taxus canadensis* Marsh.   AMERICAN YEW, GROUND HEMLOCK

W1912A JA See 34, *Pinus strobus,* W1912C.

F1933A CHJJ Chest colds. Take whole plant and boil it up in 2 qt. water. Put the patient in a chair, cool him with a blanket, and stand the steeping, hot liquid under him. Steam until perspires well.

F1938A CHJJ See 405, *Symplocarpus foetidus,* F1938C.

T. canadensis

F1939A PH See 26, *Onoclea sensibilis*, F1939A.

F1939B NL See 255, *Aralia racemosa*, F1939A.

H1973A Anyms. For when person gets numb in fingers or legs. Boil twigs up like tea.

H1973B Anyms. Put in all medicines to give them strength.

H1973C DC For colds. Makes you sweat.

PINACEAE (PINE FAMILY)

30. *Abies balsamea* (L.) Mill.   BALSAM FIR

A. balsamea

W1912A JA See 34, *Pinus strobus*, W1912C.

W1912B JA See 38, *Thuja occidentalis*, W1912A.

F1939A PH See 26, *Onoclea sensibilis*, F1939A.

FndA NL Coughs. Boil in water to "cut it." Take straight or dilute with alcohol.

FndB KD Cuts, sores, ulcers on legs. With: (1) 254, *Aralia nudicaulis*, FndA. Boil young shoots 5 in. long or just the bark. Boil until becomes mush. Make poultice.

FndC ?? Gum for cuts.

FndD NL See 38, *Thuja occidentalis*, FndA.

H1973A Anyms. See 137, *Populus tremuloides*, H1973A.

H1973B JT Colds.

H1973C Anyms. Gonorrhea.

H1973D Anyms. Colds. Heats you up, good for chest. Cook as strong as you can. Take ½ cup, feel the heat.

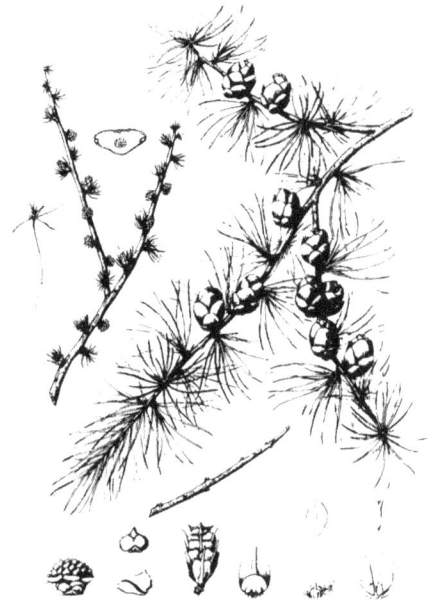

L. laricina

31. *Larix laricina* (DuRoi) Koch   TAMARACK, AMERICAN LARCH

W1912A JJ See 35, *Tsuga canadensis*, W1912E.

W1912B PT See 21, *Osmunda claytoniana*, W1912A.

W1912C KD See 35, *Tsuga canadensis*, W1912C.

W1912D JA See 34, *Pinus strobus*, W1912C.

32. *Picea* sp.   SPRUCE

W1912A JA See 34, *Pinus strobus*, W1912C.

F1938A CHJJ See 444, *Smilax herbacea*, F1938A.

H1973A JT Chewing gum, ingrown nails, cuts. Apply gum.

H1974A RB Colds. Make tea. Vomit in spring.

33. *Pinus rigida* Mill.   PITCH PINE

W1912A BB Boils. Apply pitch to boil as salve.

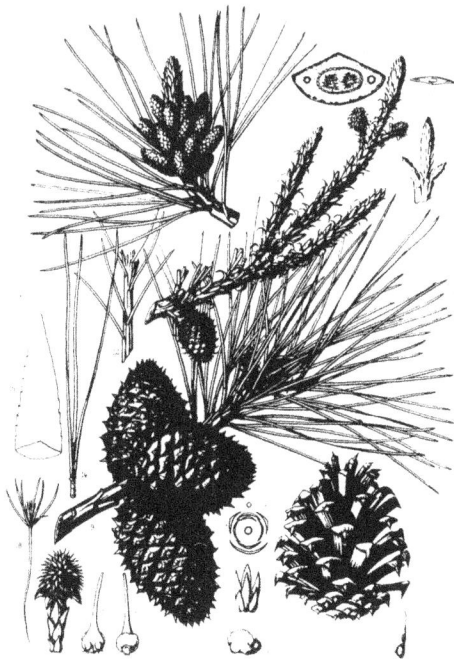

P. rigida

W1912B JJJr Laxative. Cut hole through bark, pitch will collect. Take a teaspoonful (sweetened) twice a week.

W1914A JJJr Rheumatism. Also moves bowels (see above).

F1938A CHJJ Burns.

F1938B CHJJ See 440, *Veratrum veride*, F1938B.

H1973A Anyms. To get rid of fleas. Also for boils. Smoke from burnt leaves for former, put pitch on latter.

H1973B ON Cuts in joints. Put pitch in joint.

34. *Pinus strobus* L.   WHITE PINE

W1912A KD Weakness in infants. When it does not walk for 2 or 3 yr. Boil 1 gal. of leaf clusters in 1 gal. water for ½ hr. Boil them hard. Let cool. Wash child at bedtime for 9 days. Keep child warm.

W1912B KD Difficulty in breathing in fat people. With: (1) 345, *Viburnum trilobum*, W1912D. Use double handful of bark, add handful of bark of (1) scraped downward. Boil in ½ gal. water down to 1 qt. Drink wineglass 2 hr. apart.

W1912C JA Colds, coughs, rheumatism. With (1) 31, *Larix laricina*, W1912D; (2) 32, *Picea* sp., W1912A; (3) 30, *Abies balsamea*, W1912A; (4) 35, *Tsuga canadensis*, W1912D; (5) 37, *Juniperus virginiana*, W1912B; (6) 29, *Taxus canadensis*, W1912A. Take a small handful of each, boil in 2 gal. water down 1/3. Drink warm when thirsty.

W1912D BB Powder for chafed babies. With: (1) 83, *Castanea dentata*, W1912B. Both woods must be rotten. Dry and powder. Dust on child. Can use each separately.

W1912E JJJr Cleans stomach, rheumatism, cramps, stiffness in limbs, prevents typhoid. Bark scraped from young shoots and eaten.

W1912F SH Toilet powder for babies. Powder rotten wood. Dust on spots sore from perspiration.

W1912G JJ For a kind of venereal disease when the penis gets dry and the skin cracks. With: (1) 254, *Aralia nudicaulis*, W1912B. Take pitch, break up root of (1), add beef tallow and beeswax. Put in pot, cook, stir. Strain, let cool. Apply as salve.

W1912H JJJr Consumption. Get pine knots where rest of the wood has rotted away. Take 12 knots about 13-14 in. long. Split out the red heart. Boil in 5 qt. water, down to ½. Take ½ cup 3 times per day after meals. Second day, take 3 teaspoons. Will increase appetite.

W1912I JSch Prevents sicknesses of all kinds. Leaves burned in a pail in the spring and fall. Let the smoke fill the house.

W1912J KD For broken coccyx in adults (see 110, *Polygonum aviculare*, W1912A, for children's treatment). If this is not treated, it will turn into consumption. Get gum from tree, put in pan, reduce until hard. Apply as poultice. White substances of joints of bones decays under fever. Poultice pulls it out.

W1912K KD See 369, *Eupatorium perfoliatum*, W1912F.

W1914A ?? For regular consumption. Cut 1 small knot into fine pieces. Put in 8 qt. water. Boil down about ¼. Drink bellyful, vomit. Repeat for 4 days in mornings. Then a little at a time.

W1914B DJ Heal navel on newly born. Make a flour from and tie on.

W1914C RS Ghost medicine. When people have been away for some time and come back home again they burn the young branches or shoots in a pot in the house to drive away ghosts. The smoke is allowed to go through the house. Another method is to hang up a small branch of young pine when coming home again.

W1914D JJJr See 353, *Anthemis cotula*, W1914B.

F1933A SS For emetic when someone dies and you can not forget it. Boil in a gallon of water. Drink as much as you can, drinking water afterwards for 3 mornings. Face east when vomiting.

F1933B SS See 254, *Aralia nudicaulis*, F1933A.

F1933C JS See 71, *Platanus occidentalis*, F1933A.

F1938A CRE/ET Ointment for cuts, wounds. Heat pine resin with beeswax and tallow and make into a salve.

F1938B CHJJ See 185, *Prunus serotina*, F1938C.

FndA CG Cough syrup. With: (1) 185, *Prunus serotina*, FndD. Take green end needles, add bark of (1) and steep. Make syrup, add sugar.

FndB KD Deep cuts. With: (1) 255, *Aralia racemosa*, FndB. Take a handful of shoots, add a handful of roots of (1). Boil in ½ gal. water until roots soften. Wash wound with solution. Poultice with bark and roots, keeping it moist.

FndC JS Cuts, bruises, sores, scabs on face. With: (1) *Populus* sp. Take 3/4 cup pine gum, add ½ pt. tallow, and the green buds of (1). No water. Boil. Strain and cool.

FndD SG Drawing thorns and slivers. With (1) 68, *Sanguinaria canadensis*, FndE. Poultice 1 day, take off, wash with (1).
For wounds and to drive ghosts away. Needles and branches picked up after freshly blown down by wind.
To wash eyes of a person who has seen a dead person. Burn windblown branches. Inhale smoke, let smoke in eyes to wash.
Spring emetic. Put bundle of windblown branches in 5 qt. water. Boil, drink warm 3 times in morning.
Blood tonic. Put bundle of young trees in 3 qt. water with 1 cone. Drink. Do not vomit.

H1973A Anyms. See 99, *Ostrya virginiana*, H1973B.

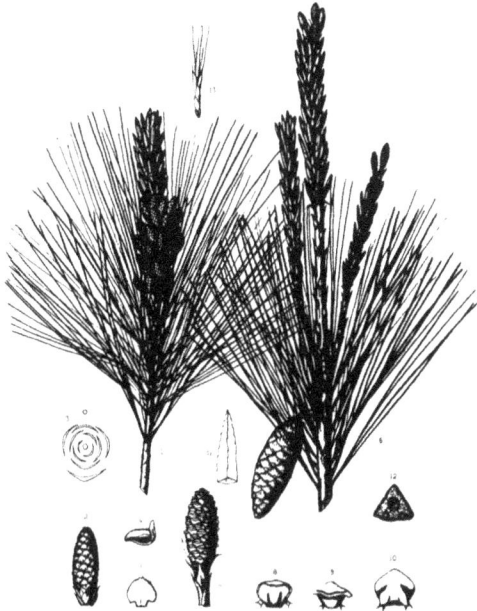

P. strobus

H1973B Anyms. See 87, *Quercus ilicifolia*, H1973B.

H1973C Anyms. Better than penicillin. For poison ivy too. Use knots, the older, the better. Make shavings. Boil. Drink.

H1973D TL Colds, sore throats, infections. Take bark from limb, roll it and boil to make steam (needles used too). Inhale steam for head cold. Pitch made into salve like "Vicks" and put on chest for chest colds (mix with honey). Salve for infections and sore throats too. Put on stick and suck.

H1973E ON Scrofula. Boil twigs, drink. Boils on horses neck. Use needles.

35. *Tsuga canadensis* (L.) Carr.   HEMLOCK, NORTHERN HEMLOCK

W1912A JA Coughs and colds.

W1912B BB Rheumatism. Put handful of leaves and bark in 6 qt. water. Boil until steams. Put hot brick in water. Put blanket over pot and sit over it

as long as desired. Then soak feet. Do 3 times a week.

W1912C KD For complaint. Blood gets bad, cold. Starts with fever and soreness. Person tired. With: (1) 93, *Betula lenta*, W1912C; (2) 450, *Cypripedium calceolus*, W1912A; (3) 31, *Larix laricina*, W1912C; (4) 38, *Thuja occidentalis*, W1912B. Take a handful of this plant, add a handful of twigs of (1) and (4), a handful of bark of (3), and a handful of roots of (2). Put in cold water, 1 gal. Boil down to ½ gal. Take 1 wineglass every hour.

W1912D JA Coughs, colds, rheumatism. See 34, *Pinus strobus*, W1912C.

W1912E JJ Venereal disease. With: (1) 31, *Larix laricina*, W1912A; (2) 222, *Cornus foemina*, W1912B. Take scraped bark, add scraped bark of (2) and pitch and rotten wood of (1). Put in pouch, pound up, and sift. Put this in a cotton bag. Place penis in bag and tie around waist.

F1933A SS See 254, *Aralia nudicaulis*, F1933A.

F1933B CHJJ Colds. Take small bundle of needles, smash. Boil in 1-2 qt. of water. Drink when thirsty.

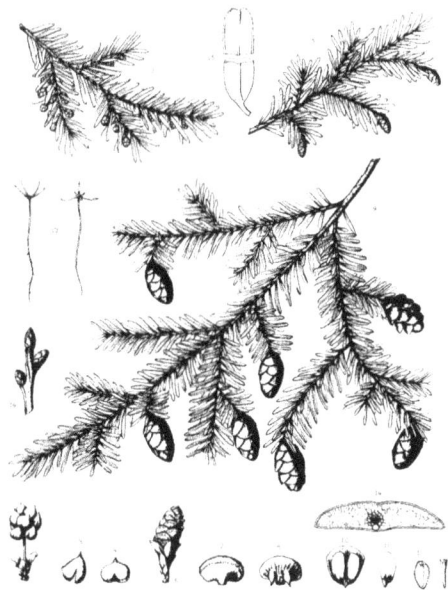

T. canadensis

F1938A CHJJ See 23, *Polypodium virginianum*, F1938A.

F1939A PH See 26, *Onoclea sensibilis*, F1939A.

H1973A DC For colds. Makes you sweat. Bundle end twigs and bark for 10 min. if chilled. Drink a lot. If mild cold with fever, drink cold.

CUPRESSACEAE (CYPRESS FAMILY)

36. *Juniperus communis* L.   COMMON JUNIPER, LOW JUNIPER, PASTURE JUNIPER

W1912A DJ Colds and coughs caused by becoming overheated and then getting chilled. The blood turns to water. Put a small bundle in 10 qt. water and boil down to 9 qt. Drink ad lib.

W1914A JJJr No use given.

J. communis

37. *Juniperus virginiana* L.   RED CEDAR, SAVIN

W1912A ED For the water, a diuretic. With: (1) 375, *Inula helenium*, W1912E; (2) 15, *Lycopodium*

J. virginiana

sp., W1912A. Boil a handful of leaves with the roots of (1) and the whole plant of (2) in 8 qt. water until 4 qt. remain. Take a tablespoon before meals.

W1912B JA See 34, *Pinus strobus*, W1912C.

W1914A JJJr No use given.

38. *Thuja occidentalis* L.   NORTHERN WHITE CEDAR, ARBOR-VITAE

W1912A JA Cuts, bruises, sprains, sores. With: (1) 30, *Abies balsamea*, W1912B. Pound and boil sprigs and leaves. Wash affected sores with solution as hot as you can stand. Use leaves as poultice.

W1912B KD See 35, *Tsuga canadensis*, W1912C.

W1912C KD Weakness in the hips due to untreated broken coccyx. With: an unspecified plant referred to as *odera* [*ohté:ra* = "root"]. Take a single handful of leaves and a double handful of roots of (1). Boil 20 min. in a gallon water. Cool to warmth of bath water. Put in pan and dash the solution on the hips.

FndA NL Rheumatism. Also for women after parturition with 30, *Abies balsamea,* FndD. Make a steam bath using the branches.

FndB PH Rheumatism. Boil tips in large tub of water. Patient puts feet in water as hot as can stand. Cover with a blanket. Take about 3 such baths.

FndC ?? A tea of cedar was used by women during confinement.

H1973A TL Colds. Leaves boiled, steam inhaled.

T. occidentalis

ANGIOSPERMS
MAGNOLIACEAE (MAGNOLIA FAMILY)

39. *Magnolia acuminata* (L.) L.   CUCUMBER TREE, CUCUMBER MAGNOLIA

M. acuminata

W1912A ED Toothache. Take inside bark from east side of tree, scraping upwards. Place handful in cup of water. Steep. "Chew the water" [out of the saturated bark?].

W1912B JJ See 227, *Euonymus europaea,* W1912B.

FndA DJn Trunk for masks, canoes.

LAURACEAE (LAUREL FAMILY)

40. *Lindera benzoin* (L.) Blume   SPICEBUSH, BENZOIN-BUSH, BENJAMIN-BUSH

W1912A HJn Colds. Steep twigs. Make tea.

W1912B JJ Gonorrhea and syphilis. With: (1) 229, *Euonymus* sp., W1912B. Take root of 1 plant with 5 in. sections of root of (1) bundled 1½ in. thick. Put in 4 qt. water, boil down to ½ qt. Drink as much as can.

W1912C JJ See 247, *Rhus* sp., W1912A.

W1914A DJ When a person has cold sweats—not very sick at first. With: (1) 403, *Arisaema triphyllum*, W1914C. Put 1 stem from the root up with 1 plant of (1) in 8 qt. water and boil quite a bit. Steam whole body, covering whole body with a blanket. Put heated stone in water too.

F1933A CHJJ Colds, measles—causes measles to erupt. Strip leaves, bundle twigs, boil to tea.

F1938A AJ See 72, *Hamamelis virginiana*, F1938A.

H1973A DC Tea. Bark off twigs boiled up.

H1973B ON Tea. Twigs put in deer meat that is being cooked, for spicing.

41. *Sassafras albidum* (Nutt.) Nees   SASSAFRAS

W1912A BB Rheumatism. With: (1) 241, *Staphylea trifolia*, W1912B; (2) 158, *Pyrola elliptica*, W1912F. Put equal amounts of each in 1 qt. whisky. Take tablespoon before breakfast and at bedtime.

W1912B HJn For fever in women [a woman] after childbirth or when [she] has come down with a cold. Steep roots 4 in. long, 1½ in. diameter in a gallon water for ½ hr. For fever, drink cold, for chills drink hot—a teacup 5 times a day.

F1933A SS Blood purifier. Peel root bark, pound. Make infusion, drink.

F1933B CHJJ For blood. Steep root bark, drink tea.

F1933C CHJJ See 225, *Celastrus scandens*, F1933A.

F1938A AJ Blood purifier. Pound root, scrape bark, put in 2 qt. water, make tea. Boil 3 min.

F1938B AJ Eye inflammation or cataracts. Remove liquid from inside stem. Add a little water. Drop mixture in eye.

FndA AJm Bathing sore eyes. Make tea, bathe eyes.

FndB DJn Blood medicine, nosebleed, high blood pressure. Leaves for wounds, cuts, bruises as poultice. Boil pith of new sprouts from ground.

FndC CRE/ET High blood pressure. Gather roots in spring, boil in water, make tea, drink.

H1973A DC See 434, *Smilacina racemosa*, H1973A.

H1973B DC Eyewash. Take white material from inner bark of new growth. Pour boiling water over this. Wash eyes with this.

H1973C Anyms. Blood pressure—thins blood.

H1973D TL Tonic.

H1974A ON Diabetes. Clears blood. Make tea.

S. albidum

ARISTOLOCHIACEAE (BIRTHWORT FAMILY)

42. *Asarum canadense* L.   WILD GINGER, ASARABACCA

W1912A LJ Colds and fevers in children. Preparation not given.

W1912B LJ Protects other medicines from contamination. Also prevents bad dreams caused by the dead. Make a tea and drink.

W1912C KD Convulsions in children. With: (1) 103, *Claytonia virginiana,* W1912C. Make flour of both roots, use teaspoon of each in 4/5 gal. of hot water that is boiling hot. Let cool. Take a teaspoon often.

W1912D HJn Fever. Steep 4 pieces of root 2 in. long in ½ qt. water for 5 min. Take ½ cup warm before meals. Lie down, cover with blankets, sweat. Repeat 2-3 times a day.

W1912E ED For headache and fever in child. Take 3 3-in. pieces of root, smash, put in 1 qt. water that is cold. Take ½ teacup when chilly.

W1912F MS Fevers, typhoid. With: (1) 375, *Inula helenium,* W1912G; (2) 301, *Mentha spicata,* W1912B. Put all in mortar and pound, dry in oven, sift. Add teaspoon of powder to 1 qt. boiling water. Put back in stove to simmer. Take teacupful when needed. Take cold if hot and hot if cold.

W1912G KD For when babies cry until they hold their breath. With: (1) 394, *Sonchus asper,* W1912A. Cut roots into small pieces, take 1 teaspoon. Take small pinch of (1). Place both in 4/5 glass of hot water, let cool. Take 1 small teaspoon, often.

W1912H BB For long lasting headache. Make tea and drink teacupful. Cover head with quilts and make patient sweat.

W1912I DJ For any kind of fever and for sweating. Plant may be added to all kinds of medicine to make them stronger. Put root of 1 plant in 1 cup cold water, mash root a little and let stand a little—about 20 min. Drink all or ½. Repeat, if neces-sary.

W1912J PJ Ghost contamination. Make a tea or decoction and wash the hands and face with this, drink some. Also sprinkle over shoulder.

W1912K KD See 437, *Trillium* sp., W1912B.

W1912L RS Urinary disorders. With: (1) dirt from an ant hill. Put 4 roots in 1½ qt. water, boil down ½. Add a little of (1). Drink as much as you can as often as you can.

W1914A DJ For when people are sick and are not to be seen. Make a tea of this, wash hands and faces and sprinkle clothes, then you can go in and see the sick. Person is not afraid when told that this medicine has been used.

W1914B RS If you have been to a funeral, you are not fit to visit the sick. This medicine makes you fit. Make a decoction, drink some, wash hands and face.
Also for lack of appetite. Smash roots of 1 plant, put in 2 qt. water, boil quite a bit. Drink all you want.

W1914C JD For when a person is not fit to see a sick person (e.g., during menses, after attending a funeral) who is using good medicine. Smash 1 root in ½ cupful water. Wash face and hands and take the sick person a little to drink.

WndA ED Antipoison charm. Chew root when in crowd and no one will poison you.

F1933A CHJJ Scarlet fever, colds, peevies. Boil roots, drink.

F1933B CHJJ See 375, *Inula helenium,* F1933B.

F1938A CHJJ For stoppage of urine. With: (1) 130, *Citrullus colocynthis,* F1938A; (2) 331, *Galium* sp., F1938A. Take 4-5 roots, mix with ½ handful of seeds of (1) and roots of (2). Boil down. Drink a lot.

F1938B HRE Fevers, colds, stimulant. Fever in summer. Boil roots of 4-5 plants in 2 qt. water a short time. Drink like water.

F1938C CHJJ See 375, *Inula helenium,* F1938B.

F1939A JG Fever. Boil root and whole plant.

F1939B JJm See 56, *Hydrastis canadensis,* F1939A.

FndA CHJJ & AJm Fever. Boil root in quart water. Drink all you can.

FndB SS Coughs and measles. Boil root in quart of water, drink all you can.

FndC KD Spring tonic for the old. Works as a physic. Pound root, steep, drink.

FndD JG See 263, *Osmorhiza claytonii,* FndB.

FndE SS See 375, *Inula helenium,* FndA.

FndF CG Colds, typhoid fever. Take a little of the root and steep it up. Drink every little while—get sweat pretty easy from that.

H1973A Anyms. Fever, no matter where, will cut it down, and chest congestion. Cut up, steep, put on where fever is or drink 2 cups. Then vomit.

H1973B Anyms. See 99, *Ostrya virginiana,* H1973B.

H1973C Anyms. See 254, *Aralia nudicaulis,* H1973C.

H1973D DC Fever, chills; makes you sweat. Cook a handful of roots, drink.

H1973E Anyms. See 434, *Smilacina racemosa,* H1973C.

H1973F Anyms. Boils, laziness. With: (1) 190, *Rubus idaeus,* H1973C; (2) 256, *Panax quinquefolius,* H1973B. No details given.

H1973G TL Fever, just tea, and a general tonic.

Used with other unspecified plants. Take when first getting cold.

H1973H ON For horses that are sick from not being used. Mix with whisky after boiling.

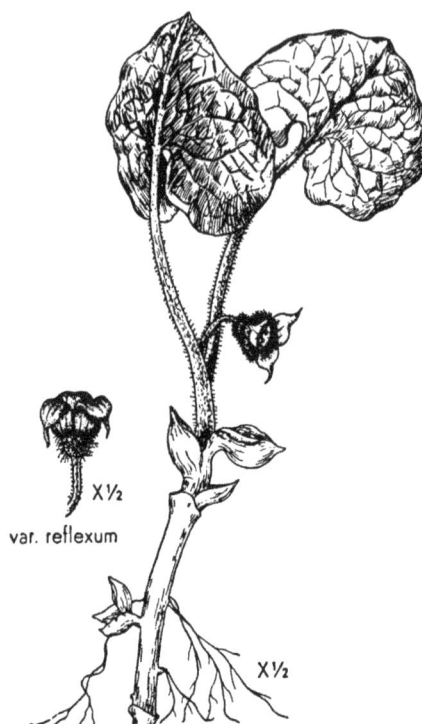

var. reflexum

X½

X½

A. canadense

NYMPHAEACEAE (WATERLILY FAMILY)

43. *Nuphar lutea* (L.) Sibth. & Sm.   YELLOW POND-LILY, COW-LILY, SPATTERDOCK, BULL-HEAD LILY

W1912A PT For swollen abdomen. With: (1) 222, *Cornus foemina,* W1912C. Boil large bunch in 1 gal. water. Cool. Wash stomach. Also drink some. If continues, apply boiled roots as poultice.

W1912B JJ Smallpox. The root has holes in it just the same as a person who has recovered from smallpox. With: (1) rattlesnake. Chop up 2 handfuls of root, put in 2 qt. water and add 2 in. of body of (1). Stir. Drink it all as quickly as possible. Guaranteed to dry up disease, no matter if person is half dead it will cure.

W1912C DJ A cause of consumption. See 414, *Scirpus tabernaemontanii,* W1912A, and 304, *Prunella vulgaris,* W1912D, for the cure for this.

W1912D KD See 57, *Ranunculus abortivus,* W1912E.

W1912E KD See 437, *Trillium* sp., W1912B.

W1912F KD See 123, *Sarracenia purpurea,* W1912A.

W1912G KD See 57, *Ranunculus abortivus,* W1912C, W1912D.

W1914A ?? Hung up in house to ward off witches. Makes the witch blind when he or she comes into the house.

W1914B RS Ghost medicine. After the death of a member of the family, the roots of this plant are sliced up and left standing in cold water—the water then being sprinkled in the person's bedroom and all the rooms of the house, also in the stable and barn, or wherever the person might go. Also, around the outside of the house. A good sized piece of root is hung up in the house. Tobacco is sprinkled when preparing this medicine, and an invocation is made stating the purpose of the medicine and asking the dead people to stay a-way.

W1914C JD To keep witches away. Hang up inside. Also for witchcraft diseases. Get fresh root and tie on sore place. Tobacco used in telling what to be used for.

F1938A CRE/ET Antiwitch. Take root, hang over door. When witches look toward your house, they will only see a pond.

FndA NL Heart trouble. Dry and scrape with grater. Put in hot water and drink a little.

FndB KD See 396, *Taraxacum officinale,* FndA.

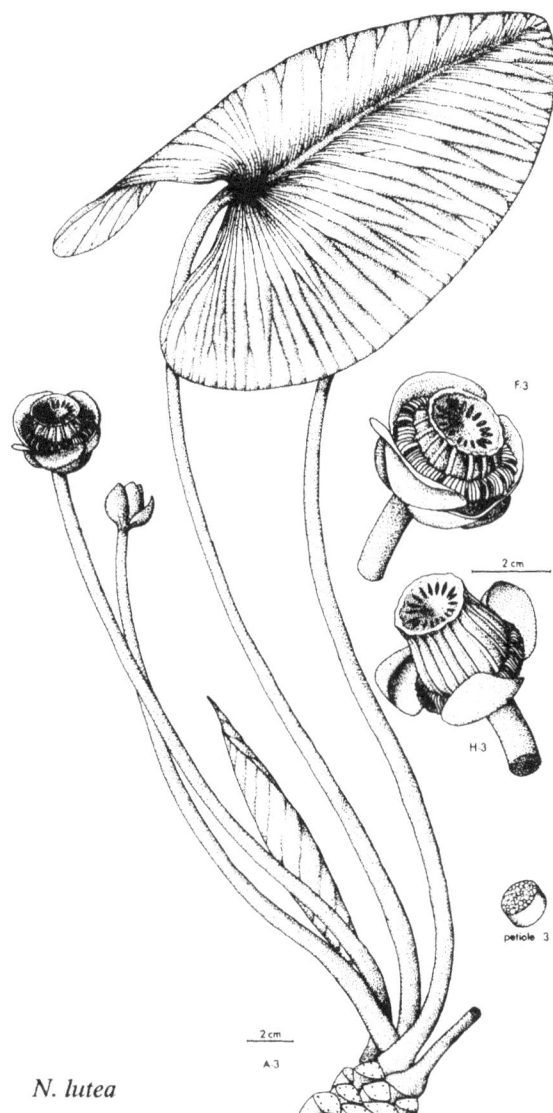

N. lutea

44. *Nuphar* sp.   POND-LILY

W1912A KD See 375, *Inula helenium,* W1912D.

RANUNCULACEAE (CROWFOOT FAMILY)

45. *Actaea pachypoda* Ell.   WHITE BANE-BERRY, WHITE COHOSH, DOLL'S-EYES

FndA SS For when a man urinates blood. Boil root in 3 qt. water for 10 min. Drink glassful 4 times a day.

46. *Actaea spicata* L. ssp. *rubra* (Ait.) Hultén   RED
BANEBERRY, RED COHOSH, SNAKE-
BERRY

W1912A KD See 91, *Alnus incana*, W1912A.

W1914A DJ For when there is something wrong
with a dog's head and he will not hunt anymore.
Put 1 root of plant in 1 qt. water. Hold dog's
mouth shut and pour down nose so that it will get
into his head.
Also for rheumatism or anything wrong with the
legs. Wash legs with the liquid.

W1914B JD See 267, *Gentiana andrewsii*, W1914A.

H1973A Anyms. For giving young men the right
sense. Drink and sprinkle on head.

47. *Anemone canadensis* L.   CANADA
ANEMONE, WIND-FLOWER

W1912A DJ To counteract witch medicine made
from 369, *Eupatorium perfoliatum*, W1912G, which
was been placed in one's liquor flask. Put the same
quantity of this plant in the liquor flask as *E.
perfoliatum*.

F1938A CHJJ For worms. Kills them inside—cuts
them up. Boil root in 1 qt. water. Drink all possible. Worms come up—can hear them squealing
inside. Something you eat they do not like and
they bother you.

48. *Anemone virginiana* L.   THIMBLEWEED,
TALL ANEMONE

W1912A DJ Antilove medicine—will take off bewitchment. With: (1) 311, *Linaria vulgaris*,
W1912A. Make 2 small bunches of the plants,
smash and steep. Take 3 qt. and vomit. Use rest to
vomit a second time.

W1912B DJ Emetic. Roots mashed and left in a
cup of water all night. Take in morning.

W1912C DJ To tell if wife is crooked (unfaithful).

A. virginiana

Put roots under pillow and you will dream about
the matter and learn the truth.

W1912D DJ Medicine for revenge against a man
who has played a trick on a man's son. Person will
go crazy or die. Get root and tell it and tobacco to
kill man who played trick. Can be undone by using
same plant and tobacco.

W1914A DJ Used as a cure for love medicine
made from 122, *Malva neglecta*, W1914A. Put 2
roots in 6 qt. water, boil quite a bit. Drink quite a
lot to vomit. If in a hurry to get better, wash head
first, then body.

W1914B DJ See 314, *Penstemon hirsutus*, W1914A.

F1933A SS Tuberculosis. Pound handful of roots.
Then cook whole plants in 2 qt. water.

FndA FSt Diarrhea. Boil roots of 2 plants in quart water. Cool it. Drink all you can.

FndB JS Love medicine for either sex. Soak stems and roots in warm water. Wash face in it. Think how you want the girl or boy to behave.

49. *Aquilegia canadensis* L.   WILD OR RED COLUMBINE, ROCK-BELLS, MEETING-HOUSES

W1912A ED For kidneys. Steep handful of roots in 1 qt. hot water. Take 1 teacupful 3 times a day before meals.

W1912B DJ See 329, *Galium aparine,* W1912A.

W1912C KD See 437, *Trillium* sp., W1912C.

C. palustris

A. canadensis

50. *Caltha palustris* L.   MARSH MARIGOLD, COWSLIP

W1912A DJ Emetic against a love charm. The charm will no longer work after this. Smash roots of 1 plant, steep a little in 6 qt. water. Take early in the morning and vomit. Repeat for 4 days.

W1914A ?? Cure for love medicine. Smash roots of 1 plant in cupful of warm water. Drink right away and vomit.

W1914B JD Emetic. Cut up roots finely, make a small bag of a white cloth, put in ½ cup water, let stand all night. Drink before sunrise, follow with warm water.

H1973A Anyms. Not given.

51. *Cimicifuga racemosa* (L.) Nutt.   BLACK SNAKEROOT, BLACK COHOSH

W1912A ED To promote flow of milk in women. Steep handful of root in gallon of water. Take 1 glass 3 times a day before meals.

F1938A HRE & SRE Rheumatism. Steep roots, bundle with blankets, and sweat over its steam.

F1938B CRE/ET (?) Backache in baby. Baby cries when on back, is cranky, irritable. Place a handful of fresh leaves in a bag. Put the bag on the baby's back as is.

F1938C CHJJ See 405, *Symplocarpus foetidus*, F1938C.

F1938D CRE/ET (?) See 444, *Smilax herbacea*, F1938A.

F1939A CG Rheumatism. Boil roots in kettle, wrap in blanket, sit over steam.

F1939B DJ Blood purifier. Use root. No further details.

FndA FSt Sore back of baby. Smash leaves and poultice.

FndB CG For back disorder. Pound leaves, spread on cloth. Wet with whiskey and apply to back.

FndC SS Rheumatism. Boil root in water. Place feet in water, wash joints and limbs.

H1973A DC Arthritis. Put in cider. No other details.

C. racemosa

C. virginiana

52. *Clematis virginiana* L.   VIRGIN'S-BOWER, TRAVELLER'S-JOY

W1912A BB For sore penis (venereal disease). Dry root, sprinkle on sores. Also steep handful of roots in 1 qt. water until thick like cream. Wash with this solution.

H1973A DC Kidney trouble—when it burns. Drink by the quart.

H1973B Anyms. Gives you strange dreams. Cut stem 2-3 in. and cook or boil 5 min. Put liquid on your face and hands.

53. *Coptis trifolia* (L.) Salisb.   GOLD-THREAD(S), CANKER-ROOT

W1912A MS For sore mouth in children. Steep 3 small roots in a teaspoon of water, wash out mouth.

W1912B ED Sore mouth in children. Dry root, powder, and apply this to the sore places.

W1912C DJ For when you vomit sour stuff. May get yellow and break out all over. Biliousness, jaundice. Smash 6 roots about 1 ft. long and steep in a cup of water. If bad, drink up right away.

Make second batch from same roots if necessary, but do this only once.

W1912D SH Sore eyes, mouth, throat. Bunch roots size of little finger, steep in ½ qt. water.

W1912E RS Trench mouth. See 7, *Parmelia conspersa,* W1912A.

W1914A DJ For little babies when they throw up often. Put a little bunch of the roots in 1 qt. water. Boil a little. Take 1 teaspoon at a time.

F1938A CHJJ Sore mouth. Chew root tendrils.

F1938B JH Jaundice; yellow skin, eyeballs, dizziness. Boil a handful of roots in gallon of water until the scum comes on top. Drink until vomit.

F1938C AJ Sore mouth in baby. Put warm water over a few dry roots. Steep. Apply to inside of mouth with teaspoon, a little at a time.

F1939A JS Sore eyes. See 56, *Hydrastis canadensis,* F1939B.

FndA FSt Sores in mouth. Make solution for root, gargle.

FndB JG See 263, *Osmorhiza claytonii,* FndB.

FndC CHJJ See 291, *Verbena hastata,* FndA.

H1973A DC See 434, *Smilacina racemosa,* H1973A.

H1973B DC Ulcers in the mouth.

H1973C ON Mouthwash.

H1974A Anyms. Blindness. See 331, *Galium* sp., H1974A.

H1974B Anyms. Causes baby to break out in a rash for 1 day, but after that you will have a healthy baby. Sickness caused by bad blood inherited from mother. Baby cries and cannot sleep. Boil a piece in just a little water. Let boil until water begins to turn yellow. Let cool. Give a drop or 2 to baby boy or girl.

54. *Hepatica nobilis* Mill. var. *acuta* (Pursh) Steyerm. SHARP-LOBED HEPATICA

W1912A DJ Prevents conception. Steep 3 plants in 3 qt. water for a strong dose. Drink the quantity as soon as possible.

W1912B JJJr Chewed by women to bewitch men and make them go crazy by affecting their hearts, leading to their weeping and chasing a woman (see 295, *Galeopsis tetrahit,* W1914B, for a cure for this). Woman uses tobacco to collect it. She chews it, then kisses the man.

W1914A DJ Childbirth inducement for middle-aged woman having first baby. Put a handful of

H. nobilis

plants in 6 qt. water and boil quite a bit. Drink as often as you like.

W1914B RS See 332, *Mitchella repens,* W1914B.

F1933A CHJJ Summer complaint in children. With: (1) 232, *Ceanothus americana,* F1933D; (2) 302, *Monarda* sp., F1933A; (3) 167, *Tiarella cordifolia,* F1933B; (4) *odikdegwadŏ'.* Boil (after dried) in 2 qt. water until boiling good. Take out roots. Get child to drink.

FndA AJm Blood purifier. Preparation not given.

55. *Hepatica nobilis* Mill. var. *obtusa* (Pursh) Steyerm.   BLUNT-LOBED HEPATICA

W1912A ED To tell fortune by roots. No details given.

W1912B JSl Sore abdomen experienced by pregnant woman caused by eating food handled by young girls in menses. Put 6 plants in 4 qt. water. Boil down to about 3 qt. Take ½ teacup 3 times a day.

W1912C KD See 437, *Trillium* sp., W1912C.

F1940A CHJJ See 152, *Epigaea repens,* F1940A.

56. *Hydrastis canadensis* L.   GOLDEN-SEAL, ORANGE-ROOT, YELLOW PUCCOON (RARE AND COMMERCIALLY EXPLOITED)

W1912A MS To kill love medicine, an emetic. With: (1) 387, *Solidago canadensis,* W1912C; (2) 256, *Panax quinquefolius,* W1912D. Steep root, add 1 plant top of (1) and 1 root of (2) in 1 gal. water until it colors the water. Drink as an emetic, wash face, head, hair.

W1912B MS For heart trouble or run down system. Pound 2 big roots into a white cloth and put into a pint of whisky. Take a tablespoon 3 times a day before meals.

W1912C HJn For fever and gall use. Dry and powder 3 roots, add ½ teacupful warm water. Cover with paper, stir, and drink before meals. Cover person.

F1933A KD For earache when it pains but will not burst. Person will not become deaf. With: (1) 291, *Verbena hastata,* F1933A; (2) 308, *Fraxinus nigra,* F1933A. Take roots, add 1 root of (1) and young growth of (2) about 4 in. long. Burn (2) until sap runs out. Pound roots and pour sap over them. Put 2 drops in ear for 5 min., then let juice run out.

F1938A CHJJ Whooping cough and diarrhea. Boil root, drink all you can.

F1938B JC Emetic for biliousness. Pinch root, steep in pint of water. When turns amber, remove root. Drink ½ cup at a time, warm, until it works.

F1938C AJ Liver trouble and fever. Breaks up fever. Upset, sour stomach, gas after eating. Scrape and grate root to powder. Put pinch in warm water. Take 1 cup a day.

F1939A JJm Tuberculosis. With: (1) 256, *Panax quinquefolius,* F1939A; (2) 68, *Sanguinaria canadensis,* F1939C; (3) 42, *Asarum canadense,* F1939B. Take small piece 1 in. long, mix with ½ in. of root of (1), 1 in. root of (2), 3-4 3-in. roots of (3). Steep and drink.

F1939B JJm Scrofula, sore eyes. With: (1) 53, *Coptis trifolia,* F1939A. Steep. Take bowl of warm water, put in ingredients. Then boil. Drop solution in eyes.

FndA FSt Fever and pneumonia. Steep root or pour boiling water over it. Let cool and drink.

57. *Ranunculus abortivus* L.   SMALL-FLOWERED CROWFOOT, KIDNEYLEAF CROWFOOT, CHICKEN-PEPPER

W1912A KD Sore eye from catching cold. Pick in spring. Take ¼ handful of roots and leaves and smash. Add 3/4 glass boiling water and pour off again. Strain. Wash eye 4 times a day and at bedtime.

W1912B KD When girls faint suddenly. Is a blood disease. When blood is too strong or weak. When foaming at the mouth accompanies fainting, use both of the ingredients given. With: (1) 43, *Nuphar lutea*, W1912D. Cut up roots, a handful. Dry root of (1) and scrape it. Put 2 teaspoons of these in a pint of water. Steep. Put in bottle and cork it. Then, let it stay in the sunlight for 4 hr. Take wineglassful anytime when head feels badly. If recovery is not immediate, soak white or yellow paper in the liquid and slap it on the forehead.

W1912C KD Epilepsy in men. With: (1) 323, *Lobelia cardinalis*, W1912E; (2) 43, *Nuphar lutea*, W1912G; (3) 15, *Lycopodium* sp., W1912B. Handful of dry root with large handful of (1), 5 tablespoons of (2), small double handful of (3). Bundle tightly. Will smell like muskrat. Pour a quart of boiling water over these and leave boiling on stove for about 20 min. Take wineglass every 20 min. for 1 hr. Then, take 4 hr. apart. When better, an unspecified emetic is taken.

W1912D KD Epilepsy in women. With: (1) 323, *Lobelia cardinalis*, W1912E; (2) 312, *Mimulus ringens*, W1912A; (3) 43, *Nuphar lutea*, W1912G; (4) 266, *Sium suave*, W1912C. Take 1 handful with 1 handful of (1) that has a little stem and some flowers, 2/3 handful of (2), and 4 tablespoons of the scraped root of (3) (the fourth plant is not mentioned). Put roots in vessel and pour 1 qt. boiling water over them. Cover and cool slowly. Drain. Take wineglassful every ½ hr. until better. This medicine is followed by an emetic made from *Ranunculus abortivus* and (1) 402, *Acorus calamus*, W1912E; (2) 43, *Nuphar lutea*, W1912G. Take a good handful, add a small handful of (1) and (2). Boil hard for ½ hr. in 1 qt. water. Drink lukewarm, 1 cup, 20 min. apart. Vomit is then buried.

W1912E JSI Sore teeth. No details given.

W1912F KD See 437, *Trillium* sp., W1912C.

W1912G KD See 114, *Polygonum punctatum*, W1912A.

W1914A DJ Stomach trouble. Rotten stomach when you do not know what the matter is. Also

R. abortivus

good for counteracting poison another has given you. Good for snakebite too. Smash handful of roots, boil a little in ½ cup water. Drink and follow with warm water. Vomit. For snakebite, wash with the liquid.

W1914B DJ Smallpox. Boil a handful of the roots in 1 pt. water. Take in a day, make more, repeat. It dries up the pox.

58. *Ranunculus acris* L.   COMMON BUTTER-CUP, TALL OR FIELD BUTTERCUP (NATUR.)

FndA KD Diarrhea. Also for pains or cold in chest. For former, steep 1 root 4 in. long in warm water until changes color. Drink 1 cup, repeat as needed. For latter, make mustard plaster by smashing whole plant.

R. acris

R. bulbosus

59. *Ranunculus bulbosus* L.   BULBOUS CROW-
FOOT OR BUTTERCUP, MEADOW-BLOOM
(NATUR.)

W1912A DJ Venereal disease and sore teeth. For
the former, dry and cut up 4 plants and boil them
down a little in 3 qt. water. Drink often until
quantity is gone. For latter, cut off little bulb and
plug the cavity. Will break up the tooth.

60. *Ranunculus hispidus* Michx. var. *nitidus* (Muhl. ex
Ell.) Duncan (*R. septentrionalis* Poir.)   SWAMP
BUTTERCUP

FndA FSt Toothache. Will break up tooth and
come out by itself. Put root in cavity.

R. hispidus

61. *Ranunculus recurvatus* Poir. ex Lam.
HOOKED OR ROUGH BUTTERCUP

W1914A DJ For sore or hollow teeth—it kills the worms. Chew the roots.
Also for 4 kinds of venereal disease. Boil down a handful of roots in 1 qt. water, a little. Take ½ cupful 4 times a day.

W1914B JD See 355, *Aster macrophyllus,* W1914A.

62. *Ranunculus* sp.   CROWFOOT, BUTTERCUP

W1914A DJ See 289, *Myosotis* sp., W1914A.

63. *Thalictrum dioicum* L.   EARLY MEADOW-RUE

W1912A KG Sore eyes from cold in head. Boil roots of 2 plants in 1 qt. water down to ½. Wash eyes with it first, then tie wet cloth over eyes.

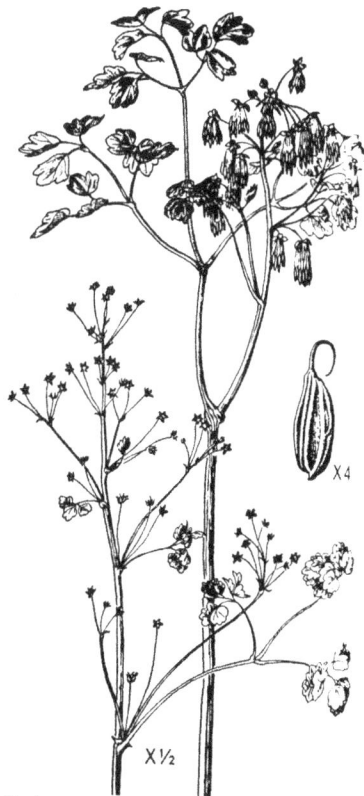

T. dioicum

F1933A SS Good for the heart—for heart palpitations. Boil a handful of the roots in 1 gal. water. Boil down to 1 qt. Drink cool as water.

F1938A CRE/ET Good for the heart, any disturbances. Put 5 short roots in 2 qt. water, bring to boil, let cool. Drink abundantly when thirsty throughout the day.

H1973A Anyms. Makes you crazy. Used the same way as 76, *Cannabis sativa,* H1973A.

64. *Thalictrum pubescens* Pursh   TALL MEADOW-RUE, FALL MEADOW-RUE, LATE MEADOW-RUE

W1912A JJJr Nosebleed. Smash, bundle up. Put handful in teacup of water, let stand 10 min. Wet the head and neck with this and rub hard.

F1938A CRE/ET Gall medicine. With: (1) 68, *Sanguinaria canadensis,* F1938F. Make tea of roots. Drink as needed.

BERBERIDACEAE (BARBERRY FAMILY)

65. *Caulophyllum thalictroides* (L.) Michx.   BLUE COHOSH, PAPOOSE-ROOT

W1912A DJ For any kind of fever. Take roots of 1 plant (2 if very bad) and boil in 8 qt. water until 7 qt. remain. Take at least 3 times a day before meals.

F1933A CHJJ See 294, *Collinsonia canadensis,* F1933A.

FndA CHJJ Not given.

FndB DJn Did not know.

FndC ?? Tonic. Roots used. No details given.

FndD FSt Good for gallstones. Smash root, put in teacup, add hot water, drink. Lie down until you get ready [to vomit?], then drink hot water and vomit.

J. diphylla

66. *Jeffersonia diphylla* (L.) Pers.   TWIN-LEAF, RHEUMATISM ROOT

W1912A DJ Diarrhea in children and adults. Also for gall. Boil 1 plant (whole) in 4 qt. water down to 3 qt. Take a bellyful right away.

67. *Podophyllum peltatum* L.   MAY-APPLE, WILD MANDRAKE, GROUND LEMON

W1912A JJJr Sweeny or atrophy of shoulder muscle of horse. Lance and make a hole inside. Put in the seeds and pulp of a ripe fruit. This will draw out the infection. Repeat if necessary.

W1912B ED Laxative, used in spring before flowers. Cut out where rootlets branch—this part is poisonous. Cut root into 4 parts an inch thick. Put in quart of warm water and leave over night. Drink 1 cup in morning, repeat 3 days.

W1912C KD Anciently used as a cathartic. Add tablespoon [of roots?] to pint of water and boil down to a cup. Roots will soften. Take 2 tablespoons 3 times daily about 2 hr. before meals.

W1912D SH Strong physic. Take root size of little finger, wash and chew up raw. Must be careful—very strong.

W1912E JJm Laxative for horse bound up on green grass. Boil 2 3-in. pieces [of root?] in 1 qt. water, down to ½. Put in bottle and give through the nose. One dose is enough.

W1912F BB Root poisonous, but can eat fruit.

W1912G DJ Increases strength so that one is able to physically punish someone suspected of bewitchment. Use not clear. With: (1) 408, *Luzula* sp., W1912A; (2) 272, *Asclepias incarnata*, W1912B. Boil 1 plant of each a little in ½ qt. water. Also rub roots on the hands.

W1912H DJ Corn medicine. Keeps away birds and worms. Root mixed with water for sprouting corn.

W1912I KD Anciently used for boils. See 354, *Arctium* sp., W1912E.

W1913A JJJr Laxative. Chew a piece (of root) the length of the first joint of your little finger.

W1914A JD Physic. One root in 1 qt. water, cold. Smash, let stand a little while. Middle of the afternoon, take 1 tablespoon, and after a little while, take another. At bedtime, take ½ teacupful.

F1933A CHJJ For constipation. Take 12 roots the width of the hand, roast them until nearly burnt. Put in 2 qt. water, boil down ½. Then cool. Drink before going to bed. Lie still. Take again in 1 day after well cleaned out.

F1933B SS Thinks it is poisonous. She hates it.

F1938A AJ Thinks it is poisonous, but can eat the apples or oranges.

F1939A HJ Physic. Knows it is poison. Learned from his grandmother how to make nonpoisonous. Cut out knobby joints of rootstock and roast what is left.

FndA JG Laxative. Did not know it was poisonous. Boil whole plant, roots and all.

FndB FSt Knew it was poison and that some drank it.

FndC AJm Laxative. Root used.

FndD SH Physic, but must be careful. Chew piece of root about length of little finger.

P. peltatum

H1973A DC Cathartic. Bake roots until just quite dry—do not burn them. Make into a powder. Use a little (about the size of a dime) and pour over boiling water in a cup. Drink.

H1973B FS Did not know use.

H1974A DC Strong cathartic. Get root, dry, pound to powder. Use only a little (less than a dime)—pour cup of boiling water over, let stand 10 min., drink with empty stomach. Take a lot of fluids later.

PAPAVERACEAE (POPPY FAMILY)

68. *Sanguinaria canadensis* L.   BLOODROOT, PUCCOON

W1912A LJ Bleeding lungs, for menses. Steep into tea.

W1912B ED Gonorrhea. Also for sickness caught from a menstruating girl. Smash roots, put in 2 qt. cold water. Drink a quart at a time.

W1912C MS Head colds, catarrh. Dried and snuffed.

W1912D BB Head colds. Powder root, snuff.
Cuts. Cook roots, smash, poultice cuts.

W1912E KG Hemorrhages in stomach and lumps. Steep 3 roots in 3 qt. water a little. Take ½ cup 4 times a day before meals and before retiring.

W1912F JSI For pain inside. Put 3 roots in ½ qt. water, boil a little and steep. Take 1 tablespoon 3 times a day before meals.

W1912G JJ See 212, *Dirca palustris*, W1912D.

W1912H DJ See 403, *Arisaema triphyllum*, W1912B.

W1912I SH See 397, *Tussilago farfara*, W1912A.

W1912J KD See 433, *Polygonatum pubescens*, W1912A.

W1914A DJ Cramps in the stomach; hiccoughs. For former, smash 1 root and boil a little in 1 qt. water. Take cupful 3 times a day.

W1914B JD See 355, *Aster macrophyllus*, W1914A.

F1933A CHJJ Cuts on foot from axe, will stop bleeding. Boil roots in 1-2 qt. water, cool off to blood temperature and apply to wound.

F1933B JS See 189, *Rubus allegheniensis*, F1933A.

F1933C CHJJ See 304, *Prunella vulgaris*, F1933A.

F1933D CHJJ See 375, *Inula helenium*, F1933B.

F1938A CRE/ET Colds and fever. With: (1) 345, *Viburnum trilobum*, F1938C. Put 1 bunch of whole plant with piece of (1) the size of a finger, cut into 4 pieces. Take ½ cup at once.

F1938B AJ Colds. Boil 2 small roots in ½ pt. water, boil down to syrup, add sugar. Take 1 teaspoon every 2 hr. for 2 days.
Cuts and boils. Split root in ½, put in water under fire until hot, not boiling. Wash wound or put on boil.
Piles. Boil roots in water, dampen clean white cloth in solution, push piles back into intestines 3-4 times.

F1938C CRE/ET Blood medicine. Gather in spring, dry root for storage. Steep in water and drink.

F1938D CHJJ Cuts and wounds. Cook roots, wash with water, apply roots as poultice.

F1938E AJ Fever and kidneys. When kidneys over or under work. Also regulates heartbeat and makes blood redder. With: (1) 345, *Viburnum trilobum,* F1938D. Mix 2 roots the size of ½ cup and boil in 1 gal. water. Cool. Drink like water.

F1938F CRE/ET See 64, *Thalictrum pubescens,* F1938A.

F1938G AJ See 186, *Prunus virginiana,* F1938A.

F1938H FJ Poison ivy. Crush and apply juice to blisters.
See 253, *Impatiens pallida,* F1938A.

F1938I AJ See 72, *Hamamelis virginiana,* F1938A.

F1939A JG Sore eyes. Powder root, put in table-spoon boiling water, wash eyes when cooled.
Cough, fever, wounds, sores with inflammation. Take ½ qt. water and ½ teaspoon of powder and boil a short time. Drink tablespoon at a time. Make stronger for wounds.

F1939B CG For sores on legs. Poultice leaves on sores.
Asthma. Steep 3-4 roots in 2 qt. water, boil down to 1 qt., add sugar.
Cuts, sores. Rub on root.

F1939C JJm See 56, *Hydrastis canadensis,* F1939A.

F1956A SJ Colds and sore throat. Also for sores and cuts. Chew.

FndA SS Coughs and colds. Steep roots in water. Drink warm, ad lib.
Emetic. Take larger amount.

FndB AJm Cough medicine. Use roots.

FndC AJ(?) Wound medicine. Steep in hot water to extract juice, cool, drink every day.

S. canadensis

FndD SS Cuts. Mash root, put in water, put cut in water.
TB. Used as a drink.

FndE SG See 34, *Pinus strobus,* FndD.

FndF ?? Venereal disease and cuts. Pound stem.

H1973A JT Intestinal trouble—like a tonic. With unspecified plants. Preparation not offered.

H1973B Anyms. See 254, *Aralia nudicaulis,* H1973C.

H1973C DC See 434, *Smilacina racemosa,* H1973A.

H1973D Anyms. Ulcers or for women that are ugly (because it cuts the woman's strength down). Crack up dried roots in cloth, put in cup, pour boiling water over it, take spoonful. Works in 10-15 min. Use until gone.

H1973E Anyms. Swellings above the waist. Powder the root, put in cup, pour hot water over. Drink 1 big spoonful.

H1973F TL Skin infections. No details.

H1973G Anyms. See 75, *Ulmus rubra*, H1973B.

H1974A RB Panacea.

H1974B Anyms. Ulcers. Boil small piece of root in gallon water. No details.

FUMARIACEAE (FUMITORY FAMILY)

69. *Corydalis sempervirens* (L.) Pers.   PINK CORY-DALIS, ROCK HARLEQUIN, PALE CORY-DALIS

1914A DJ For piles. With: (1) 116, *Rumex crispus,* W1914B. Put 1 plant of each in 6 qt. water. Boil quite a bit. Take as often and as much as you like. May wash piles too.

70. *Dicentra cucullaria* (L.) Bernh.
DUTCHMAN'S-BREECHES or, perhaps,
Dicentra canadensis (Goldie) Walp.
SQUIRREL-CORN, TURKEY-CORN

W1912A KD Medicine for distance runners, strengthens limbs. See 399, *Alisma plantago-aquatica.* Eat root before running and immediately after.

PLATANACEAE (PLANE-TREE FAMILY)

71. *Platanus occidentalis* L.   SYCAMORE, PLANE-TREE, BUTTONWOOD

W1912A ED See 308, *Fraxinus nigra,* W1912B.

F1933A JS Skin eruptions, scabs. With: (1) 34, *Pinus strobus,* F1933C; (2) 133, *Populus balsamifera,* F1933A; (3) deer tallow. Scrape down a bushel of bark. Boil having ground it up in several gallons of water until 1 remains.

F1938A CRE/ET Eczema. With: (1) sheep oil; (2) deer tallow; (3) sweet butter. Powder 3 tablespoons bark. Boil 2 tablespoons of (1) or (2) and the same of (3). Mix these together and make a salve, pour into a mold. Rub on skin, skin peels off like bark of tree.

HAMAMELIDACEAE (WITCH-HAZEL FAMILY)

72. *Hamamelis virginiana* L.   WITCH-HAZEL

W1912A HJn Taken when one can not eat—an emetic. Scrape bark upward from a 6 ft. long stick. Boil for ½ hr. in 6 qt. water. Drink before breakfast and vomit. Do twice a week.

W1912B MS Bloody dysentery, cholera. Steep handful of bark of twigs in 2 qt. water. Drink as much as you like.

W1912C BB (cf. 249, *Zanthoxylum americanum,* W1912G) Toothache. No details.

W1912D JJJr Cold around the heart. Take handful of leaves and twigs, make a bundle, add 2 qt. water, boil down to ½. Take ½ cup in morning before breakfast, 4 times a day totally. Make ½ the quantity again.

W1912E JJ See 186, *Prunus virginiana,* W1912B.

W1914A DJ For when you catch cold and have heaves. Put 1 branch of second growth in 6 qt. water. Make a poultice and place over affected body part.

W1914B JJJr For when a pregnant woman has fallen or got hurt in any way and fears child will be born prematurely. Put a good handful of new growth shoots in 4 qt. water. Boil pretty well. Drink as often as you like.
Also for bruises on body. Same as above, also apply bark as poultice.

F1933A SS Blood purifier. With: (1) 157, *Chimaphila umbellata,* F1933F. Make bundle of tips, sprouts of succours and combine with stem and root of (1) — leaves removed. Boil in 1 gal. water. Drink any amount, anytime.

F1933B CHJJ See 375, *Inula helenium,* F1933B.

F1938A AJ Panacea. With: (1) 40, *Lindera benzoin,* F1938A; (2) 256, *Panax quinquefolius,* F1938C; (3) 68, *Sanguinaria canadensis,* F1938I; (4) *odadeo?dasta?*; (5) *o?tcohgwa:?*; (6) 129, *Viola*

sp., F1938A. Use 1 stick cut up into 4 in. lengths with root of (1)-(6). Bundle and boil in kettle down to 1 gal. Put away in cool place. Drink 1 glass, 3 times a day.

F1938B CHJJ Emetic. Peel bark upward, boil, drink all you can early in the morning before breakfast.
Venereal disease. With: (1) 91, *Alnus incana*, F1933A. Prepare same as above, but add (1).

F1938C HRE Emetic. Scrape bark upward. Boil handful in 10 qt. water until it goes down a little. Drink lukewarm early in the morning.

F1938D AJ TB—when you catch cold and stop coughing. No appetite. With: (1) 217, *Cornus alternifolia*, F1938A. Scrape bark from both until you get a cupful. Boil in 2 gal. water, about 10 min. Boil hard. Drink as much as possible until it (i.e., vomit) comes up.

F1938E CRE/ET See 97, *Corylus americana*, F1938A.

F1938F AJ & CRE See 186, *Prunus virginiana*, F1938A.

F1939A NL Tea for medicine for coughs and colds. Bundle young branches and boil for a long time. Take hot.

F1939B PH See 26, *Onoclea sensibilis*, F1939A.

FndA AJm Astringent. Use bark. No further details.

H1973A DC Chest colds. Boil a fistful new growth twigs in 2 qt. water. Strain, drink like water.

H1973B TL Arthritis. See 133, *Populus balsamifera*, H1973B.

H1973C TL Asthma, TB. Mixed with other unspecified plants.

H1973D ON Regulates kidneys. Peel bark, boil in water, poultice. Also bundle twigs, boil, drink.

H1974A DC For young person with TB or lung trouble. Scrape bark up, cook big bunch of shavings in gallon or 2 of water for 10 min. Use liquid to drink for vomiting, drink lukewarm, drink ½ of 5 qt. first time—vomit. If cannot vomit, use feather to tickle throat. Get all liquids out. Then use second 5 qt. [*sic*] the same day. Do this for 3 days. Can also drink by mouthful to heal spots and scars in lungs. Do all of this on an empty stomach before breakfast.

ULMACEAE (ELM FAMILY)

73. *Celtis occidentalis* L.   HACKBERRY, SUGARBERRY

W1912A NH Did not know use.

W1912B DJ See 93, *Betula lenta*, W1912B.

W1912C DW Woman's medicine, regulates menses. Decoction of the bark.

W1914A DJ For suppressed menses in young girls caused by working in the sun. With: (1) *gasi'sût*. Wrap tobacco in bright red ribbons and bury in ground between roots of both trees. Cut bark downward off the east side. Put small handful of bark in 10 qt. water and boil down to 1 qt. Take a good drink every time person is dry.

C. occidentalis

74. *Ulmus americana* L.   AMERICAN ELM, WHITE ELM

W1912A HJn Excessive menstruation. Scrape bark of roots from east side, 2 handfuls. Steep 1 hr. in 1 gal. water.

W1912B MS Piles caused by drinking after a girl who has menses. Take bark from east side, preserving inner bark. Steep 2 big pieces in 3 qt. water. Boil down until dark. Drink ad lib.

W1912C KD See 353, *Anthemis cotula*, W1912C.

W1912D KD See 91, *Alnus incana*, W1912A.

W1914A DJ To drink for broken bones. With: (1) 86, *Quercus bicolor*, W1914B. Take 2 in. long strip of bark from a scar on each tree. Boil in 6 qt. water. Drink as often as you like.

FndA JG & SG See 75, *Ulmus rubra*, FndA.

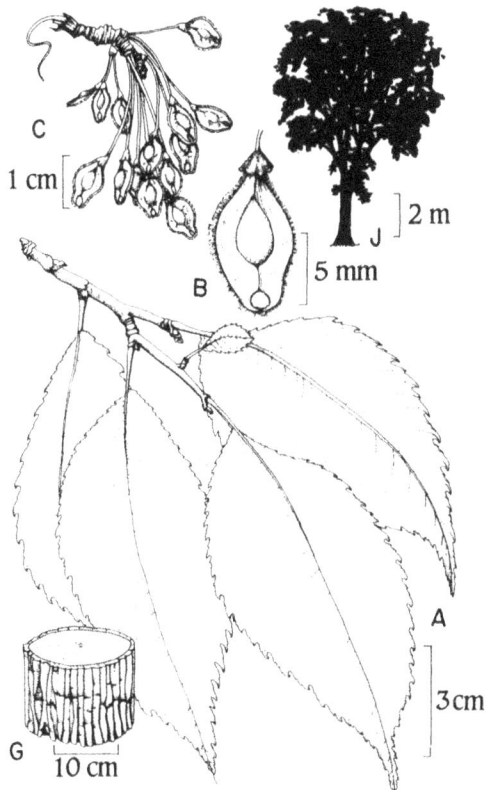

U. americana

75. *Ulmus rubra* Muhl.   SLIPPERY ELM, RED ELM

W1912A SH Infected and swollen glands said to be first stage of tubercular glands. With: (1) 403, *Arisaema triphyllum*, W1912E. Find pole about 2 in. in diameter and scrape bark from east side. Boil mashed, inside bark in 1 qt. water. Smash and poultice (1) first to draw out, then when it breaks, put poultice of bark, wetting it with the liquid.

W1912B JJJr Infected and swollen glands (see above). With: (1) 81, *Juglans cinerea*, W1912G. Boil mashed, inside bark in 1 qt. water. Poultice first with (1) — using the root bark taken about noon away from the sun. Then poultice with *Ulmus rubra*.

W1912C DJ Good when paralyzed—when you want to lie down and sleep, feel drowsy all the time and do not want to get up. With: (1) 185, *Prunus serotina*, W1912G. Find a tree with a black center and remove 4 chips about 7-8 in. long out of the heart of this tree. Strip bark on east side of (1) at about head height. Boil these in 6 qt. water. Take ½ cup anytime.

W1912D DJ Sore eyes. Take a strip of inside bark about 2 in. long and put in ½ cup warm water. Let stand 10 min. Wet cloth and put a couple of drops in eye. Then wash eye with clean water. Use solution 3 times in space of ½ hr.

W1912E ED See 86, *Quercus bicolor*, W1912A.

W1914A ?? For sleepiness and weakness (sits with palms up). With: 185, *Prunus serotina*, W1914A. Same as 1912C, above, except ingredients are put in 10 qt. water and boiled down to 2 qt. Drink enough to make vomit. Do for 4 days, then drink small amounts for 2 more days.

W1918A JJJr See 224, *Cornus sericea*, W1918A.

F1933A SS See 254, *Aralia nudicaulis*, F1933A.

F1938A AJ Dry birth. Steep 2 ft. section of bark in gallon of water. Drink 1 qt. or more.

FndA JG & SG Chew inner bark to facilitate childbirth. For parturition, use with: (1) 96, *Carpinus caroliniana*, FndA; (2) 74, *Ulmus americana*, FndA. Mix bark with chips of bark of (1). Boil until water is sticky. Let woman drink. Then mix bark with bark of (2) after birth to prevent inflammation.

H1973A DC Infected kidneys. Use bark for 2 weeks. No further details.

H1973B Anyms. To clean stomach. With: (1) 68, *Sanguinaria canadensis*, H1973G. Cook bark until hard like gravy. Drink, then take ½ cup of drink made from (1).

H1973C ON Sore throats. Chew inner bark.

H1973D FK Biliousness. Grandmother would boil inner bark in spring.

H1974A DC Kidneys. Boil bark, drink quite a bit.

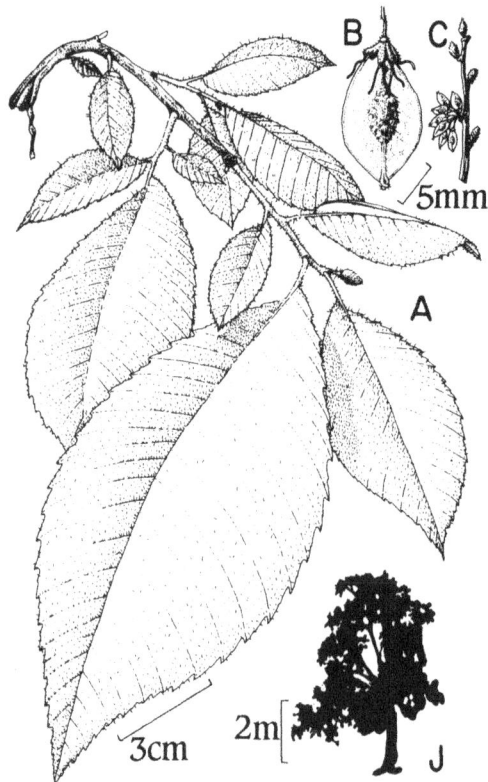

U. rubra

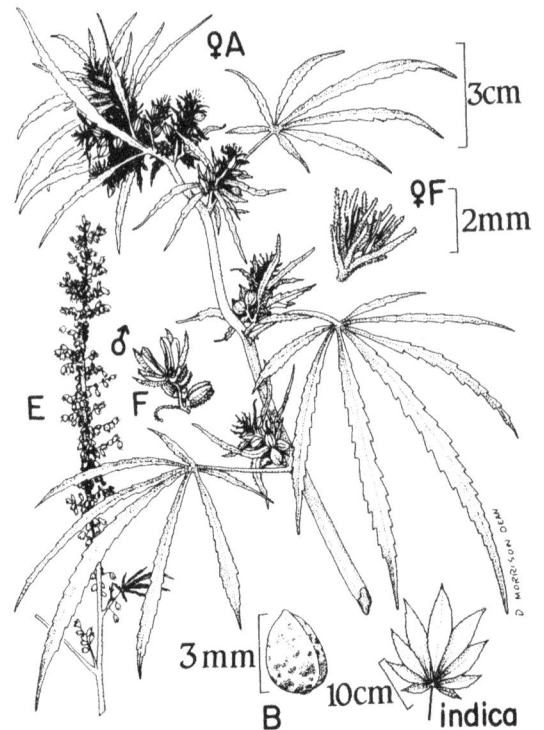

C. sativa

CANNABACEAE (HEMP FAMILY)

76. *Cannabis sativa* L.   HEMP, MARIJUANA (ILLEGALLY CULTIVATED IN THE WILD & RARELY ESCAPING)

H1973A Anyms. Used when the patient gets well but has not gotten it into his head that he is recovered. This plant will get him going.

URTICACEAE (NETTLE FAMILY)

77. *Laportea canadensis* (L.) Wedd.   WOOD-NETTLE, GIANT STINGING NETTLE

W1912A ED Facilitates childbirth. Smash handful of roots, put in a teacup of warm water. Drink immediately. Drink twice.

W1914A DJ To counteract poison made of menstrual blood mixed with fruit. Smash the roots of 2 plants, put in 6 qt. water, boil quite a bit. Drink

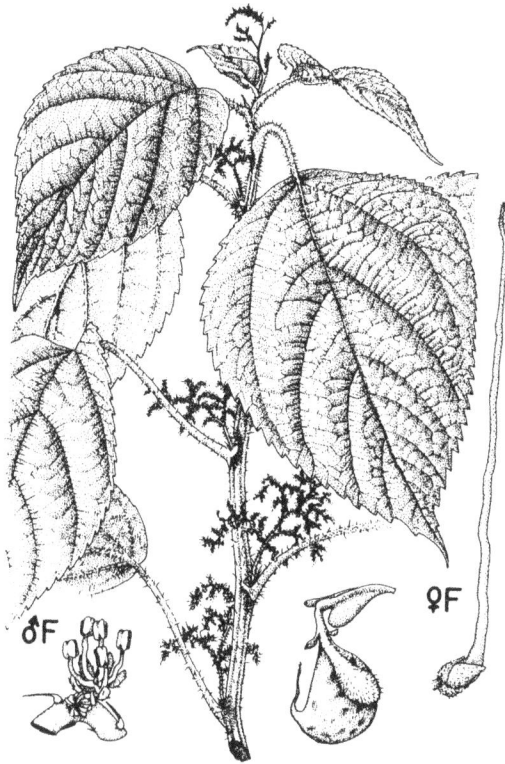

L. canadensis

U. dioica

quite a lot, early in the morning. Vomit for 4 mornings.

W1914B ?? (cf. 85, *Quercus alba,* W1914A) In this case, the wood nettle has been called by a name that is also used for the white oak. This may be the result of a misidentification. However, the use given is consistent with the other uses given for the nettles: namely, witchcraft or counterwitchcraft.
When your woman goes off and will not come back. Made by relatives of person who is left. Counteracts loneliness. Take all the bark off of 2 trees (suggesting oak, not nettle) that touch—just the bark where they touch. Put in 6 qt. water, boil down to 1 qt. Drink all you can and vomit.

W1914C JD Neutralizes love medicine. Put 2 roots in 5 qt. water, boil quite a bit. Drink about ½ of it and try to vomit. Make a hole in the ground deep enough to hold the vomit. Get a good-sized white stone in fields. Heat in fire, vomit in hole a second time, then place heated stone in quickly. Cover

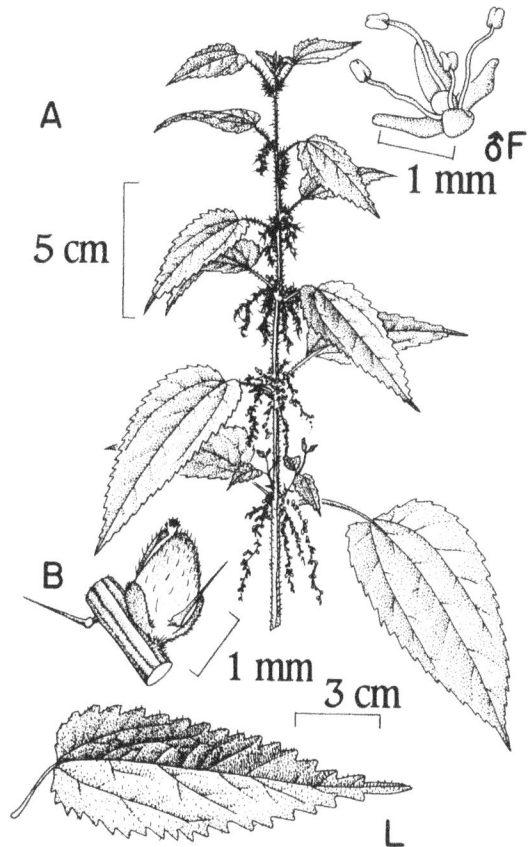

over. The person who caused the trouble will go crazy.

F1933A CHJJ See 214, *Epilobium angustifolium,* F1933B.

78. *Pilea pumila* (L.) Gray   RICHWEED, CLEARWEED, COOLWORT

H1973A Anyms. Sinus problems. Squeeze water out of stem and inhale.

79. *Urtica dioica* L.   STINGING NETTLE, GREAT NETTLE

H1973A DC No use.

H1973B Anyms. Witching medicine. With: (1) 224, *Cornus sericea,* H1973A. Cut a staff of (1) about

3-4 ft. long. Cut this staff in two and make a cross. Secure cross with strip of bark. Sharpen one end of cross and stick in ground. Find a snake (any kind), pick it up, holding it between thumb and index finger. It will wrap around your wrist. When around wrist, cut its throat . . . it will stop moving. When no longer moving, remove from wrist, it will maintain a curled position. Place snake on cross by curling its body around the crossbar of the cross. Let the blood from the snake continue to drip as it hangs from the cross. Let it drip into a white cup or container. Let the blood dry. When dried, mix nettle in with dried blood of the snake and the dried blood of the person you want to witch. Then burn tobacco and say what you want to happen to the person you are witching. If it works, what you have said will happen within 1 month.

JUGLANDACEAE (WALNUT FAMILY)

80. *Carya ovata* (Mill.) Koch   SHAGBARK HICKORY, SHELLBARK HICKORY

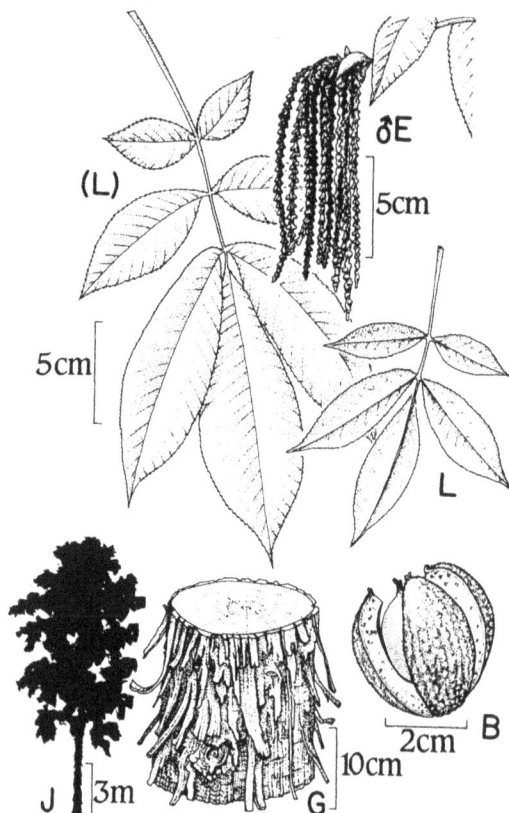

C. ovata

W1912A KD To kill worms in adults. With: (1) 133, *Populus balsamifera,* W1912A; (2) 81, *Juglans cinerea,* W1912E. Take the white from the inside bark off of a tree about 3 in. in diameter, about a single handful. Add a single handful of the bark of (1), and a double handful of the bark of the young growth of (2). Put these in ½ gal. water. Boil down to 1 pt. and drain. Add 3 lb. white sugar and boil to a syrup. Give 2 teaspoons, 1 before breakfast and 1 at night before supper. When bowels loosen, stop. Repeat in 4 days if necessary.

FndA SG Torch fuel. Tie to end of stick.

H1974A RB Arthritis. Bundle bark from east side. Boil 2 bundles at same time. Drink 1 bundle, poultice with the other.

81. *Juglans cinerea* L.   BUTTERNUT, WHITE WALNUT

W1912A ED See 11, *Lycopodium digitatum,* W1912A.

W1912B MS To stop bleeding. Smash bark, steep or chew bark and apply to wound.

W1912C KD For yellow skin, too much gall, skin becomes thin. With: (1) 84, *Fagus grandifolia,* W1912C. Cut into small pieces leaving some wood on. Put 1 gal. of chips along with a handful (of chips?) of (1) in a 2 gal. kettle. Add water and 2 lb. white sugar. Boil down to 3/4 gal. Take 1 table-spoon before retiring and before breakfast.

W1912D SH Ulcers in the mouth. With: (1) 134, *Populus balsamifera,* W1912A. See also 26, *Onoclea sensibilis,* W1912B. Take a single handful of the buds of each and steep in 1 qt. water. Wash mouth with this.

W1912E KD See 80, *Carya ovata,* W1912A.

W1912F EH See 243, *Acer pensylvanicum,* W1912A.

W1912G JJJr See 75, *Ulmus rubra,* W1912B.

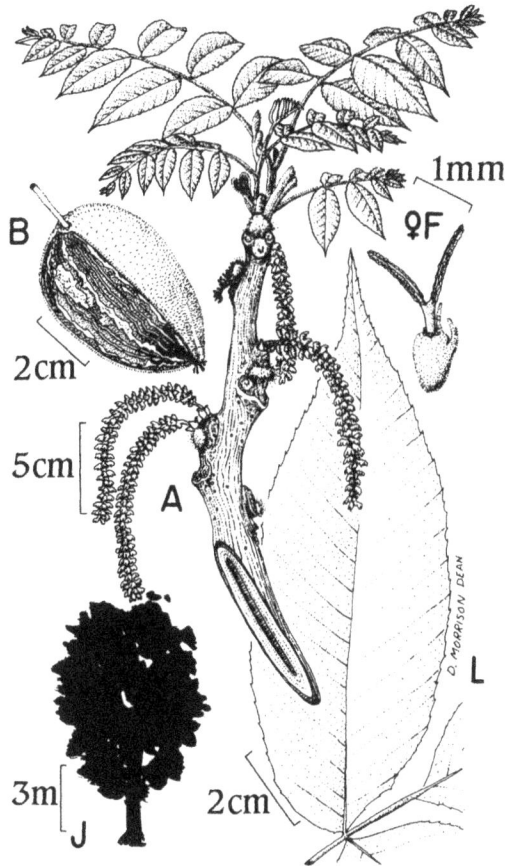

J. cinerea

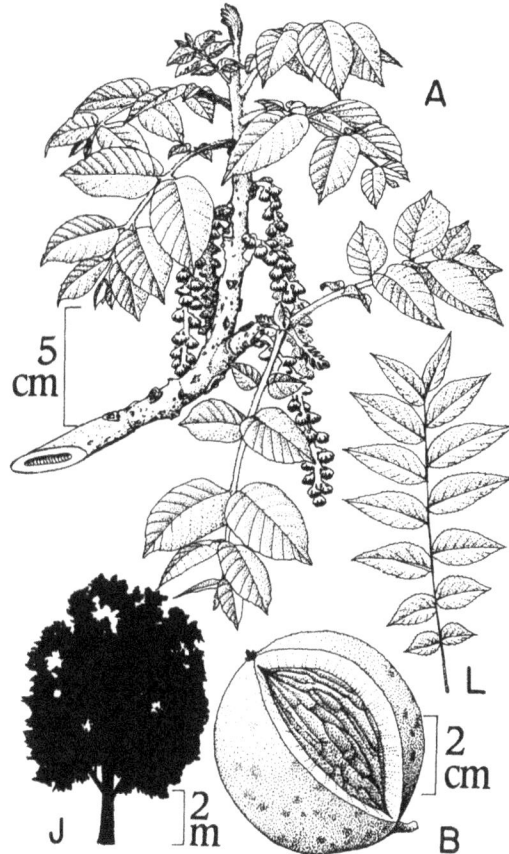

J. nigra

W1912H KD See 114, *Polygonum . punctatum,* W1912A.

W1914A DJ See 82, *Juglans nigra,* W1914A.

F1938A CHJJ Physic. Scrape bark down, boil, drink cupful once.

FndA AJm Edible nut, shucks for dye and for coloring new steel traps.

FndB CHJJ Juice will stop toothache.

FndC JG Laxative, and for venereal disease. Scrape 3-4 shoots downward, put in 2 qt. water. Boil 3 min. Let cool. Take ½ cup 3 times before meals.

FndD JG See 263, *Osmorhiza claytonii,* FndB.

FndE SS See 340, *Triosteum aurantiacum,* FndB.

H1973A JT Physic. Boil in water, make like tea. Too much makes very strong physic. Drink 4 tablespoons.

H1973B DC Good cathartic. Cook 1½ qt. of bark chips in an iron kettle all day (7-8 hr.). Keep adding cold water. Boil down to 3 qt. Take a spoonful before breakfast.

82. *Juglans nigra* L.   BLACK WALNUT

W1912A BB Medicine for rain. Take bark that has been hit by lightning and place it in a cup of water. Leave for a couple of minutes. It will rain in 2 days. Save bark for future use.

W1912B ED For the blood. With: (1) 425, *Asparagus officinalis,* W1912A. Take a handful of bark with a handful of the roots of (1). Boil in 12 qt. water down to 4 qt. Add brandy and 5 lb. brown sugar. Take 1 tablespoon before meals.

W1912C ED For craziness or headache. Take bark of young growth from east side, scraping upward. Wet it with warm water, add a pinch of salt. Apply as a poultice.

W1912D PS See 137, *Populus tremuloides,* W1912C.

W1914A DJ Laxative. With: (1) 81, *Juglans cinerea,* W1914A; (2) 133, *Populus balsamifera,* W1914A. Scrape a yard strip of each bark downward. Put in 5 qt. water, boil down to 3 qt.

F1933A SS See 254, *Aralia nudicaulis,* F1933A.

F1948A CHJJ Edible.

FndA JG Brown dye. Boil down 3 large shoots in 2 qt. water, scrape bark off, let boil down until brown in color.

FAGACEAE (BEECH FAMILY)

83. *Castanea dentata* (Marsh.) Borkh.
AMERICAN CHESTNUT (BLIGHTED)

W1912A HJn Worms in young dogs. Scrape bark and include it in the dog's food.

W1912B BB See 34, *Pinus strobus,* W1912D.

F1933A CHJJ See 96, *Carpinus caroliniana,* F1938A.

84. *Fagus grandifolia* Ehrh.   AMERICAN BEECH, BEECHNUT

W1912A DJ See 120, *Tilia americana,* W1912C.

W1912B JSl See 86, *Quercus bicolor,* W1912B.

W1912C KD See 81, *Juglans cinerea,* W1912C.

F1933A SS See 254, *Aralia nudicaulis,* F1933A.

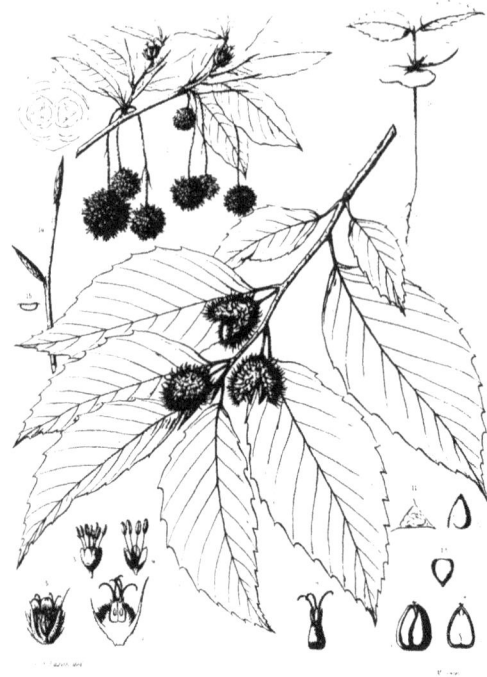

F. grandifolia

H1973A JT Abortions—only when mother is suffering. Use bark.

85. *Quercus alba* L.   WHITE OAK

W1912A JSl See 86, *Quercus bicolor,* W1912B.

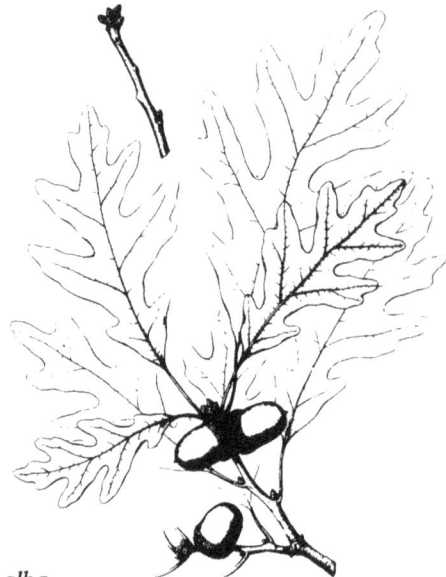

Q. alba

W1914A DJ No use given.

W1914B See name given for 77, *Laportea canadensis,* W1914B.

F1939A NL Bark used for horses having distemper.

86. *Quercus bicolor* Willd.   SWAMP WHITE OAK

W1912A ED Catarrh. With: (1) 75, *Ulmus rubra,* W1912E. Take leaves of both, dry them and smoke them, making the smoke come out through the nostrils.

W1912B JSl Consumption. With: (1) 85, *Quercus alba,* W1912A; (2) 84, *Fagus grandifolia,* W1912B; (3) 96, *Carpinus caroliniana,* W1912B; (4) 99, *Ostrya virginiana,* W1912C. Take 3 chips of bark of each off the north side. Add to 6 qt. water, boil down to 3 qt. Drink a teacup at a time beginning 2 days before the new moon.

W1914A DJ For when wife runs around, takes away lonesomeness. No preparation given.

W1914B DJ See 74, *Ulmus americana,* W1914A.

F1938A CHJJ See 23, *Polypodium virginianum,* F1938A.

87. *Quercus ilicifolia* Wang.   SCRUB OAK, BEAR OAK

H1973A DC For female troubles. Wash inwardly, drink.

H1973B Anyms. Sugar medicine. With: (1) 34, *Pinus strobus,* H1973B. No details given.

88. *Quercus macrocarpa* Michx.   MOSSY-CUP OAK, BUR OAK, NORTHERN OVERCUP OAK

W1912A BB Diarrhea. Steep bark chips from east side in 2 qt. water. Drink teacupful as required.

W1912B BB See 345, *Viburnum trilobum,* W1912C.

Q. macrocarpa

F1938A CHJJ See 96, *Carpinus caroliniana,* F1938A.

89. *Quercus rubra* L.   RED OAK

Q. rubra

W1914A DJ For ruptured or not properly healed navels. Scrape bark off lump (where branch has broken off and partly grown back—it looks like a navel), powder, and tie on navel.

F1938A CHJJ See 96, *Carpinus caroliniana*, F1938A.

90. *Quercus* sp.   OAK

H1973A JT Sore throats that will not heal. Poultice of inner bark.

H1973B Anyms. See 99, *Ostrya virginiana*, H1973B.

BETULACEAE (BIRCH FAMILY)

91. *Alnus incana* (L.) Moench ssp. *rugosa* (DuRoi) Clausen   SPECKLED ALDER, HOARY ALDER, HAZEL ALDER

W1912A KD Internal hemorrhage. With: (1) 74, *Ulmus americana*, W1912D; (2) 120, *Tilia americana*, W1912D; (3) 246, *Acer spicatum*, W1912A; (4) 46, *Actaea rubra*, W1912A; (5) 393, *Solidago* sp., W1912A. Take a handful of twigs, and the same of the root of (5). Add 2 handfuls of the root of (4), the outer and inner bark of (3) from about 3-4 ft. up the east side, 2 handfuls of the young growth twigs about 5 in. long of (2), and a smashed handful of the twigs of (1). Boil down to ½ in a gallon water. Take a wineglass every 2 hr., then to 3 hr. as you improve.

W1912B PS For when it burns while urinating. Make a bundle, 1 handful, of bark. Steep in about 2 qt. water. Drink a quart at a time. Drink all as soon as possible.

W1912C KD See 26, *Onoclea sensibilis*, W1912F.

W1912D KD See 25, *Pteridium aquilinum*, W1912A.

F1933A CHJJ Venereal disease. With: (1) 447, *Corallorhiza maculata*, F1933A, and other unspecified plants (see F1938A, below). Dry barks and grind. No further details.

F1933B SS See 192, *Rubus odoratus*, F1933A.

F1938A CHJJ See 72, *Hamamelis virginiana*, F1938B.

FndA DJn Physic and spring emetic. Take lots of bark from the young shoots, boil to good brown water in 2 gal., enough to fill yourself 3 times. Keep drinking until it comes up itself or urge it with a chicken feather or finger. Drink 3 times in same day in quick succession. Keep the water lukewarm.

FndB ?? Hunting charm. Make decoction and paint a trap or bow and arrow as a charm to get game.

92. *Betula alleghaniensis* Britt.   YELLOW BIRCH

F1933A SS Scc 254, *Aralia nudicaulis*, F1933A.

F1938A CHJJ See 96, *Carpinus caroliniana*, F1938A.

H1973A JT Lactation. Boil up.

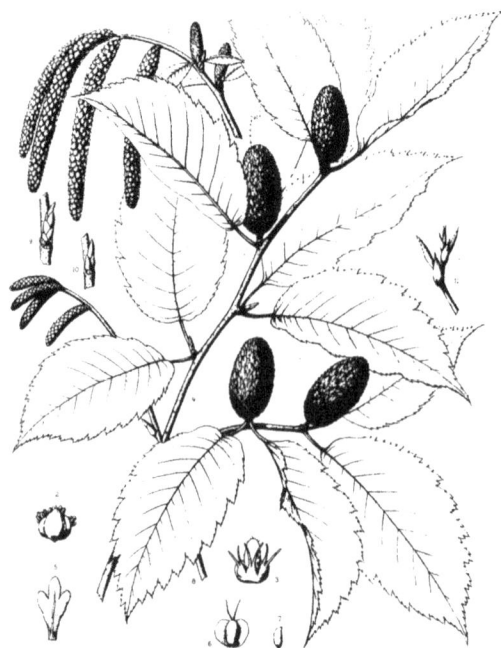

B. alleghaniensis

93. *Betula lenta* L.   SWEET BIRCH, CHERRY BIRCH, BLACK BIRCH

W1912A HJn For woman who has had gonorrhea and is about to have a baby—will help mother and baby. With: (1) 217, *Cornus alternifolia,* W1912A. From a young tree 2 in. through, take bark ½ way around the tree, cutting a strip 2-3 in. long. From (1) cut a strip 1 in. wide and 2-3 ft. long. Boil together in 2 qt. water until 1 remains. Then add 1 tablespoon of powder alum. Stir well. Take 1 tablespoon before meals.

W1912B DJ For women when they catch cold with the menses. With: (1) 73, *Celtis occidentalis,* W1912B; (2) *ga?ciset* or *dusisaa,* cf. 91, *Alnus incana,* W1912B. Put enough bark of each into 3 qt. water to make suitable strength. Take any amount, anytime.

W1912C KD See 35, *Tsuga canadensis,* W1912C.

F1933A JC This is a highly valued medicine because it sustains the deer, the mainstay of life.

B. papyrifera

94. *Betula papyrifera* Marsh.   PAPER BIRCH, CANOE BIRCH, WHITE BIRCH

FndA KD Womb shriveler. Bark burned to ashes, no further details.

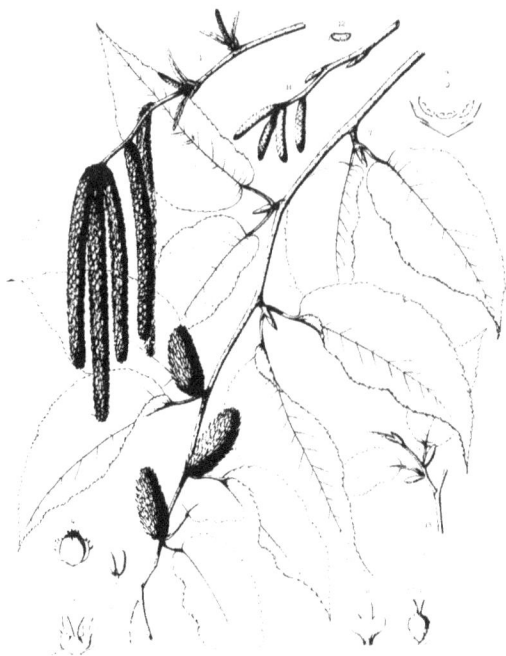

B. lenta

B. populifolia

95. *Betula populifolia* Marsh.　GRAY BIRCH, FIRE BIRCH, OLD FIELD BIRCH, WHITE BIRCH

H1973A JT Bleeding piles. Take 7 strips of bark from east side, boil and drink.

96. *Carpinus caroliniana* Walt.　HORNBEAM, BLUE BEECH, IRONWOOD WATER BEECH, MUSCLEWOOD

W1912A RS Diarrhea in babies caused by a nursing mother's notions of infidelity. With: (1) 194, *Spiraea* sp., W1912C; (2) 338, *Lonicera* sp., W1912B. Find 2 limbs where they touch and take off chips. Take 1 shoot of (1) and break it up. Put both in 5 qt. water and boil. Let cool and put in corn pounder. Then hold child over corn pounder and pour some on its head, letting it run down. Try to make it cry, then splash some on its face. Put child to bed. Then make small bundle of vine of (2) about 1 in. through. Put in ½ qt. water, steep, let it cool. Give teaspoonful to baby when cries.

W1912B JSI See 86, *Quercus bicolor*, W1912B.

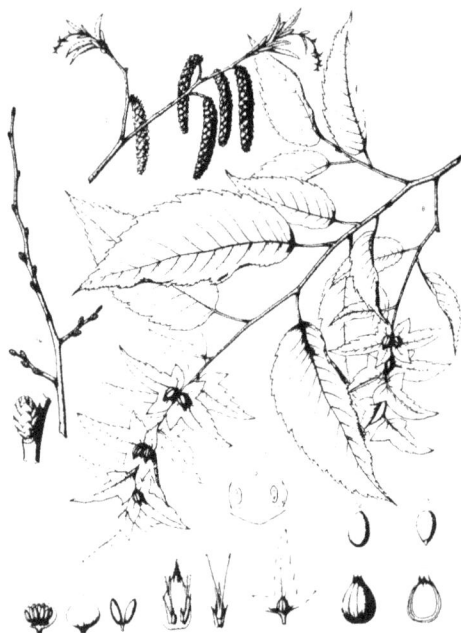

C. caroliniana

F1938A CHJJ Italian itch [scabies?]. With: (1) 88, *Quercus macrocarpa*, F1938A; (2) 400, *Sagittaria latifolia*, F1938A; (3) 89, *Quercus rubra*, F1938A; (4) 83, *Castanea dentata*, F1938A; (5) 92, *Betula alleghaniensis*, F1938A; (6) 245, *Acer saccharum*, F1938A; (7) 185, *Prunus serotina*, F1938E; (8) 218, *Cornus amomum*, F1938B; (9) swamp maple; (10) 120, *Tilia americana*, F1938B; (11) 156, *Vaccinium* sp., F1938A. Place 3 or 4 leaves of each plant in 2 gal. water. When boiled down ½, add alum the size of the end of the finger. Bathe the affected parts in it.

F1938B CHJJ For brooms.

FndA SG & JG See 75, *Ulmus rubra*, FndA.

H1974A RB For big injuries as a substitute for 120, *Tilia americana*, or basswood.

97. *Corylus americana* Walt.　HAZELNUT, AMERICAN FILBERT, HAZEL

W1912A KD For when baby's teeth are coming in. With: (1) 174, *Fragaria* sp., W1912B. Boil roots and handful of (1) without roots for ½ hr. in 1 pt. water. Take 2 tablespoons 3 hr. apart. Make necklace of young shoots of hazel and put on child when given medicine.

W1912B KD See 353, *Anthemis cotula*, W1912C.

F1938A CRE/ET Hemorrhage after childbirth may be prevented. With: (1) 345, *Viburnum opulus*, F1938B; (2) 72, *Hamamelis virginiana*, F1938E; (3) 273, *Asclepias* sp., F1938A; (4) 186, *Prunus virginiana*, F1938B; (5) *deganongwiyonta*. Take 2 stalks 2 ft. high, add the same of (4), then add 2 stalks 2 ft. high of young shoots of (1), (2), and (5). Add 1 whole plant of (3). Bunch all stalks together. Boil ingredients in 2 gal. water until 1½ remain. Cool. Drink like water twice a month after 7 months pregnancy.

F1938B CHJJ Hay fever. With: (1) 301, *Mentha spicata*, F1938B; (2) 441, *Iris versicolor*, F1938B; (3) possibly 225, *Celastrus scandens*, F1938B; (4) 230, *Ilex verticillata*, F1938C. Take ½ dozen buds, add root of (2) and 2/3 handful of (1), 5-6 berries

Co. americana

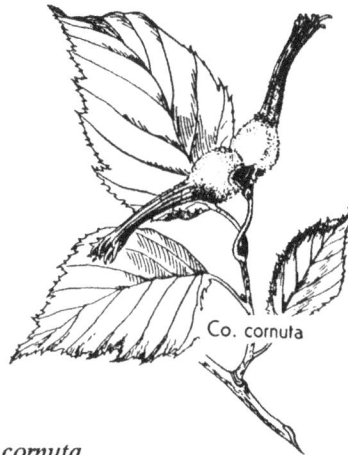

Co. cornuta

of (3), and an unspecified amount and part of (4). Put all in ½ qt. water, bring to boil, take off. Drink a cupful 3 times per day.

F1938C CRE See 186, *Prunus serotina,* F1938A.

FndA ?? Hay fever, childbirth hemorrhage, prenatal strength. Gather nuts when first frost comes, shells loosen. Eat raw.

98. *Corylus cornuta* Marsh.   BEAKED HAZEL

W1914A DJ For loneliness: "He wants to go somewhere." With: (1) 138, *Salix discolor,* W1914A; (2) 132, *Populus alba,* W1914A; (3) 368, *Eupatorium maculatum,* W1914A. Scrape bark upward, do same for (1). Put — qt. water, boil a little. Take a good drink and vomit. Wash body with liquid made of (2) and (3) afterwards. Repeat entire procedure 3 times.

FndA KD For children teething. Put ¼ handful of bark scraped down in cup of boiling water. Let cool. Wear necklace of shoots around neck. This brings on teeth without fever.

99. *Ostrya virginiana* (Mill.) Koch   HOP HORN-BEAM, IRONWOOD

W1912A SH Catarrh coughs. Boil 1½ qt. of red heart chips in 1 gal. water, down to 1 qt. Take wineglassful 4 times per day, anytime.

W1912B BB Peach stone game medicine. With: (1) 136, *Populus heterophylla,* W1912A. Take a small

O. virginiana

piece of inside bark of both trees, the second ones encountered. Take the chips in the mouth and chew them, but do not let anyone see them or you chewing them. Spit on the hands and rub them over the peach stones.

W1912C JSl See 86, *Quercus bicolor,* W1912B.

H1973A JT Cancer of rectum. Boil bark on side where sun shines. No further details.

H1973B Anyms. TB. With: (1) 34, *Pinus strobus,* H1973A; (2) 90, *Quercus* sp., H1973B; (3) 42, *Asarum canadense,* H1973B; (4) 254, *Aralia nudicaulis,* H1973B. No details given.

H1973C Anyms. Tea for swellings. No details.

PHYTOLACCACEAE (POKEWEED FAMILY)

100. *Phytolacca americana* L.   POKE, POKE-WEED, POKEBERRY, SCOKE, PIGEON-BERRY

W1912A BB Love medicine. Tie in a poplar tree, then place among roots (?).

W1912B DW Sprains and bruises. Root boiled and applied as poultice.

F1933A CHJJ Bunions. With beeswax and deer tallow. Undried roots mixed to make a salve, apply to bunions.

F1938A CHJJ See 198, *Baptisia tinctoria,* F1938A.

FndA ?? Swollen joints. Dye for baskets. Make poultice of roots and apply as poultice to swollen area.

FndB BB Expectorant, emetic, cathartic, bewitchment. No details.

FndC SS Bruises. Crush roots, apply to bruise as poultice.

FndD FSt Lumps on skin. Rub berries on lumps.

H1973A DC For vomiting. No details.

C. album

H1973B Anyms. Chest colds. Boil stem, take by spoonful.

CHENOPODIACEAE (GOOSEFOOT FAMILY)

101. *Chenopodium album* L.   LAMB'S-QUARTERS, GOOSEFOOT, PIGWEED (NATUR.)

W1912A DJ When a child dies and a woman is bothered by a flow of milk. With: (1) 314, *Penstemon hirsutus,* W1912A; (2) 160, *Lysimachia thyrsiflora,* W1912A; and (3) mud from horse's hoof. Smash whole plant of each, boil well in 1½ qt. water. Add (3). Wash with the liquid. Apply pulp as poultice.

F1938A CHJJ See 185, *Prunus serotina,* F1938C.

FndA KD Diarrhea. Take small bundle of whole plant and steep in 1 qt. water for about 5 min. Cool.

FndB SS Diarrhea. No details given.

A. retroflexus

AMARANTHACEAE (AMARANTH FAMILY)

102. *Amaranthus retroflexus* L.   PIGWEED, GREEN AMARANTH, WILD BEET, REDROOT AMARANTH (NATUR.)

W1914A ?? To kill a woman who is using you badly. With: (1) 172, *Crataegus submollis,* W1914A; (2) 445, *Smilax hispida,* W1914A. Use tobacco. Take roots like legs, fix up like a doll. Take thorns from (1), stick in doll where you want person to be sick, then boil quick [quite?] a bit if you wish it to act quickly. Person will break out like cancer.

PORTULACACEAE (PURSLANE FAMILY)

103. *Claytonia virginica* L.   SPRING-BEAUTY, MAYFLOWER

W1912A PT Convulsions in a little child. Powder 3 roots, put powder in wineglass containing 1 finger of water, add tablespoon of whisky and teaspoon sugar. Drink all of it.

C. virginica

W1912B DJ See 428, *Erythronium americanum,* W1912A.

W1912C KD See 42, *Asarum canadense,* W1912C.

104. *Portulaca oleracea* L.   PURSLANE, PUSSLEY, PULSEWEED (NATUR.)

W1914A DJ Good as medicine to cure you if someone has given you some bad medicine, such as a bloodsucker.

F1938A CHJJ Burns. Pull up bunch, put on stove and squash. Apply as poultice.

F1938B CRE/ET Bruises. Squash plant to mush. Apply as poultice. Bandage until dry. Repeat until cured.

FndA ?? Bruises. Use entire plant as poultice. Burns. Mash entire plant on stone (?).

CARYOPHYLLACEAE (PINK FAMILY)

105. *Agrostemma githago* L.   CORN-COCKLE, CORN-ROSE, PURPLE COCKLE (NATUR.)

W1914A DJ See 431, *Maianthemum canadense,* W1914A.

106. *Cerastium arvense* L.   FIELD CHICKWEED, MEADOW CHICKWEED (NATUR.)

FndA SS Injuries and miscarriage. Astringent, stops bleeding and will keep child from passing through uterus. Take a handful of plant, boil in 3 qt. water for about 10 min. Use soft water, but if hard water is used, let stand for a few days. Drink all you can 4 times a day.

107. *Stellaria media* (L.) Vill.   COMMON CHICK-WEED (NATUR.)

F1938A CHJJ Cuts and wounds. With: (1) 77, *Polygonum aviculare,* F1938A. Put together, roll and squash. Apply raw and wrap.

POLYGONACEAE (BUCKWHEAT FAMILY)

108. *Fagopyrum esculentum* Moench   BUCK-WHEAT (ESCAPE)

W1914A JJJr When baby is sick because of mother's adultery. Put 1 small plant in ½ cup water. Boil a little. Give baby a little in spoon or on finger.

W1918A JJJr For baby whose mother is running around, making baby sick. A decoction is made and the mother takes it.

109. *Polygonum arenastrum* Jord. ex Boreau DOORWEED, KNOTWEED, KNOTGRASS, SIDEWALK-WEED (NATUR.)

W1914A DJ When the butter from a young heifer is used for frying and is burned while cooking, the heifer will begin to dry up. This medicine restores milk. Put small handful of plant in a patent pailful of water. Boil quite a bit. Put corn husks in the pail of water and feed to the cow.

W1914B JJJr Love medicine. Select root of plant with a good circle of stems. Dry the root, powder it, keep in packet and place a little in other person's tea, syrup, etc. In a crowd this makes you seem like the best looking.

P. aviculare

FndA SS Miscarriage injuries, lame back. Decoction from whole plant.

110. *Polygonum aviculare* L.   KNOTWEED, KNOTGRASS (NATUR.)

W1912A ?? For when baby's coccyx breaks. With: (1) 317, *Verbascum thapsus,* W1912E. Pinch off young plant 1 in. from root, mash. Use 5 plants of (1). Take handful of each and boil in 1 qt. water. Reduce to 1 pt. Place in 2 teacups water that is boiling. Drink. Poultice leaves of (1) over fracture.

F1938A CHJJ See 107, *Stellaria media,* F1938A.

111. *Polygonum hydropiper* L.   SMARTWEED, WATER-PEPPER (NATUR.)

W1912A DJ Fever and chills. Boil a small handful in 3 qt. water. Take any amount, anytime. Best take frequently.

W1914A DJ For fever when you are continually cold and cannot sweat. Put a small handful of the plant in 4 qt. water. Boil quite a bit. Take 1/3 teaspoon, 3 times per day.

W1914B JD For fever, when you get cold. Put 1 small handful of plant in 1 qt. water with 1 pt. brandy. Boil down a little. Take a tablespoon (for an adult) about 3 times a day. Makes you sweat.

F1938A CRE/ET Indigestion. Pour boiling water over a small piece. Drink like tea.

FndA KD For swollen stomach in children. Use whole plant. No details.

112. *Polygonum pensylvanicum* L. PINKWEED, SMARTWEED

W1912A SH Horse colic, when the urine is bound up. Put a big handful of plants in a small pot of boiling water for 2 min. First dose, give a lot, then another in 15-20 min.

113. *Polygonum persicaria* L. LADY'S-THUMB, HEART'S-EASE, SMARTWEED (NATUR.)

W1912A BB To keep flies away from horses. Rub over animal.

FndA FSt Rheumatism. Boil 1 plant in 1 gal. water. Soak feet in it. Bandage. Take off in morning. Repeat.

FndB SS Heart trouble. No details given.

FndC SS Rheumatism. Make bundle 2 hands big. Cover with enough water just covering it (the bundle). Add 1 qt. water, 1 qt. vinegar, and 1 tablespoon salt. After it has boiled. Put legs in hot water, cover with blankets, go to bed.

114. *Polygonum punctatum* Ell. WATER SMARTWEED, DOTTED SMARTWEED

W1912A KD For loss of senses during menses. With: (1) 444, *Smilax herbacea*, W1912A; (2) 57, *Ranunculus abortivus*, W1912G; (3) 81, *Juglans cinerea*, W1912H. Take ½ handful of the ends of the plant, mix with a handful of the root of (1), 1½ handfuls of (2), and 5 5-in. pieces of (3) about 1 in. in diameter and peel off bark. Put with well-worn horseshoe. Add 1 qt. water. Boil down to 1 pt., cool, draw. Take ½ teacupful 1 hr. apart until pint is finished. Repeat if necessary.

115. *Rumex acetosella* L. SHEEP SORREL, RED SORREL, SOURGRASS (NATUR.)

W1912A JE Food. Used as red dye.

116. *Rumex crispus* L. CURLY DOCK, YELLOW DOCK, SOUR DOCK (NATUR.)

W1912A LJ Yellow fever (because root is yellow), and for cramps or pains in the abdomen. Mash roots, apply to swollen body as poultice. Also tea: Boil roots of 1 plant in 2 qt. water. Take ½ teacupful as often as want.

W1912B KG Upset stomach, when you cannot eat. Put roots of 2 plants in 2 qt. water and boil down to ½. Take a tablespoon 3 times per day after meals.

W1912C KG Kidney trouble. Put 3 plants in 2 qt. water and boil down ½. Drink 3 tablespoons at a time, twice a day, early in the morning and again at bedtime.

W1912D DJ Emetic before running or playing lacrosse. Makes muscles strong. With: (1) 410, *Carex oligosperma*, W1912A. Take 1 plant, add 2 plants of (1). Boil a little in 6 qt. water. Drink ½ right away. Take a little more a second time, leaving enough to wash the entire outside of the body.

W1912E KD See 348, *Achillea millefolium*, W1912D.

W1912F ED See 11, *Lycopodium digitatum*, W1912A.

W1914A JJJr For swellings anywhere caused by blood catching cold. Smash a double handful of young plants, wet, apply as poultice. Drink some of the liquid first—if drink too much, will act as physic.
Also for love medicine, if root is forked. Large plants used for this. Put root in ½ qt. water, boil quite a bit. Wash face and hands, and sprinkle clothes. Used by either a man or woman.

W1914B DJ See 69, *Corydalis sempervirens*, W1914A.

F1933A SS Physic for general bowel trouble, colds in intestines. Slice 3 roots, 4 in. long. Boil in 1 qt. water.

F1938A HRE/SRE See 255, *Aralia racemosa*, F1938A.

FndA DJn See 185, *Prunus serotina*, FndB.

FndB FSt See 450, *Cypripedium calceolus*, FndB.

H1973A JT Gonorrhea. With: (1) 174, *Fragaria* sp., H1973B; (2) 256, *Panax quinquefolius*, H1973C. No details offered.

H1973B Anyms. Gonorrhea, but it takes a long time. No details offered.

H1973C DC Blood purifier. Use root.

H1973D EP & BC Good for all illnesses.

H1973E ON Strained muscles. No details.

H1974A ON Stops bleeding.

117. *Rumex obtusifolius* L.   BITTER-DOCK, RED-VEINED DOCK, BROAD-LEAVED DOCK (NATUR.)

F1933A JS Whooping cough. Pound root, cook 1 hr. in 1 gal. water. Fix every day, give to child. Will not stop cough, will stop whoop.

F1938A HJk Contraceptive, stops menses also. Details not given.

F1938B CRE/ET Blood disorder. With: (1) 157, *Chimaphila umbellata*, F1938B; (2) 255, *Aralia racemosa*, F1938B. Use 2 roots of each per gallon water. Boil down to 1 qt. Drink all in 1 day.

F1939A HJ See 255, *Aralia racemosa*, F1939B.

CLUSIACEAE (MANGOSTEEN FAMILY)

118. *Hypericum ellipticum* Hook.   PALE ST. JOHN'S-WORT

W1914A DJ Used after a remedy for the suppression of menses. With: (1) 254, *Aralia nudicaulis*, W1914B. Put 3 stems along with 2 roots of (1) in 3 qt. water. Boil. Take a cupful 3 times a day.

119. *Hypericum perforatum* L.   ST. JOHN'S-WORT (NATUR.)

FndA RSn Prevents sterility. Add 2 roots to blood medicine.

FndB JG Fever. Preparation not given.

TILIACEAE (LINDEN FAMILY)

120. *Tilia americana* L.   BASSWOOD, WHITE-WOOD, LINDEN

W1912A JA For baby that does not walk, but should. Boil bunch of branches of the young growth in 3 gal. water, down to 1 gal. Bathe child twice a day, repeat if necessary.

W1912B HJn Panacea for hurts anywhere. With: (1) 214, *Epilobium angustifolium*, W1912C. Take young growth twigs, add a small bunch of roots of (1). Steep ½ hr. in 1 gal. water. Take ½ cup frequently.

W1912C DJ For burns or scalds. With: (1) 84, *Fagus grandifolia*, W1912A. Take equal amounts of leaves of both and boil them until well cooked. Mash, tie on as poultice.

W1912D KD See 91, *Alnus incana*, W1912A.

W1912E JJ See 194, *Spiraea* sp., W1912B.

W1912F KD See 369, *Eupatorium perfoliatum*, W1912F.

F1938A CHJJ Use not given.

F1939A PH See 308, *Fraxinus nigra*, F1939A.

H1973A DC When you can not urinate. Strip bark downward, make bundle. Cook in 2 qt. water.

H1973B Anyms. See 217, *Cornus alternifolia*, H1973A.

H1973C FS Use not given.

H1974A RB For severe injuries. Make tea.

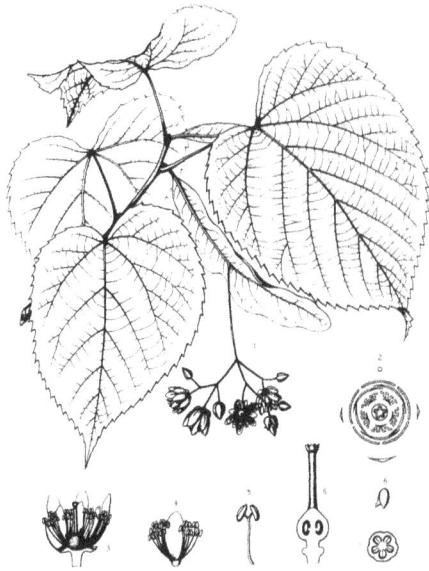

T. americana

H1974B ON Picks you up when you are run down. Bundle shoots, steep.

H1974C DC When you stop urinating too much. See H1973A, above.

MALVACEAE (MALLOW FAMILY)

121. *Malva moschata* L.   MUSK-MALLOW (NATUR.)

M. moschata

F1938A CHJJ For chills and lassitude—for winter use. Steep whole plant. Drink the warm water anytime.

122. *Malva neglecta* Wallr.   CHEESES, MALLOW (NATUR.)

W1912A KD Swellings of all kinds. With: (1) 339, *Sambucus canadensis,* W1912D. Pound double handful of whole plant, same for leaves of (1). Soak in cold water just enough to wet it. Then scrape bark of (1) downward and wet it with hot water. When poultice removed, rub with bark of (1).

W1912B KD When baby's stomach swells up from cow's milk and improper food. Take a tumbler full of "cheeses" and 2 glasses of onions (Dutch set), with 1 cup milk and 1 cup lard. Boil down a little. Spread this over a cotton sheet big enough to cover the baby's stomach, and tie it on well. Leave it on all night. Also apply it similarly to the back. Also, melt a teaspoon of lard and give it to the baby—add a little sugar.

W1912C KD See 369, *Eupatorium perfoliatum,* W1912F.

W1914A DJ Cure for a love medicine. If you stay with a girl a short time and then go away, if she wants to keep you she gets the "cheeses" and cuts them up and puts them in your food (because they are always found around the house). For a cure, put 1 whole plant in 6 qt. warm water, mash up the plant, let stand a little, and drink a lot. Then vomit. For an alternate cure, see 48, *Anemone virginiana,* W1914A.

M. neglecta

F1933A KD Poultice for baby's swollen stomach. With: (1) Dutch set onions. Pick the "cheeses" and poultice.

SARRACENIACEAE (PITCHER-PLANT FAMILY)

123. *Sarracenia purpurea* L.   PITCHER-PLANT, SIDE-SADDLE PLANT

W1912A KD Recurring chills followed by fever. With: (1) 302, *Monarda* sp., W1912D; (2) 43, *Nuphar lutea*, W1912F. Mix a double handful of finely cut leaves with 2 handfuls of whole plant of (1) and 3 tablespoons of flour from root of (2). Put in gallon of water, boil to ½. Drink as much as possible when chill comes.

W1912B JA High fever and shakiness. Make tea from dried leaves. Take when fever comes.

F1933A CRE/ET Basket medicine and love charm.

F1933B CHJJ Medicine for thirst. Love medicine. No details.

F1933C JS Whooping cough. Boil whole plant in water for 15 min. Do not boil too much. Strain through deer leather. Drink cold, throw away unused medicine.
Also a love and lacrosse medicine. Sprinkle on person, or burn and make request.

F1933D SS Love and basket medicine. No details.

F1933E CHJJ See 332, *Mitchella repens*, F1933C.

F1938A AJ Peddling medicine, love medicine. No details.

F1938B CHJJ See 198, *Baptisia tinctoria*, F1938A.

F1939A NL Fever. Steep it, drink.

F1939B HJ Basket medicine.

FndA FSt & SS Basket medicine. No details.

FndB WS Whooping cough. No details.

S. purpurea

H1973A JT Pneumonia. No details offered.

H1973B Anyms. Fever. No details offered.

VIOLACEAE (VIOLET FAMILY)

124. *Hybanthus concolor* (Forst.) Spreng.   GREEN VIOLET

W1912A ED For when a mare with foal gets hurt in the belly. Take a single handful of the roots and stem and steep them in 1 gal. water. Serve a quart at a time in the feed—the colt will then come alright.

125. *Viola pubescens* Ait.   YELLOW VIOLET

W1914A DJ To cure facial eruptions that may turn to worms and are caused by going with a woman during her menses. Her breath on your face will do it. Add a plant to 2 qt. water and boil quite a bit. Drink ½ qt. and wash with the rest.

126. *Viola pubescens* Ait. var. *eriocarpa* (Schwein.) Russell   SMOOTH YELLOW VIOLET

W1912A DJ See 158, *Pyrola elliptica*, W1912B.

127. *Viola sagittata* Ait.   ARROW-LEAF VIOLET

W1912A KD See 437, *Trillium* sp., W1912B.

128. *Viola striata* Ait.   CREAM VIOLET, PALE VIOLET, STRIPED VIOLET

W1914A DJ (Hunters in the woods used to take a flower each and interlock them. The hunter whose flower held the longest was to take all the game.) When a young man who has been refused by the parents of a young girl has taken the girl's hair as bait and placed it with some roots in a hole in a tree and plugged up the hole. When the hair moves, the girl becomes sick and half crazy, and she may even drown herself. Take a small bunch of whole plants, mash them, put in ½ qt. warm water. Wet the head with them.

129. *Viola* sp.   VIOLET

W1912A MR For wounds. Smash and poultice.

W1912B SH When baby becomes sick as a result of mother's infidelity. With: (1) "grass from swamp." Put roots of 1 plant with whole plant of (1) in ½ qt. water and steep. Mother and baby must drink.

V. sagittata

V. striata

F1938A AJ See 72, *Hamamelis virginiana*, F1938A.

H1973A DC Eyewash. Roots used.

H1973B ON Boils on horse's head. Crush root, poultice.

CUCURBITACEAE (GOURD FAMILY)

130. *Citrullus colocynthis* (L.) Schrad. (*C. vulgaris* Schrad.)   WATERMELON (RARE ESCAPE)

F1938A CHJJ See 42, *Asarum canadense*, F1938A.

131. *Sicyos angulatus* L.   BUR CUCUMBER, STAR CUCUMBER

W1914A ?? See 20, *Osmunda cinnamomea*, W1914A.

FndA JG Venereal disease. Take the whole vine, bundle it up. Let the root come with it. Boil in 2 qt. water. Drink that as often as she can. Eat no oily or greasy foods like meat (it spoils the medicine). Take nothing salty. Remain on diet until cured.

P. alba

SALICACEAE (WILLOW FAMILY)

132. *Populus alba* L.   WHITE POPLAR, SILVER-LEAF POPLAR (NATUR.)

W1914A DJ A body wash after medicine taken for loneliness. See 98, *Corylus cornuta*, W1914A for the antilove medicine. Used with: (1) 368, *Eupatorium maculatum*, W1914A. Put 2 branches about 3 ft. long with the roots of (1) in 8 qt. water and boil down to 1 qt. Wash body, head first. Repeat entire procedure 3 times.

H1973A FS Tonic. Also cleans you out after a cold. Make small bundle of inner bark. Drink all you can, gag up, bring up phlegm. Repeat.

133. *Populus balsamifera* L.   BALSAM POPLAR, TACAMAHAC, HACKMATACK

W1912A ?? See 80, *Carya ovata*, W1912A.

W1914A DJ See 82, *Juglans nigra*, W1914A.

F1933A JS See 71, *Platanus occidentalis*, F1933A.

F1939A PH For worms in horses or people. Scrape bark of young shoots upward. Boil in 2 qt. water. Give 1 qt. of this to a horse, 2-3 tablespoons for people.

F1939B KD Worms. Scrape bark from young growth, branches about 4-5 in. long. Boil in water.

S. angulatus

P. balsamifera

P. grandidentata

Make drink, then take castor oil. Worms come out in stool.

H1973A DC See 434, *Smilacina racemosa,* H1973A.

H1973B TL Arthritis. With: (1) 72, *Hamamelis virginiana,* H1973B. Notch downward toward east, pick up chips and mix with (1). No further details.

134. *Populus balsamifera* x *deltoides* = *P. xjackii* Sarg. BALSAM POPLAR (BALM-OF-GILEAD)

W1912A SH See 81, *Juglans cinerea,* W1912D. See also 26, *Onoclea sensibilis,* W1912B. Note: There is some doubt about the identification of this plant.

135. *Populus grandidentata* Michx.   BIG-TOOTH ASPEN, LARGE-TOOTHED ASPEN

W1912A JJJr Itch. Take the dust off the bark early in the spring when the catkins are on. Rub this on affected parts.

136. *Populus heterophylla* L.   SWAMP COTTON-WOOD, BLACK COTTONWOOD (RARE)

W1912A BB See 99, *Ostrya virginiana,* W1912B.

137. *Populus tremuloides* Michx.   QUAKING ASPEN, TREMBLING ASPEN

W1912A MS Pleurisy. Scrape bark and wet it with hot water. Apply as poultice.

W1912B HJn For when baby cries, but is not sick. Steep bark of young trees 1 in. through in 1 teacup warm water. Drain and cool. Take 1 teaspoon every hour.

P. heterophylla

P. tremuloides

W1912C PS (possibly 133, *Populus balsamifera*) Measles. Scrape up bark, make into bundle the size of a walnut. Put in teacup cold water, squeeze with fingers. Take a little at a time.

W1912D BJ Cramps caused by worms. Scrape bark, steep in 1 pt. of water.

W1912E ED Syphilis with chancres. With: (1) 138, *Salix discolor*, W1912B; (2) 185, *Prunus serotina*, W1912F. Scrape down the bark of young growth of each. Put handful in 1 qt. warm water. Apply as poultice, keeping it wet.

W1912F JJ See 212, *Dirca palustris*, W1912D.

W1914A DJ Worms in horses and humans. Take a double handful (twice) of bark. Put in 6 qt. water and boil down 1 qt. Mix with feed.

FndA CG Worms in children. Scrape bark off shoot toward the tip. Put in ½ cup warm water. Let stand. Let child drink a little.

FndB SS Worms in children. Take 2 young shoots, break up, bring to boil in 1 qt. water. Let child drink ad lib.

FndC AJm Worms in horses. Bark used. No further details.

FndD AJ Fits in dogs caused by worms. Also for cats. Cut a piece of bark about 6-8 in. long by 1 in. thick. Then cut it up into 1 in. pieces. Put in 2 qt.

clear water and boil it for 20 min. Then cool. When real cold, give it to the dog or cat—all that it can drink.

H1973A Anyms. To stop bed wetting. With: (1) 30, *Abies balsamea*, H1973A. No details given.

138. *Salix discolor* Muhl.   PUSSY-WILLOW, GLAUCOUS WILLOW

W1912A JJ See 194, *Spiraea* sp., W1912B.

W1912B ED Syphilis with chancres. See 137, *Populus tremuloides*, W1912E.

W1914A DJ See 98, *Corylus cornuta*, W1914A.

FndA CHJJ Bleeding piles. Steep bark.

S. discolor

139. *Salix nigra* Marsh.   BLACK WILLOW, SWAMP WILLOW

W1912A JA See 247, *Rhus* sp., W1912I.

140. *Salix sericea* Marsh.   SILKY WILLOW

F1939A KD For abscesses in the mouth and throat.

S. nigra

S. sericea

F1939B CHJJ [Informant] does not use this as medicine because worms appear in the stalk when it grows old.

141. *Salix* sp.   WILLOW

W1912A JJ See 194, *Spiraea* sp., W1912B.

BRASSICACEAE (MUSTARD FAMILY)

142. *Armoracia rusticana* (Lam.) Gaertn., Mey. & Scherb. (*A. lapathifolia* Gilib.)   HORSE-RADISH (NATUR.)

W1912A ED For the blood. Smash roots, steep handful in 5 qt. water. Take a teaspoonful 3 times a day before meals.

H1973A JT Sugar diabetes. No details offered.

143. *Brassica napus* L.   TURNIP (TURNIP-GREENS)

WndA ED Sores, boils. Dry leaves, heat, apply.

A. rusticana

144. *Cardamine bulbosa* (Schreb. ex Muhl.) BSP.
SPRING CRESS

W1914A DJ Poison to kill anyone. Roots are mixed with anything.
See 300, *Mentha canadensis,* W1914A, for a cure for this.

145. *Cardamine concatenata* (Michx.) Schwarz
CUT-LEAF TOOTHWORT, PEPPER-ROOT

W1912A PS Mesmerizing plant. No further details.

W1912B DJ For attracting women and girls. Cut up the roots fine and put in your pocket. At sundown go out and whistle and call her name. She will come and do whatever you want.
Also a hunting medicine. Cut up roots and rub on the gun or traps. Whistle a little when you come to the right place.

W1912C JSI Headache. Smash root and mix with cold water. Tie on head as poultice early in morning. Use twice in 24 hr., then make a new poultice if needed. Dampen with water.

W1912D RS For divination. It is floated in still water in a vessel. Tobacco is burnt and root is told what it is to be used for.
Also a fishing medicine. Smash roots, wet with water. Tie the line and hook up in this and leave all night. A little of the liquid is also put on fish worms.

W1914A DJ A kind of love medicine if placed in the mouth. The person whistles and the person he or she wants will hear and come. Mention person's name.

146. *Cardamine diphylla* (Michx.) Wood   TWO-LEAVED TOOTHWORT, PEPPERWORT

W1912A LJ Colds; to stimulate appetite and regulate stomach. Preparation not given.

W1912B JJ Heart palpitations from fright or remorse or anger. Gather root in fall, let stand in water overnight. Eat lots of it anytime.

W1912C DJ When the heart jumps and the head goes wrong; when love medicine is too strong. Mash 3 or 4 small roots, place in ½ cup warm water, cover and let stand. Drink up right away. Apply roots as poultice for swelling.

W1912D KD See 375, *Inula helenium,* W1912D.

W1914A DJ Heart disease. Dry the root, make into a flour (a good handful), put in ½ qt. water, boil quite a bit. Take in 2 drinks on the same day. Make more if required.

FndA ?? For gas on the stomach. Eat root raw.

FndB JG Fever, summer complaint. Take large plant, bundle in 1 qt. boiling water. Let steep. Drink when cooled.

147. *Cardamine douglassii* Britt.   PURPLE CRESS

W1912A ED Counteracts any kind of poison, and is also used for divining the perpetrator of witchcraft. Get it at night and at the last quarter of the moon. Smash roots, put in small wooden cup and cover. Let stand for ½ hr. Face of witch will appear. Stick a needle in the image's eye and it will also make his eye sore.

C. douglassii

R. sylvestris

148. *Cardamine* sp.   CRESS

F1938A CHJJ See 301, *Mentha spicata*, F1938C.

149. *Rorippa sylvestris* (L.) Besser   CREEPING
YELLOW-CRESS (NATUR.)

W1912A RS Fever in baby which may run into
cough or cold. Make a bundle 2-3 in. across, add
to ½ qt. water. Boil, cool, put in bottle and mix
1 pt. sweet milk with it. Mother drinks all of this
she can. In old days, juice from boiled corn bread
used instead of milk.

150. *Sisymbrium officinale* (L.) Scop.   HEDGE-
MUSTARD (NATUR.)

W1914A DJ See 175, *Geum strictum*, W1914A.

151. *Thlaspi arvense* L.   FIELD PENNYCRESS,
FANWEED, FRENCH-WEED, MITHRIDATE
MUSTARD (NATUR.)

FndA KD Sore throats. Pick in June. Make infusion of whole plant. Bundle plants and smash.

S. officinale

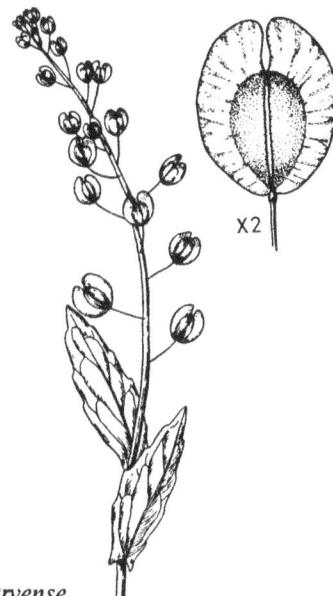

X2

Th. arvense

ERICACEAE (HEATH FAMILY)

152. *Epigaea repens* L.   TRAILING ARBUTUS, MAYFLOWER, GROUND-LAUREL

F1938A SRE Kidney medicine. With: (1) 192, *Rubus odoratus,* F1938A. Strip bark from roots. Use with stalks and leaves. Bundle size of circled thumb and forefinger. Boil 15-20 min. in 1 gal. water. Drink like water.

F1938B AJ See 157, *Chimaphila umbellata,* F1938A.

F1940A CHJJ Kidney medicine. With: (1) 157, *Chimaphila umbellata,* F1940A; (2) 153, *Gaultheria procumbens,* F1940A. Take 2-3 of each and boil until water turns pink. Take all you can in a day.

F1940B DJn Labor pains in parturition. With: (1) 332, *Mitchella repens,* F1940A; (2) 55, *Hepatica nobilis* var. *obtusa,* F1940A. Preparation not given.

FndA CHJJ Kidney medicine. Whole plant used. No further details.

FndB SS Kidney medicine. Cook whole plant until boils in 1 gal. water for 10 min. Drink all you can.

H1973A Anyms. For digestion. Cut leaves, boil, drink a spoonful.

153. *Gaultheria procumbens* L.   WINTERGREEN, CHECKERBERRY, TEABERRY, MOUNTAIN-TEA, IVORY-LEAF

F1933A SS Blood purifier. Boil to make an infusion. Drink like tea.

F1933B CHJJ Tea. Put leaves in hot water.

F1940A CHJJ See 152, *Epigaea repens,* F1940A.

FndA CRE/ET Colds or just as tea. Put leaves in boiling water and drink hot.

FndB JG See 263, *Osmorhiza claytonii,* FndB.

H1973A Anyms. Beverage. With: (1) 174, *Fragaria* sp., H1973A.

H1973B DC See 434, *Smilacina racemosa,* H1973A.

H1973C DC Kidney trouble. Make tea.

H1974A DC Kidney trouble, rheumatism, arthritis. No preparation offered.

154. *Kalmia latifolia* L.   MOUNTAIN LAUREL, CALICO-BUSH, SPOON WOOD, IVY-BUSH

F1938A CRE & SRE Not used for medicine.

K. latifolia

155. *Vaccinium stamineum* L.   DEERBERRY, SQUAW-HUCKLEBERRY

?? ?? Not used as a medicine.

156. *Vaccinium* sp.   DEERBERRY, BLUEBERRY

W1912A ED See 418, *Elymus canadensis,* W1912A.

F1938A CHJJ See 96, *Carpinus caroliniana,* F1938A.

PYROLACEAE (SHINLEAF FAMILY)

157. *Chimaphila umbellata* (L.) Bart. ssp. *cisatlantica* (Blake) Hultén   PIPSISSEWA, PRINCE'S-PINE

?? ?? Diuretic. Use leaves and stems. No details given.

W1912A SH When a pregnant woman feels feverish and drowsy. Woman is not sick, baby is. Make a small bundle about 1 in. thick from whole plant. Put in ½ qt. water, steep. Take cupful until used up, 4 times a day.

W1912B ED See 11, *Lycopodium digitatum*, W1912A.

W1912C SH See 354, *Arctium* sp., W1912D.

W1912D ED See 12, *Lycopodium lucidulum*, W1912A.

WndA ED Worms in babies. Steep small bunch of plants in 2 qt. water. Take ½ cup before meals, 3 times a day.

F1933A SS Cancer of the stomach. With: (1) *tcadwaweye^{n}ho.* Boil stalk and root along with smashed root of (1) in 1 gal. water. Take cupful before meals and going to bed.

F1933B SS For blood chill. With: (1) 221, *Cornus florida*, F1933A. Take stem and roots and add bark of (1) scraped downward. Boil in 1 gal. water. Drink ad lib.

F1933C SS See 254, *Aralia nudicaulis*, F1933A.

F1933D CHJJ See 332, *Mitchella repens*, F1933C.

F1933E CHJJ See 418, *Elymus canadensis*, F1933A.

F1933F SS See 72, *Hamamelis virginiana*, F1933A.

F1938A AJ Rheumatism. With: (1) 152, *Epigaea repens*, F1938B. Boil ¼ lb. of each in a gallon of water for ½ hr. Cool. Drink cupful before meals.

C. umbellata

F1938B CHJJ See 117, *Rumex obtusifolius*, F1938B.

F1938C HRE See 255, *Aralia racemosa*, F1938A.

F1939A HJ See 254, *Aralia nudicaulis*, F1939A.

F1939B HJ See 255, *Aralia racemosa*, F1939B.

F1939C JJm Dropsy. With: (1) 273, *Asclepias syriaca*, F1939A.

F1940A CHJJ See 152, *Epigaea repens*, F1940A.

FndA JG See 263, *Osmorhiza claytonii*, FndB.

FndB FSt See 192, *Rubus odoratus*, FndA.

FndC SS See 340, *Triosteum aurantiacum*, FndB.

H1973A ON Medicine strengthener. Mix with any medicine, drink.

158. *Pyrola elliptica* Nutt.   SHINLEAF, WILD LILY-OF-THE-VALLEY

W1912A KG Fits or epileptic seizures. Take 2 roots and leaves of 2 plants and add 2 qt. boiling

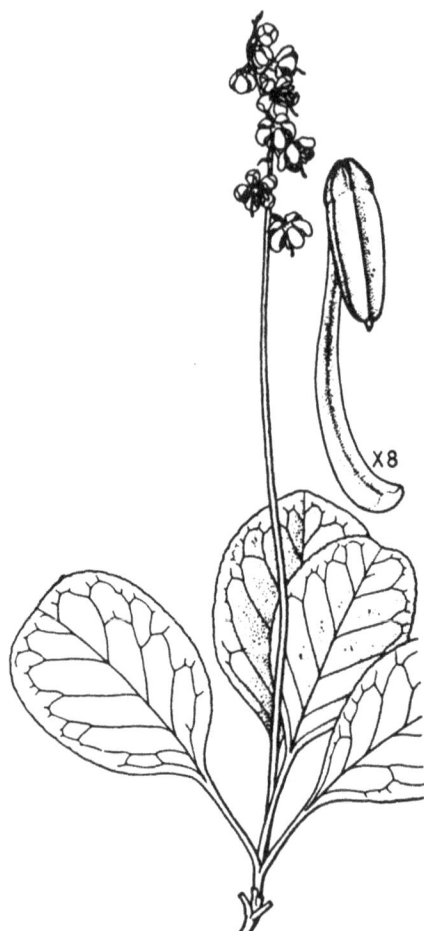

P. elliptica

water. For baby, give 1 teaspoon, 3 times a day. For adults, take 4 times a day.

W1912B DJ Indigestion. With: (1) 126, *Viola eriocarpa*, W1912A. Boil 4 whole plants of each in 8 qt. water until 4 remain. Drink as much as possible when thirsty.

W1912C JJJr Sore eyes. No details given.

W1912D JSl Sore eyes—sty, inflamed lids, etc. Put a small bunch of the whole plant in ½ qt. water. Boil down ½. Wet small cloth and drop much of it into the eye.

W1912E ED See 12, *Lycopodium lucidulum*, W1912A.

W1912F BB See 41, *Sassafras albidum*, W1912A.

W1914A DJ Sore legs (if a man is a good worker and someone wants to spoil him). With: (1) 239, *Polygala paucifolia*, W1914A. Put 1 plant with a good handful of (1), smash up with a little warm water and tie on as poultice.

PRIMULACEAE (PRIMROSE FAMILY)

159. *Lysimachia quadrifolia* L.   WHORLED LOOSESTRIFE

F1933A JS Emetic. Pull up the root, wash, put in hot water, and steep. Drink it and throw up at once.

160. *Lysimachia thyrsiflora* L.   TUFTED LOOSE-STRIFE

W1912A DJ See 101, *Chenopodium album*, W1912A.

GROSSULARIACEAE (CURRANT FAMILY)

161. *Ribes americanum* Mill.   WILD BLACK CURRANT

W1912A DJ See 298, *Lycopus virginicus*, W1912A.

W1914A DJ Fortune telling or divination. With: (1) 163, *Ribes rotundifolium*, W1914A. Scrape bark off from newly grown twigs about 12 in. long. Put

R. americanum

R. hirtellum

R. rotundifolium

in 3 qt. water, boil down to 1 qt. Add $.25 piece. Drink quite a bit pretty near bedtime. Put tobacco and a personal possession of the sick person under your pillow and you will dream of the correct medicine.

F1938A CHJJ See 380, *Polymnia uvedalia*, F1938B.

FndA ?? Overactive kidneys. Peel root. No further details.

H1973A Anyms. For when a woman hurts inside from lifting. Put a handful of leaves in 1½ qt. water, boil ½ hr., drink.

162. *Ribes hirtellum* Michx. NORTHERN GOOSE-BERRY, CURRANT

WndA KD See 224, *Cornus sericea*, WndA.

163. *Ribes rotundifolium* Michx. ROUND-LEAF CURRANT, WILD GOOSEBERRY

W1914A DJ See 161, *Ribes americanum*, W1914A.

CRASSULACEAE (SEDUM FAMILY)

164. *Sedum telephium* L. LIVE-FOREVER (NATUR.)

W1912A BB Given to horses as diuretic. Steep.

F1938A CHJJ See 198, *Baptisia tinctoria*, F1938A.

FndA ?? For when babies cry; bruises. Preparation not given.

FndB HRE For injuries resulting from witching, e.g., sprained ankle. Pound stalks and leaves, poultice swelling.

H1974A ON Paralysis. Rub on paralyzed face.

SAXIFRAGACEAE (SAXIFRAGE FAMILY)

165. *Mitella diphylla* L. COOLWORT, MITER-WORT

W1914A DJ Counteracts bad luck brought about by throwing partridge entrails or feathers in fire. Used as an emetic, and as a body and gun wash. Put small bunch of whole plants in 8 qt. water and boil considerably. Drink some and vomit. Wash body and rifle. If this fails, burn the largest root of the plant and make a charred line above and below the eye.

FndA KD For sore eyes. Crumble 1 plant in hand and put in ½ cup of lukewarm water. Cool. Strain. Drop 3 drops in eye.

166. *Saxifraga pensylvanica* L. SWAMP SAXI-FRAGE, WILD BEET

F1933A CHJJ See 315, *Scrophularia lanceolata*, F1933A.

M. diphylla

F1938A CHJJ See 301, *Mentha spicata,* F1938C.

FndA ?? Ingredient in Little Water Medicine ritual. Referred to in song 11, first group of Little Water Medicine ritual.

FndB HRE Belongs in blood medicine. No details.

FndC FSt For weak kidneys. Steep root in 1 qt. water. Drink all you can. Hold as long as you can, then urinate.

FndD DJn See 185, *Prunus serotina,* FndB.

167. *Tiarella cordifolia* L.   FOAMFLOWER, FALSE MITERWORT

W1912A MR For wounds. Smash roots, put on wounds.

W1912B DJ For hunting pheasants or partridges. With: (1) 297, *Lycopus americanus,* W1912B.

Steep 2 of each plant in 5 qt. water. Drink and wash whole body with liquid. Partridges [and pheasants] will come close.

W1912C JSl To fatten little children. A tonic. Put roots and leaves of 2 plants in 2 small cups of water and steep a little. Give 6 times a day, 3 teaspoons at a time.

W1912D ED See 27, *Polystichum acrostichoides,* W1912A.

W1914A JJJr Hunting medicine for cleaning rifle. Put a double handful of whole plant in 3 qt. water, boil a little. Wash rifle all over. Drink quite a bit to vomit, and wash or wet body.

W1914B JD For sore mouth in baby. Smash 2 roots in ½ cup warm water, cover, let stand 5 min. Give a teaspoon at a time, or put it in mouth.

S. pensylvanica

Ag. gryposepala

F1933A KD For sore eyes when pustules break out on edge of eyelids. Dry and rub 7 leaves in hands. Put tablespoons of this in river or soft water. Let sit for 4 hr., drain. Drop solution in eyes.

F1933B CHJJ See 54, *Hepatica nobilis* var. *acuta*, F1933A.

ROSACEAE (ROSE FAMILY)

168. *Agrimonia gryposepala* Wallr.   AGRIMONY, COCKLEBUR, HARVEST LICE, STICKSEED, TALL HAIRY AGRIMONY

W1912A LJ Diarrhea, summer complaint, vomiting in children. Put roots of 1 plant in 1 pt. water, steep. Give wineglassful when thirsty.

W1912B ED Diarrhea. With: (1) 175, *Geum aleppicum*, W1912A. Steep both roots in 2 qt. water. Drink as much as you like.

W1912C SH Bloody diarrhea. Boil roots of 5 plants in 2 qt. water down to 1½ qt. Take ad lib until cured.

F1933A CHJJ Diarrhea. Preparation not given.

F1933B HJm Diarrhea and emetic for summer complaint. An alternate medicine for 304, *Prunella vulgaris*, F1933B. Make a bunch from 2 plants, boil in 1 qt. water.

FndA HRE Basket medicine, or anything you want to sell. Bundle roots and flowers, steep in hot water, sprinkle on.

169. *Agrimonia* sp.   AGRIMONY

W1912A KD Too much gall. With: (1) 310, *Chelone glabra*, W1912B. Take 1 big handful of roots of both and boil in ½ gal. water down to 1 pt. Take ½ glassful 3 hr. apart.

170. *Amelanchier arborea* (Michx. f.) Fern.   SHAD-BUSH, SERVICEBERRY

W1912A HJn Gonorrhea. Steep bark.

171. *Crataegus punctata* Jacq.   HAWTHORN

W1914A ?? Cure for *uske^{n}na'dau* caused by 102, *Amaranthus retroflexus*, W1914A. Take the bark from the east side of the plant at about the height of your head. Put in 6 qt. water and boil quite a bit. Drink as much as you like.

FndA KD (plant may be 178, *Malus coronaria*) Stop menstrual flow (cf. 24, *Adiantum pedatum*, FndA). With: (1) sweet orchard apple, *Pyrus malus*. Use handful of young shoots and bark from 1 apple branch, add cut bark from young shoots

A. arborea

upward of (1). Boil in ½ gal. water down to 1 qt. Drink as water.

172. *Crataegus submollis* Sarg.   HAWTHORN

W1914A DJ See 102, *Amaranthus retroflexus,* W1914A.

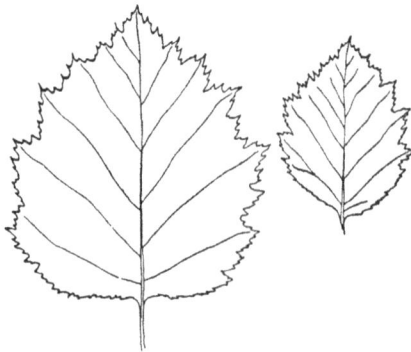

C. submollis

173. *Dalibarda repens* L.   FALSE-VIOLET, ROBIN-RUN-AWAY, DEWDROP

F1933A CHJJ Use not given.

FndA JG See 263, *Osmorhiza claytonii,* FndB.

174. *Fragaria* sp.   STRAWBERRY

W1912A KD See 348, *Achillea millefolium,* W1912D.

W1912B KD See 97, *Corylus americana,* W1912A.

W1912C JJJr See 189, *Rubus allegheniensis,* W1912A.

H1973A Anyms. See 153, *Gaultheria procumbens,* H1973A.

H1973B JT See 116, *Rumex crispus,* H1973D.

H1973C Anyms. Regulating women. Make tea from whole plant. Drink 1 cup.

H1973D TL Stroke. With an unknown plant. No preparation offered.

H1973E Anyms. Colic in babies.

H1974A DC Chancre sores, sties. Boil roots for 5 min., wash eye or mouth.

175. *Geum aleppicum* Jacq.   YELLOW AVENS (NATUR.)

W1912A ED See 168, *Agrimonia gryposepala,* W1912B.

W1914A DJ For horses cramps. Horses kick and look toward abdomen. With: (1) 150, *Sisymbrium officinale,* W1914A. Put 2 plants of each in 3 qt. water, boil quite a bit. Wet oats or other food with some and give some through nose with a bottle.

W1914B JJJr Cure for love medicine brought about by 396, *Taraxacum officinale,* W1914B. Put a double handful of the roots in 5 qt. water and boil pretty well. Drink before sunrise to make vomit.

G. alleppicum

G. canadense

Do for 4 days, covering up the vomit where no one will see. A part of (1) is taken each time along some old road not often used. The material is thrown to one side and told to go away and never come back.

FndA SS Diarrhea and high fever with convulsions. Put 2 whole roots in 1 qt. water. Boil hard for 5 min. Let stand and cool. Drink until fever and diarrhea stops.

176. *Geum canadense* Jacq.   WHITE AVENS

W1914A DJ Prevents snakes from biting. With: (1) cinnamon (optional). Smash and mix roots of 1 plant with (1) — no water. Put in small bottle. Rub on shoes and pants. If rubbed on stick, and touched to snake's head, he will die.

F1938A CHJJ See 301, *Mentha spicata*, F1938A.

FndA JG Love medicine. Decoction from whole plant.

177. *Geum rivale* L.   PURPLE AVENS, CHOCOLATE-ROOT, WATER AVENS

W1912A DJ Snake bite preventative. Mash 2 plants and let stand ½ hr. in 1 pt. of cold water.

Put on clothes and shoes and carry a stick covered with the medicine. Also may mix plant with perfume. When snake smells it, it will die.

?? ?? Fever. Will stop diarrhea too. Steep roots.

G. rivale

178. *Malus coronaria* (L.) Mill.   AMERICAN CRAB, SWEET-CRAB

W1912A JJJr Black eye. Scrape off bark from branches about 2 ft. long. Boil a little. Poultice on bark, keep wet with liquid.

W1912B JJ See 186, *Prunus virginiana,* W1912B.

W1914A DJ For sore eyes (from snow blindness) and black eyes. Scrape bark either way, add cold water and make tea. Wash eyes. For black eyes, tie on wet cloth, letting a little get in the eye.

W1914B DJ For difficult birth. With: (1) 187, *Rosa acicularis,* W1914A. Scrape bark off downward from 20 in. long twig and from 16 in. long twig of (1). Put in ½ qt. cold water, let stand for 3 min. Drink all of it.

W1914C JD Suppressed menses. Two quarts water added to second growth of roots. Boil a little. A small string of wampum is taken along and a single bead is taken off and placed by a root that points in the direction of the sun at 12 o'clock. Drink quite often.

F1939A SG Sore eyes, bruises or inflammation of eyes. Take bark from 1 young shoot, put in teacup, pour boiling water over, cover, let steam for 10-15 min. Wash eyes with solution.

FndA JG Dye for burden strap. Boil outer bark down in 2 qt. water. Reduce to color of scarlet. Put strings in water.

179. *Malus pumila* Mill.   COMMON APPLE, PARADISE APPLE (NATUR. & ESCAPING)

W1912A JSl Earache. Use second growth of apple tree sucker shoots. Scrape bark upward off of 1 yr's growth. Boil in 2 tablespoons of water for 5 min. Drop some in ear, then plug with a rag.

W1912B ED See 187, *Rosa acicularis,* W1912A.

F1938A CHJJ Bruises, black eye. With: (1) 247, *Rhus* sp., F1938D. Squash apple peelings with bark or the bark of (1). Soak in water, bandage.

P. opulifolius

180. *Physocarpus opulifolius* (L.) Maxim.   NINE-BARK

W1914A JD See 241, *Staphylea trifolia,* W1914A.

181. *Porteranthus trifoliatus* (L.) Britt.   INDIAN PHYSIC, BOWMAN'S-ROOT

W1912A JJJr See 27, *Polystichum acrostichoides,* W1912E.

F1938A HRE Grippe. Boil 2 roots in 1 qt. water. Do not make too strong. Drink 1 cup at a time. Makes you sweat.

FndA FSt Cold, fever, chills from fever, sore throats from colds. Pound root to powder. Pour on hot water. Put teacupful of water over a pinch of powder for strong medicine.

FndB SS Physic. No details.

182. *Potentilla canadensis* L.   DWARF CINQUE-FOIL, FIVE-FINGERS

F1933A SS Diarrhea. Pound root and steep.

P. canadensis

183. *Prunus americana* Marsh.   HEDGE-PLUM,
WILD PLUM, SLOE

H1973A Anyms. Diarrhea. Dry berries. No further
details offered.

184. *Prunus pensylvanica* L. f.   PIN-CHERRY,
FIRE-CHERRY, WILD RED CHERRY, BIRD-
CHERRY

F1938A CHJJ See 185, *Prunus serotina,* F1938C.

185. *Prunus serotina* Ehrh.   BLACK CHERRY,
WILD CHERRY, RUM-CHERRY

W1912A BB Consumption or "old cough." Fix
same as above. Take every 2 weeks for 3 mornings.

W1912B HJn For cuts. Bark placed over wound.

W1912C HJn Headaches. Scrape 2 handfuls of the
bark of young shoots and steep a little. Remove
switches and apply as poultice to head.

W1912D LJ Colds, fevers, soreness, cough, inflammation of lungs, consumption. Steep handful

(double) of bark in 4 qt. water, boil down to 2 qt.
Add 1 pt. good whisky and sugar.

W1912E LJ Bronchitis—"catching heavy cold in
babies." Scrape 2 handfuls of bark, boil in 2 qt.
water. Keep it hot after removing from heat.
Bundle baby, place over steam. Let drink another
batch made with whisky in 1/3 qt. water.

W1912F ED See 137, *Populus tremuloides,*
W1912E.

W1912G DJ See 75, *Ulmus rubra,* W1912C.

W1914A DJ See 75, *Ulmus rubra,* W1914A.

WndA BB For too much gall. Scrape bark down
from 4 switches of good size, young growth. Steep
and boil down to ½ in 4 qt. water. Use for 3
days.

F1933A JS For sores all over the body caused by
bad blood. With: (1) 354, *Arctium* sp., F1933A; (2)
396, *Taraxacum officinale,* F1933A. Scrape handful
of bark downward, add root of (1) and 4 roots of
(2). Boil. Drink.

F1933B CRE/ET Blood medicine. Scrape off bark
and boil immediately. Drink.
Inhalant for colds. Dried bark put in teacup, boiling water poured over, add salt. Cool to lukewarm.
Snuff liquid.

F1938A AJ Headache. Scrape bark downward, put
on cloth, apply to forehead.
Colds, sore throats. Take bark from year-long
shoots, put in teacup, add boiling water. Let cool,
take mouthful every ½ hr.

F1938B CHJJ Cough syrup. Remove bark, dry in
sun. No further details.

F1938C CHJJ For burns. With: (1) 184, *Prunus
pensylvanica,* F1938A; (2) 33, *Pinus rigida,*
F1938B; (3) 101, *Chenopodium album,* F1938A;
(4) deer tallow; (5) beeswax. Squash roots of the 4
plants, mix with (5) until soft like water, then add
(4) until like lard. Keep adding (4) until soft. Stir
over fire. Apply to burn. Will heal in 3-4 days.

F1938D CRE/ET Headache and head cold. Scrape bark off young shoots. Sprinkle cold water on it. For headache, apply bunch of it to forehead and nape of neck. For colds, inhale the water sprinkled on the bark in the morning, during the day, and before bed. Clears the head.

F1938E CHJJ See 96, *Carpinus caroliniana,* F1938A.

F1939A PH See 26, *Onoclea sensibilis,* F1939A.

FndA SG & JG Headache. Scrape bark off young shoots, upward. Boil, drink.
Colds. Preparation not given.
Gonorrhea. Scrape root bark from crotch of forked root.

FndB DJn Blood purifier. With: (1) 166, *Saxifraga pensylvanica,* FndD; (2) 354, *Arctium* sp., FndA; (3) 116, *Rumex crispus,* FndA; (4) 282, *Ipomoea pandurata,* FndA; (5) 5, *Peltigera* sp., FndA; (6) *gahgawaya?s.* Take bark off of second growth shoot about 2 in. long. Add roots and some leaves of (1), root of (3), and unspecified parts and amounts of (2), (4), and (6). Steep ingredients in 3 gal. water.

FndC FSt See 192, *Rubus odoratus,* FndA.

FndD CG Cough syrup. See 34, *Pinus strobus,* FndA.

H1973A DC Cough medicine.

H1973B FS Tea, cough syrup. Peel bark down to green dry for making tea. Dried and smashed for cough syrup.

H1973C TL Colds. Take soft part of bark and boil—very young growth.

H1973D ON Headache, colds. Use bark off shoots.

H1973E FK Cough medicine. Scrape bark, dry.

186. *Prunus virginiana* L.   CHOKE-CHERRY

W1912A JJJr Stops diarrhea in horses. Take

branches, leaves, and berries, boil a handful in 1 qt. water down to ½. Pour dose through the nose.

W1912B JJ Consumption. With: (1) 338, *Lonicera* sp., W1912A; (2) 72, *Hamamelis virginiana,* W1912E; (3) 247, *Rhus* sp., W2923J; (4) 178, *Malus coronaria,* W1912B. Scrape roots upward, mix equal parts. Put a big tablespoon in 2 qt. water, boil. Drink all you can.

F1938A CRE Blood purifier, for prenatal strength. With: (1) 72, *Hamamelis virginiana,* F1938F; (2) 345, *Viburnum trilobum,* F1938E; (3) 225, *Celastrus scandens,* F1938C; (4) 68, *Sanguinaria canadensis,* F1938G; (5) 97, *Corylus americana,* F1938C. Take 2 young shoots, add same of (1), (2), and (5), with root of (3) and slice of root of (4). Cook in 1 gal. water for ½ hr. Cool. Drink after sixth month, 1 cup a day until birth.

F1938B CRE/ET See 97, *Corylus americana,* F1938A.

F1938C CHJJ See 23, *Polypodium virginianum,* F1938A.

FndA HH Food, prenatal care, blood purifier.
Hemorrhages. Use stalk.
Diarrhea. Use bark.
Wounds. Remove inner bark downwards.

H1973A DC Cough syrup. Bundle, boil down, make syrup. Add 3 tablespoons of sugar. Take by teaspoon.

H1974A DC Diarrhea. No preparation offered.

187. *Rosa acicularis* Lindl.   PRICKLY ROSE (RARE)

W1912A ED For blindness. That is, when eye turns white. With: (1) 245, *Acer saccharum,* W1912A; (2) 179, *Malus pumila,* W1912B. Take rose leaves and bark scraped down, add bark from east side of (2), cutting it upward, and the bark scraped downward of (1) — a handful of each. Put in gallon water, steep. Squeeze drops in eyes 3 times a day.

R. acicularis

W1912B DJ Black magic. With: (1) 445, *Smilax hispida*, W1912B; (2) 354, *Arctium* sp., W1912B. Take 1 of each plant and boil a little in 1 qt. water. They first take (2) and bundle it like a doll, make a pointed stick, and stick it into the doll where you wish to affect the person you wish to injure (that [*sic*] 'you are angry at'). This may be cured by using 289, *Epilobium angustifolium*, W1912B.

W1914A DJ See 178, *Malus coronaria*, W1914B.

188. *Rosa rubiginosa* L.   SWEETBRIER, EGLAN-TINE (NATUR.)

W1914A JD For difficult urination caused by cop-ulation with a woman during menses. This medicine will open [you]. With: (1) 236, *Parthenocissus quinquefolia*, W1914B. Put a double handful of each plant in 5 qt. water, boil quite a bit. Both plants are "witch." Drink 1 qt. at a time.

189. *Rubus allegheniensis* Porter ex Bailey NORTHERN BLACKBERRY, SOW-TEAT BLACKBERRY, HIGHBUSH BLACKBERRY

W1912A ED To make dogs good hunters and to ensure them from theft. Smash secondary rooting, put in 1 qt. water and let stand for 1 hr. Give to dog anytime.

R. allegheniensis

W1912B JJJr Blood remedy for all ages. With: (1) 174, *Fragaria* sp., W1912C. Cut up and boil 2 roots about 1 ft. long in 3 qt. water, down to ½. Add root of 3 plants of (1). Take ¼ cup 3 times a day before meals. Patient will recover by the time the medicine is gone. Dilute if too strong.

F1933A JS TB. With: (1) 68, *Sanguinaria canadensis*, F1933B; (2) 214, *Epilobium angusti-folium*, F1933D. Smash and boil all roots in 1 gal. water. Drink anytime.

F1938A CRE/ET If baby's navel is red and sore after birth. Squash the root and make into a fine mash. Bandage navel until healed.

F1938B CRE/ET Coughing, thin, weak, fever every afternoon, sweats. When you catch cold, do not take care of yourself, and it turns to TB. With: (1) *deganōōgwíyōta?*; (2) *deyagoninowanyégowa*. Take 3 roots, add 3 roots of (1) and 1 stalk and root bunch that fits between fingers and thumb or (2). Boil thoroughly in 1½ gal. water for a few minutes. Drink like water. Drink no other liquid. Drink until cough stops.

F1938C CHJJ See 437, *Trillium* sp., F1938A.

F1938D CHJJ See 440, *Veratrum viride*, F1938C.

F1939A DJn Summer complaint. Scrape root up and down. No further details.

FndA AJm For summer complaint, diarrhea, blood purifier. Preparation not given.

190. *Rubus idaeus* L. ssp. *sachalinensis* (Levl.) Focke
WILD RASPBERRY

W1912A BB For tea. Steep young twigs.

W1912B KD See 348, *Achillea millefolium*, W1912D.

W1912C ED See 12, *Lycopodium lucidulum*, W1912A.

W1912D PT See 21, *Osmunda claytoniana*, W1912A.

F1938A CRE/ET For kidneys. Gather leaves, boil them, drink the water.

F1938B CRE/ET For burning and pain when passing water. Removes bile from the bladder—acts as emetic and physic. Boil about 12 of the top leaves only in 2 qt. water for 5-6 min. Cool, drink 1 qt. all at once. Rest.

FndA SS Blood purifier. For low or high blood pressure. Boil 2 root tips in 2 gal. water. Boil down to 3 qt. Cool, drink like water. Take 2 gal. a week.

H1973A JT For ladies who are run down because of sickness of period. With: (1) "snakeroot" (42, *Asarum canadense*?). Take 2 big roots, add 2 tablespoons of (1). Put in ½ qt. water, boil for ½ hr. Drink ½ cup at a time.

H1973B JT Tonic. No details offered.

H1973C Anyms. See 42, *Asarum canadense*, H1973F.

191. *Rubus occidentalis* L.   BLACK RASPBERRY, BLACK-CAP, BLACK THIMBLEBERRY

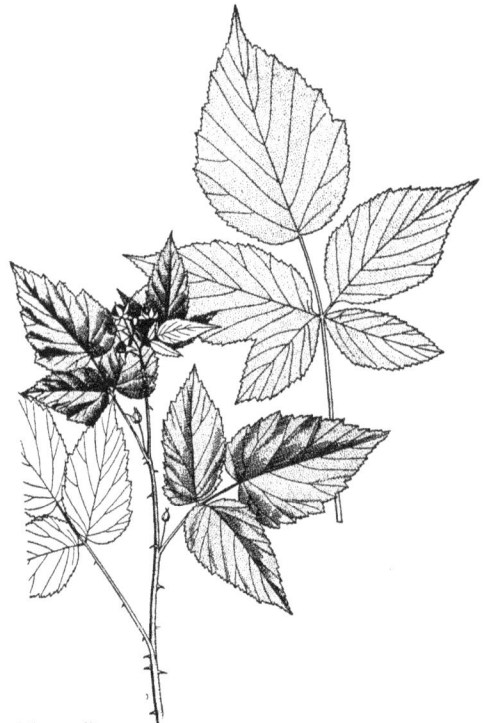

R. occidentalis

W1912A KD See 348, *Achillea millefolium*, W1912D.

W1912B PT See 21, *Osmunda claytoniana*, W1912A.

W1914A DJ A medicine for when a man is hunting and his wife is fooling around. Take some of the roots at both extremities without killing the bush. Put in 1 qt. water, boil a little, take drink yourself, wash body. Give woman a little. She will then stay home no matter if he stays away a year.

F1939A SS Whooping cough in children. Boil bundle size of hand of roots, stalks, and leaves in 3 qt. water, strain with cheesecloth, drink 4 glasses a day.

FndA AJJ & CHJJ Food, emetic, physic. Spring medicine for removing bile. Top leaves for medicine, no details given.

H1973A DC Diarrhea. No details offered.

R. odoratus

192. *Rubus odoratus* L.   PINK THIMBLEBERRY, PURPLE FLOWERING RASPBERRY

W1912A MS For kidneys or gonorrhea or indigestion. Remove leaves, boil bundle of twigs or switches—6 of them—in 1 gal. water. Take drink anytime—more at night and morning. Make strong for gonorrhea.

W1912B KG Blood medicine for newly born babies and for the bowels. Take berry and add 4 pieces of root and stem about 7-8 in. long and put in 3 qt. of water. Boil down ½. Take 6 times a day, and ½ cup before meals.

W1913A PJ To counteract diarrhea. Bark is scraped and made into decoction.

F1933A SS Venereal disease. With: (1) 91, *Alnus incana,* F1933B. Take whole root along with 2 in. sprigs of (1). Boil in 1 gal. water. Wash the parts, especially the chancres and sores. Drink the rest ad lib.

F1933B SS See 254, *Aralia nudicaulis,* F1933A.

F1938A CRE/ET See 152, *Epigaea repens,* F1938A.

F1939A NL Food.

FndA FSt For when woman has miscarriage. With: (1) 185, *Prunus serotina,* FndC; (2) 157,

Chimaphila umbellata, FndB. Take leaves off of plant, add inner bark of (1) and whole plant of (2). Boil and steep all in 1 gal. water. Drink.

H1973A JT Settle stomach. Cut up limbs, make tea.

H1973B DC Diarrhea. Boil 2 roots for 10 min., let cool, drink.

H1973C FK Colds. Use roots.

193. *Spiraea alba* DuRoi   NARROW-LEAVED MEADOW-SWEET

W1912A JSl See 267, *Gentiana andrewsii,* W1912B.

194. *Spiraea* sp.   MEADOW-SWEET

W1912A ED Antidote for love medicine. Scrape bark of young twigs upward, put handful in teacup

S. alba

W. fragarioides

of warm water and let stand for ½ hr. Drink first thing in morning and immediately return to bed. Take 14 qt. warm water, vomiting 3 times. Repeat once a day for 3 days.

W1912B JJ Consumption, for initial stages only. With: (1) 141, *Salix* sp., W1912A; (2) 224, *Cornus sericea*, W1912C; (3) 249, *Zanthoxylum americanum*, W1912F; (4) 120, *Tilia americana*, W1912E; (5) 343, *Viburnum lentago*, W1912B; (6) 138, *Salix discolor*, W1912A. Scrape bark upward from all plants, same quantities of each. Place in a box, then take 1 handful to 5 qt. water and boil. Drink until vomit, mornings and evenings.

W1912C RS See 96, *Carpinus caroliniana*, W1912A.

W1914A JD To make children stronger so that they will walk sooner. Scrape 1 handful of bark, put in 1 qt. warm water, let stand. Wash body.

195. *Waldsteinia fragarioides* (Michx.) Tratt.   BARREN STRAWBERRY, FALSE STRAWBERRY

W1912A ED Blood remedy. With: (1) 359, *Aster* sp. Mix handful of (1) roots with the strawberry plants. Boil in 8 qt. water until 4 remain. Take ½ pt. in morning before meals.

W1912B DJ Snake bite. Smash, wet a little, apply as poultice.

W1914A DJ Makes hair grow after losing it due to fever. With: (1) 238, *Vitis riparia*, W1914A. Put 1 plant with 1 sprig of (1) in 2 qt. water. Boil down to ½ qt.

CAESALPINIACEAE (CAESALPINIA FAMILY)

196. *Cassia hebecarpa* Fern.   WILD SENNA, PARTRIDGE-PEA

?? KD Worm remedy.

W1912A KD See 369, *Eupatorium perfoliatum*, W1912A.

H1973A DC See 434, *Smilacina racemosa*, H1973A.

FABACEAE (BEAN FAMILY)

197. *Amphicarpea bracteata* (L.) Fern.   HOG-PEANUT

W1915A DJ See 323, *Lobelia cardinalis*, W1915A.

198. *Baptisia tinctoria* (L.) Vent.   WILD INDIGO, RATTLEWEED

F1938A CHJJ Liver sickness. With: (1) 123, *Sarracenia purpurea*, F1938B; (2) 100, *Phytolacca americana*, F1938A; (3) 368, *Eupatorium maculatum*, F1938A; (4) 164, *Sedum telephium*, F1938A; (5) $ge^nde^nmde^nsta?gowa$; (6) $utge\check{s}u^nta$; (7) *gamunskuta*. Put whole roots of each in water. Cook until color of water has turned red.

FndA SS For cramps in the arms, legs, stomach. Medicine for athletes. Boil a handful of the root in 2 qt. water for 5 min. Rub liquid on legs. Bathe before game to run fast.

FndB HJ It draws bile from one point in you from where it does not belong. Steep and drink. Concentrates bile. Then take physic to eliminate it.

FndC CHJJ Liver and rheumatism medicine. With: (1) 368, *Eupatorium maculatum*, FndB; (2) 25, *Pteridium aquilinum*, FndE; (3) 255, *Aralia racemosa*, FndA. No preparation given.

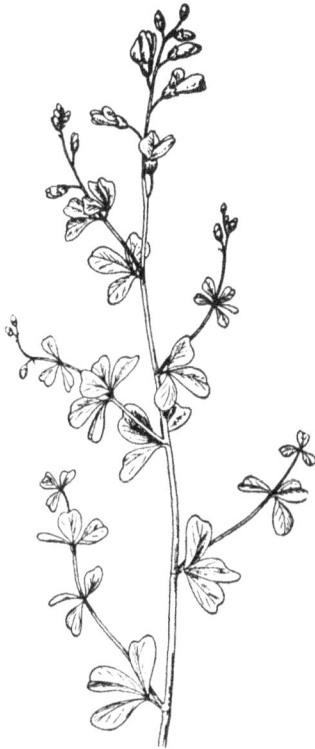

B. tinctoria

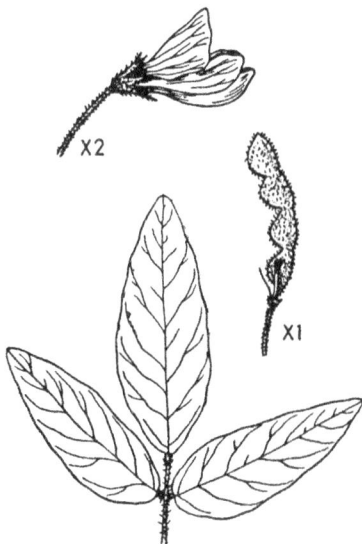

D. canadense

199. *Desmodium canadense* (L.) DC.   GIANT TICK-CLOVER OR TREFOIL, SHOWY TICK-TREFOIL, BEGGAR-TICKS

F1938A CHJJ Biliousness. Boil root in 1½ qt. water. Drink all you can.

200. *Desmodium glutinosum* (Muhl. ex Willd.) Wood STICKY TICK-CLOVER OR TICK-TREFOIL

W1912A ED Basket medicine. Take root, smash it, putting a handful in a teacupful of cold water. Sprinkle on baskets and customers will pay the price you ask.

201. *Lespedeza* sp.   BUSH-CLOVER

W1912A DJ See 228, *Euonymus obovata,* W1912B.

202. *Melilotus alba* Desr. ex Lam.   WHITE SWEET-CLOVER (NATUR.)

W1912A DJ See 332, *Mitchella repens,* W1912A.

D. glutinosum

S. helvola

203. *Robinia pseudo-acacia* L.   BLACK LOCUST, FALSE ACACIA (NATUR.)

FndA ?? Avoid this tree. If this tree invades your property, you will not own that property long.

204. *Strophostyles helvola* (L.) Ell.   WILD BEAN (probably *Apios americana* Medic. if in Seneca territory)

F1933A SS For poison ivy and warts. Rub leaves on affected parts.

205. *Trifolium hybridum* L.   ALSIKE CLOVER (NATUR.)

T. hybridum

W1914A JJJr For scarcity of milk in women and cows. Put a whole plant in a little cold water, let stand a little, then wash breasts.

206. *Trifolium pratense* L.   RED CLOVER (NATUR.)

F1938A CRE/ET Blood medicine. Boil flowers in 1-2 qt. water. Drink like water, ½ cup, 3 times a day.

F1938B HRE For change of life in women. Steep blossoms (which may be dried and stored for winter use) in hot water. Do not boil. Put 2-3 blossoms in 2-3 qt. water. Drink cold like water.

207. *Trifolium repens* L.   WHITE CLOVER, LAWN-CLOVER, DUTCH-CLOVER (NATUR.)

W1912A KD See 396, *Taraxacum officinale*, W1912B.

WndA KD Paralysis of the eye. Mash to paste in warm water. Poultice.

208. *Vicia americana* Muhl. ex Willd.   AMERICAN OR PURPLE VETCH, TARE

V. americana

V. sativa

W1912A DJ Love medicine for women. Put 3 roots in ½ teacupful warm water for 15 min. Sprinkle on breasts, shoulders, face, and hands. Carry root in pocket, brush against man, he will follow.

209. *Vicia sativa* L. ssp. *nigra* (L.) Ehrh. (*V. angustifolia* L.)   COMMON VETCH, SPRING VETCH, TARE (NATUR.)

W1914A ?? When a woman's external sexual organs swell and she has had no sexual connections, urinary passage becomes sore and menses are suppressed. Put small handful of plants in 4 qt. water, boil quite a bit. Take a cupful 4 times per day.

W1914B ?? Love medicine. Crush a few plants in cold water, wash hands with it and shake hands with girl or touch her in some way. Give sweetened with sugar, this is better or more effective.

HALORAGACEAE (WATER MILFOIL FAMILY)

210. *Myriophyllum verticillatum* L.   WATER MIL-FOIL

W1912A DJ See 17, *Equisetum fluviatile*, W1912A.

W1914A DJ Good for little children when they lie

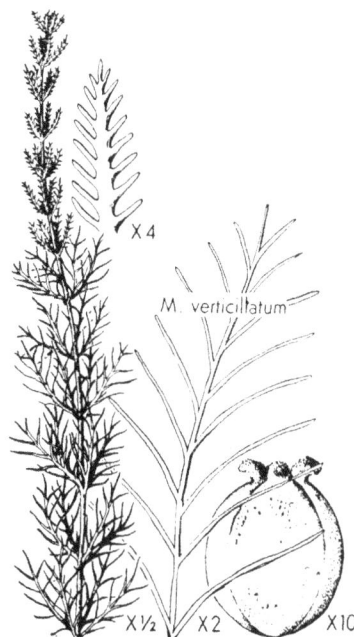

M. verticillatum

very quiet and never move—something is wrong—they will not suck, etc. Take plants from moving water (cleaner). Put whole plant in 6 qt. water, boil quite a bit. Give just a little at a time, also wash body. The child will then move as plant does in water.

Also a snowsnake medicine. Put 2 plants or branches in 4 qt. water, boil a little. Put medicine in a pail and let snowsnakes stand with points in it all night. In morning, wash whole length. Take balance of medicine and sprinkle along path of the snowsnakes (when nobody is looking). The other snakes will go in an uneven manner as the plant does in water. Body is also washed before the game with the same preparation.

LYTHRACEAE (LOOSESTRIFE FAMILY)

211. *Lythrum salicaria* L.   PURPLE LOOSE-STRIFE, SPIKED LOOSESTRIFE (NATUR.)

?? ?? Fever and sickness caused by the dead. With: (1) *nigunaza*. Decoction of whole plant(s). Drink until well.

FndA JG No use given.

THYMELAEACEAE (MEZEREUM FAMILY)

212. *Dirca palustris* L.   LEATHERWOOD, MOOSEWOOD, WICOPY, ROPE-BARK

W1912A JA Swellings on the leg or limbs. Cut branches into 2-3 in. lengths, use no leaves. Boil to a pulp next to cooking potatoes. Then peel bark and pound to pulp. Liquid not used. Apply pounded bark as poultice. Will reduce swelling.

W1912B MS Sciatica or strained back. Cut up 2 roots and steep in 4 qt. water. Drink cupful as often as [you] like. One must keep out of sight when taking this medicine.

W1912C JJ To take venereal germs out of the blood. Cut off whole root, put it under some warm ashes and let it fry. When cooked enough, take off the bark and place in 1 pail water and boil down to 1 qt. Divide liquid into 4 parts. Drink every night.

W1912D PJ Gonorrhea and syphilis—purifies the blood by loosening the bowels. With: (1) *kau^nwia*; (2) 137, *Populus tremuloides*, W1912F; (3) 68, *Sanguinaria canadensis*, W1912G. Cut off 5 lb. root and scrape. Do same for (1) and (2), but use 2½ lb. of root of (3). Dry all, pound, sift, pound again (i.e., make into a powder). Put 1 tablespoon in 2 qt. water and steep a little. Drink all you can.

W1912E KG Typhoid fever. Smash as fine as possible 3 pieces of root about 4 in. long. Put in 2 qt. water, lukewarm. Take 3 tablespoons a day.
Also a laxative. Use same root (see above), boil down to ½.

W1912F ED See 11, *Lycopodium digitatum*, W1912A.

W1912G EH See 243, *Acer pensylvanicum*, W1912A.

W1912H SH See 396, *Taraxacum officinale*, W1912F.

W1912I ED See 12, *Lycopodium lucidulum*, W1912A.

W1914A DJ For when there is something wrong with the joints of the back (lumbago). Put 1 small bush's roots in 10 qt. water, warm. Drink all you can.

F1939A HJ See 255, *Aralia racemosa*, F1939B.

F1939B CG See 214, *Epilobium angustifolium*, F1939B.

FndA CHJJ Consumption. Take whole plant, root and bark. No further details given.

FndB AJm Good for internal inflammation. Root used.

H1974A Anyms. Makes strong all over. Also an aphrodisiac. Boil whole stems. No further details offered.

ONAGRACEAE (EVENING PRIMROSE FAMILY)

213. *Circaea lutetiana* L. ssp. *canadensis* (L.) Aschers & Magnus   ENCHANTER'S NIGHTSHADE

F? ?? Wound cure.

F1938A CHJJ See 301, *Mentha spicata*, F1938C.

214. *Epilobium angustifolium* L.   WILLOW-HERB, FIREWEED, WICKUP

W1912A ED Burning while urinating. Steep handful of roots in gallon water. Drink in a day at any time—continue for 3-4 days.

W1912B DJ For pain anywhere in the body. Scrape and steep a little of the bark from 6 sticks in 1 cup water. One tablespoon should be enough, but take more if needed. Use wet chips as poultice and wet with liquid.

W1912C HJn See 120, *Tilia americana*, W1912B.

W1912D DJ See 354, *Arctium* sp., W1912B.

W1914A JD For internal injury from lifting, etc., or a hurt. Put a small handful of roots in 3 qt. water. Boil. Drink as much as you can.

E. angustifolium

F1933A CHJJ Consumption. With: (1) 336, *Lonicera dioica,* F1933A. Use all roots of both plants. Smash them. Boil in as much water as you can drink before it spoils. Drink whenever you can.

F1933B CHJJ TB. With: (1) 223, *Cornus rugosa,* F1933B; (2) 77, *Laportea canadensis,* F1933A; (3) 447, *Corallorhiza maculata,* F1933B. Smash roots of all plants. Do not quite boil them. Cool and drink 3-4 times a day.

F1933C JS TB. Preparation not given.

F1933D JS See 189, *Rubus allegheniensis,* F1933A.

F1939A NL & PH Use not known.

F1939B CG For kidneys and when the male can not urinate. With: (1) 212, *Dirca palustris,* F1939B. Steep handful of root along with inner bark of (1). Drink
Also for swollen knee. Smash root and poultice.

FndA SS Basket medicine. Take whole plant and put in water until warm. Sprinkle on basket.

215. *Oenothera biennis* L.   COMMON EVENING PRIMROSE

W1914A JD Piles. With: (1) an unspecified mint, *uⁿhia?'ã gawáhunta?,* cf. 300, *Mentha canadensis,* W1914B; (2) 304, *Prunella vulgaris,* W1914B. Put 1 root of each in 1 qt. water, boil down ½. Drink cupful twice, use rest for wash.

O. biennis

O. perennis

F1938A CHJJ Provides athletes with great strength—for lacrosse, snowsnake, wrestling. Chew roots and spit on hands and rub on arms and all over muscles.

H1973A Anyms. Used for boils and laziness once, but did not work good. Instead, used 42, *Asarum canadense,* H1973F.

216. *Oenothera perennis* L.   SUNDROPS

W1914A DJ Paralysis—"dead on the body." Make bunch of whole plants and put in 8 qt. water. Boil a little. Take a cup twice a day.

CORNACEAE (DOGWOOD FAMILY)

217. *Cornus alternifolia* L. f.   GREEN OSIER, PAGODA DOGWOOD

W1912A HJn See 93, *Betula lenta,* W1912A.

W1912B DJ To heal navel. Scrape bark towards middle. Dry, powder, wet. Tie to navel. A bullet used to be flattened and tied on top of this powder. The application "draws."

W1912C DJ See 251, *Geranium maculatum,* W1912B.

W1912D DJ & JJJr See 251, *Geranium maculatum,* W1912C.

F1933A JS Cough. Once treated son of bronchial cough. Scrape handful of bark downward in 1 qt. water. Boil (?). Drink. Yellow vomit will come up and cough will be loosened.

F1933B CHJJ Scrape bark upward and boil for an emetic.

F1938A AJ See 72, *Hamamelis virginiana,* F1938D.

FndA HRE Eyewash. Preparation not given.

H1973A Anyms. For swollen areas. With: (1) 120, *Tilia americana,* H1973B. Steep bark, poultice.

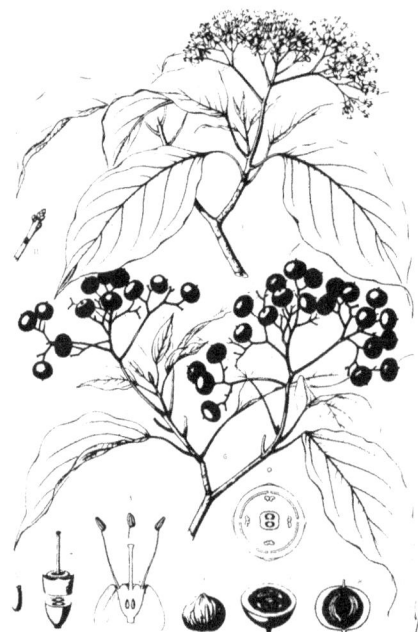

C. alternifolia

218. *Cornus amomum* Mill.   SILKY DOGWOOD, KINNIKINNIK

W1912A HJn Gonorrhea. Baby will break out red on face if mother has it. Scrape double handful of bark upward and steep in 1 qt. water. Bottle it. Wash sores first thing in morning and at bedtime. Can also dry and powder bark and rub it on sores.

W1912B EH Makes baby sleep. Scrape bark, steep in water, wash baby.

F1938A CHJJ Emetic. Leaves used as a tobacco substitute. Boil 2 handfuls of bark in all the water you can drink. Take in the morning on rising before breakfast.

F1938B CHJJ See 96, *Carpinus caroliniana*, F1938A.

F1939A CG (cf. names for 220, *Cornus drummondii*) Chest congestion, emetic. Steep handful of bark in big kettle. Drink all you can and vomit.

F1939B KD Goiter—"swollen throat." Take bark upward with leaves and tops where green. Smash and work into paste. Put silver half dollar on throat, face down (heads down). Bandage with poultice.

FndA SS See 340, *Triosteum aurantiacum*, FndB.

219. *Cornus canadensis* L.   BUNCHBERRY, DWARF CORNEL, DWARF DOGWOOD, CRACKERBERRY, PUDDING BERRY

F1933A CHJJ No use given.

FndA SS Cough, TB, fever. Make small bundle of whole plant. Put in 1 gal. water until boils. Drink.

220. *Cornus drummondii* Meyer   ROUGH-LEAF DOGWOOD (RARE)

W1912A BB Gonorrhea. Bundle 3 switches in 2 qt. water, [boil] until ½ remains. Will cure in 3 days.

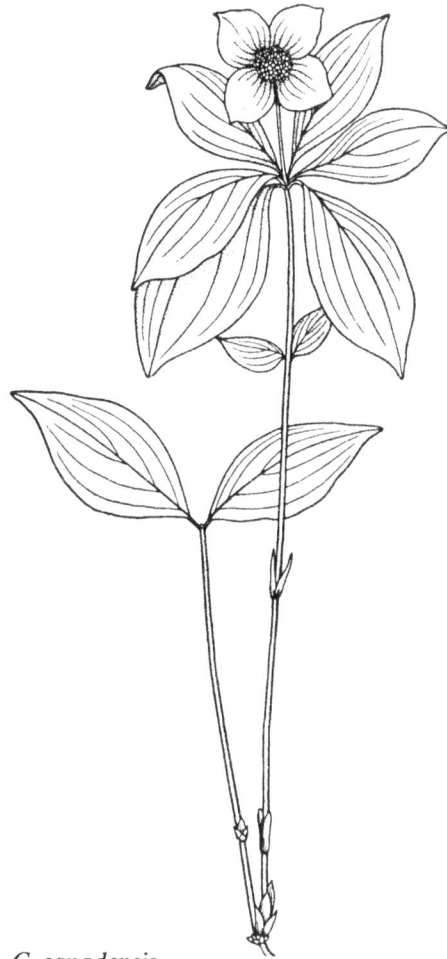

C. canadensis

C. drummondii

C. florida

221. *Cornus florida* L.   FLOWERING DOG-
WOOD

F1933A SS See 157, *Chimaphila umbellata*,
F1933B.

222. *Cornus foemina* Mill. ssp. *racemosa* (Lam.) J.
Wilson   PANICLED DOGWOOD, GRAY
DOGWOOD, STIFF DOGWOOD

W1912A PT Cuts in man or horses. Take bark
from 6 switches or stems 4-5 in. long. Add boiling
water. Sprinkle water on cut, apply bark as
poultice.

W1912B JJ See 35, *Tsuga canadensis*, W1912E.

W1912C PT See 43, *Nuphar lutea*, W1912A.

W1914A ?? Use not given.

W1914B DJ For woman after baby is born who
has swollen legs. For any swellings. With: (1) 344,
Viburnum recognitum, 1914A. Scrape bark, add
warm water, and apply as a poultice to swellings of
any kind.

W1914C DJ Peach stone and other games—as a
medicine. The gall is used. Take worm out of gall
and mix with red dye from berries. Rub on palms
and on handkerchief.

223. *Cornus rugosa* Lam.   ROUND-LEAF DOG-
WOOD

W1912A BB Emetic. Scrape bark off of 4 sticks of
young growth and put in 3-4 qt. boiling water. Stir,
let stand for few minutes. Drink and vomit. Use
another day if [you] like.

F1933A CHJJ See 418, *Elymus canadensis*,
F1933A.

F1933B CHJJ See 214, *Epilobium angustifolium*,
F1933B.

F1933C CGJJ General cathartic; sometimes an
emetic. Scrape handful of bark upward. No further
details.

224. *Cornus sericea* L. (*C. stolonifera* Michx.)
RED OSIER, RED DOGWOOD, AMERICAN
DOGWOOD

W1912A BB To whip children with.

W1912B JJJr Hemorrhage of mouth or nose.
Also, cough and headache. Scrape from end to
center off of 3 sticks 3 in. long. Steep in 6 qt. warm
water. Cover, let stand ½ hr. Take a cup before
meals and upon retiring.

W1912C JJ See 194, *Spiraea* sp., W1912B.

W1914A PJ No use given.

W1914B DJ Good medicine—especially when mixed with 276, *Nicotiana rustica,* W1914B. Bark is smoked.

W1914C DJ Emetic in winter. Not good in summer, as strangers may see it and the medicine will not work.

W1915A JJJr For magical bows and arrows. Must be gotten from high ground, not from swamps.

W1918A JJJr To kill snakes. A bow about 4 ft. long and 3/4 in. thick is made. With: (1) 276, *Nicotiana rustica,* W1918A; (2) 272, *Asclepias incarnata,* W1918B; (3) 75, *Ulmus rubra,* W1918A. A bow is made from the branch or stem, with a string of the bark of (3) and some sprigs of (1) tied to the bow near the larger end. The slippery elm (3) string is twisted with the smooth side out and the snake is struck with it. The tobacco (1) sprigs are tied on with swamp milkweed (2).

WndA KD Emetic. Scrape bark upward and boil.

F1938A CHJJ Smoke. Mix with tobacco. Also, when he knows he is going to have possessions (?). Scrape bark upward and put in warm water. Drink.

F1938B CHJJ Craziness. With: (1) 230, *Ilex verticillata,* F1938B. Scrape bark upwards from both. Boil a few handfuls of each. Drink a cupful 3 times a day.

F1939A NL No use given.

F1939B JG Ball player's emetic. Take about 3 of the young shoots, second growth, and scrape the bark up. Boil the bark in 5 qt. water about 5 min. Let cool. Drink all you can until you are filled up. Then vomit. Do this 3 mornings before sunrise.

F1939C See 218, *Cornus amomum,* F1939A.

FndA FSt Emetic and for pains in the chest. Preparation not given.

FndB JG Spring emetic. Also lacrosse and runner's medicine. Scrape bark up from about 6 shoots, young. Put shavings in pot and add 5 qt. water. Boil 3 min., let cool. Drink as much as possible in morning before the sun rises. Vomit in river or stream and it will carry the sickness away. Repeat for 3 mornings.

FndC FSt For whipping children. Mentioned in *káiwi:yo:h* (Good Message). Get 12 whips of young plants. Hit child once with each, putting each aside. If does not say will be good before used up, continue. Burn tobacco when taking shoots.

FndD KD See 162, *Ribes hirtellum,* FndA.

H1973A Anyms. See 79, *Urtica dioica,* H1973B.

H1974A Anyms. Colds. Causes vomiting.

H1974B DC Eyewash. No details offered.

CELASTRACEAE (STAFF-TREE FAMILY)

225. *Celastrus scandens* L.   AMERICAN BITTERSWEET, WAXWORK, STAFFVINE, CLIMBING OR FALSE BITTERSWEET

?? KD To put on the lips of bad children. Rub on root. Also may be put on lips and gums when teething.
For young girls who catch cold and can not menstruate. Boil whole root in ½ gal. water down to 1 pt. Very powerful.

W1912A LJ Regulator for women. Diuretic. Causes free flow of urine stopped by soreness of lifting with child. Reduces fever and soreness caused by pregnancy. Fist-size bundle of leaves and stem steeped in 2 qt. water until of sufficient strength. Take ½ cup at a time.

W1912B NH For kidney trouble following childbirth. Tea made from root.

W1912C KD See 26, *Onoclea sensibilis,* W1912D.

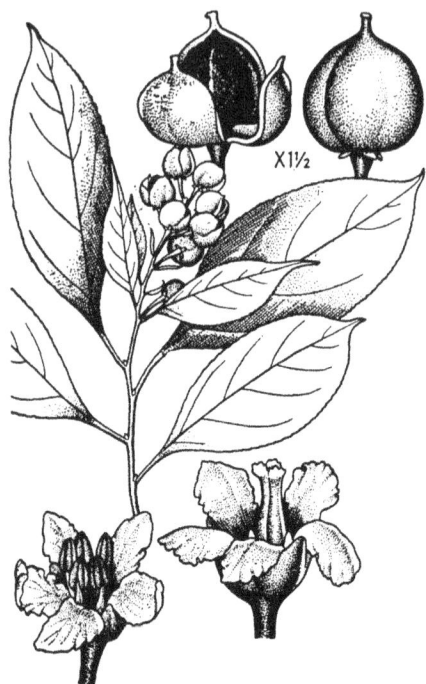

C. scandens

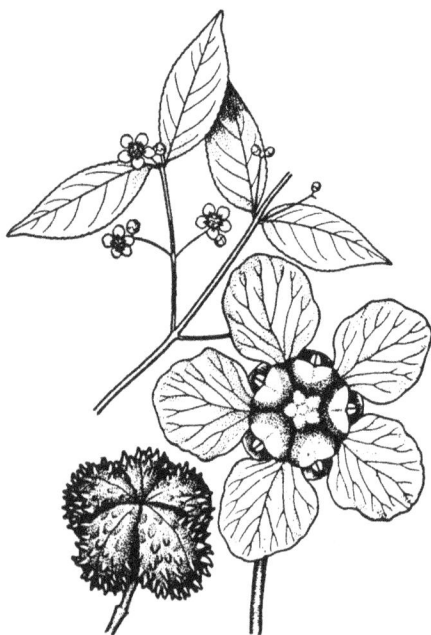

E. americana

F1933A CHJJ Dropsy or watery blood. Swellings on shins and calves, soft to the touch, leaves dent when probed with the finger. With: (1) 255, *Aralia racemosa*, F1933C; (2) 41, *Sassafras albidum*, F1933C; (3) "swamp root"; (4) *oshewe^n?do:do?*; (5) "high, has rattles on it." Take 1 root of (3) and bark of (4) and any amount of roots of others. Smash up, boil in 2 qt. water to a good boil. Drain off liquid from the first 2 boilings and the dregs, and combine those 2 a third time to make a third batch. Drink anytime.

F1938A CRE/ET Kidneys—too much fluid. Use root.

FndA NL If you eat the berries, it will wring your neck—i.e., it is poisonous. Name also derived from fact that this plant winds around itself as it climbs.

226. *Euonymus americana* L.   STRAWBERRY-BUSH, BURSTING-HEART (RARE)

W1912A KG Suppressed menses. Make bundle of 4 plants and their roots and boil down in 3 qt. until 2 remain. Take ½ cup 4 times daily, anytime.

W1912B JJJr Excessive menstrual flow. Take handful of vine, boil down to ½ in 2 qt. water. Let cool. Drink immediately and finish.

W1914A DJ To stimulate suppressed menses caused by a girl having mustard, pepper, or soda. Put small bundle of vines in 6 qt. water. Boil quite a bit until 1 qt. remains. Drink all they want—every hour is best. Do not take when pregnant because it will induce abortion.

W1914B JD Difficult urination from too much gall. With: (1) 235, *Parthenocissus inserta*, W1914A. Put a good handful of each in 5 qt. water, boil quite a bit. Drink 1 qt. at a time—bedtime is best. Hold back urinating as long as possible.

227. *Euonymus europaea* L.   SPINDLE-TREE (RARE ESCAPE)

W1912A MS Will stimulate the appetite or physic. Steep root and drink cupful before meals.

E. europaea

W1912B PJ To take worms out of a man's body caused by solid food taboo (a kind of venereal disease). With: (1) what may be 342, *Viburnum lantanoides,* W1912B; (2) 39, *Magnolia acuminata,* W1912B; (3) *skáuk;* (4) *nu'gä:äeⁿ*. Take small quantity of each, put in pail of water and boil down to 1 qt. Drink at night, and eat soft foods only.

W1912C RS Bloody urine in males caused by intercourse with [a] woman during menses. Make 2 bunches about 2 in. thick of root, put in 6 qt. water, boil down to 2 qt. Drink all you can but hold back when you wish to urinate. Then go when you cannot hold it anymore. Have nothing to do with women for 1 month. [A] woman may get the same illness—use same medicine.

F1938A CHJJ Against worms. Peel bark down. Take bark from root that runs east, toward the rising sun. Put in water. Do not quite boil. Take off heat, drink anytime you want.

FndA SS Worms in children. Scrape the bark downward off of 1 stake 4 in. long. Put in 1 qt. water, boil, let cool.

228. *Euonymus obovata* Nutt.   RUNNING STRAWBERRY-BUSH

W1912A ED Good for people who are bewitched. Used with an unspecified other plant. Bundle into a handful, steep in 3 qt. water until 2 remain. Take tablespoon every 20 min. Take all day until finished. Wash yourself morning and night with it. This plant is thought to move a little every spring. Do not know why.

W1912B DJ For stricture caused by something wrong with the blood. With: (1) 201, *Lespedeza* sp., W1912A. Boil 1 whole plant of each in 6 qt. water down to 5 qt. Drink often as much as desired, finishing what is taken out each time. None must be put back in.

W1914A DJ See 226, *Euonymus americana,* W1914A.

W1914B RS For difficult urination. Put a small handful of vines in 2 qt. water, boil quite a bit. Drink warm as much as possible. Do not urinate until obliged so as to let medicine work.

229. *Euonymus* sp.   WAAHOO

W1912A DJ See 308, *Fraxinus nigra,* W1912C.

W1912B JJ See 40, *Lindera benzoin,* W1912B.

F1933A CHJJ See 315, *Scrophularia lanceolata,* F1933A.

E. obovata

AQUIFOLIACEAE (HOLLY FAMILY)

230. *Ilex verticillata* (L.) Gray   BLACK ALDER, WINTERBERRY

F1938A HRE Physic or emetic. Take bark with berries on it, make bundle 2 in. in diameter between finger and thumb. Boil in 4 qt. water down to 1 qt. Take tablespoon before meals and retiring.

F1938B CHJJ For craziness. Scrape 2 handfuls up, boil them good. Drink cupful 3 times a day.

F1938C CHJJ See 97, *Corylus americana*, F1938B.

F1938D CHJJ See 224, *Cornus sericea*, F1938B.

FndA HW Retains vigor. Also for biliousness. Use in spring and fall. Once prepared (preparation not given) take on an empty stomach and lie still for 1 hr.—then will vomit.

FndB CHJJ Emetic for craziness. Take bark with berries on it. Make into bundles size of finger and thumb encircled. Boil down in gallon to make 1 qt. Take a tablespoon before meals and before retiring.

EUPHORBIACEAE (SPURGE FAMILY)

231. *Chamaesyce glyptosperma* (Engelm.) Small (*Euphorbia glyptosperma* Engelm.)   SPURGE, EYEBANE, MILK PURSLANE

C. glyptosperma

C. americanus

W1914A DJ For goiter. With: (1) 286, *Echium vulgare*, W1914A. Put 2 stems of each in 5 qt. water. Boil down 1 qt. Wash outside and drink ½ cup once a day.

FndA SS Stimulates lactation. Whole plant used, no details given.

RHAMNACEAE (BUCKTHORN FAMILY)

232. *Ceanothus americanus* L.   NEW JERSEY TEA, REDROOT

W1914A DJ When women have difficulty in urinating caused by catching cold during menses. Put 3 stems with roots in 6 qt. water. Boil, drink all you can. Can use same plants twice, but boil down to a quart the second time.

W1914B JD Good for suppressed menses from catching cold. Put 1 root about 3 or 4 in. long in 2 qt. water, boil quite a bit. Drink quite a lot or when feeling like it.
Also an abortive, i.e., when a pregnant woman gets hurt within 2 or 3 months of term. Put 1 root in

1 qt. water, boil quite a bit. Drink as soon as possible.

F1933A SS Good for the blood, colds. Steep root until warm. Drink in winter when cold is approaching.

F1933B CHJJ Diarrhea. With: (1) 304, *Prunella vulgaris*, F1933C; (2) 368, *Eupatorium maculatum*, F1933A. Collect plants in fall when roots are hard. Smash all 3 plants, boil in 2 qt. water for a long time. Cool, drink any amount.

F1933C CHJJ See 417, *Cinna arundinacea*, F1933A.

F1933D CHJJ See 54, *Hepatica nobilis* var. *acuta*, F1933A.

F1938A CHJJ For sores. Use bark from roots, peel, grind into powder. Dust on sores.

F1938B AJ & CRE Sore roof of mouth. Pour boiling water over bark. Cool, wash mouth in it.

FndA HJ Venereal disease. Take bark off root and grind into a powder and dust it on open sores. Then cut root into chunks 6-7 in. long, then cut up into shorter pieces. Boil 2 qt. of water enough for 2-3 days. Drink like water, a cup at a time.

233. *Rhamnus alnifolia* L'Her.   ALDER-LEAF BUCKTHORN

W1912A SH Tonic, physic, blood purifier for babies. General medicine for children. Put small handful of bark scraped downward into ½ qt. water. Steep. Give 3 teaspoons 3 times a day. If too effective, reduce dose.

W1912B DJ See 298, *Lycopus virginicus*, W1912A.

W1912C PT See 21, *Osmunda claytoniana*, W1912A.

W1914A ?? No use given.

W1914B ?? Good for little children when peevish and something is wrong with the back. Scrape off just a little bark and put it in 1 qt. warm water to make a tea (not too strong). Give a little on your finger, also wash the body.

234. *Rhamnus* sp.

W1914A DJ Emetic. Scrape bark upward from twig about 3 in. long. There must be no one around. Use early in spring.

VITACEAE (GRAPE FAMILY)

235. *Parthenocissus inserta* (Kerner) Fritsch VIRGINIA CREEPER, WOODBINE

W1914A JD See 226, *Euonymus americana*, W1914B.

236. *Parthenocissus quinquefolia* (L.) Planch. ex DC. VIRGINIA CREEPER, WOODBINE

W1912A DJ See 174, *Fraxinus nigra*, W1912C.

W1914A DJ See 312, *Mimulus ringens*, W1914A.

W1914B JD See 188, *Rosa rubiginosa*, W1914A.

P. quinquefolia

V. labrusca

V. riparia

H1973A DC Poison, not used.

H1973B Anyms. Bunches on wrists. Cut vine where soft, cook, poultice.

237. *Vitis labrusca* L.   FOX GRAPE, SKUNK GRAPE

W1912A JJJr To make a colt come when a mare is a "hard catch," i.e., to assist conception. Cut roots in 6 pieces 6 in. long. Put in 5 qt. water, boil down to ½ or less. Divide the liquid. Mix halves with oats and give at successive feedings. Mare must not go out on the road during the week preceding the stud.

238. *Vitis riparia* Michx.   FROST GRAPE, RIVER-BANK GRAPE, DUNE GRAPE

W1912A BB For the hair. Sap collected in April when sap is running. Head is washed on going to bed.

W1912B JJJr Hiccoughs. Vines chewed.

W1912C KD See 26, *Onoclea sensibilis*, W1912D.

W1914A DJ See 195, *Waldsteinia fragarioides*, W1914A.

F1939A PH Hair salve to keep hair in place. Use sap from vines in spring when sap is rising in the vines.

H1973A DC Kidney trouble—when it burns while urinating. Boil up 2 qt., drink all day instead of water.

H1973B RB For small children having trouble passing water; general stomach trouble. Soak (unspecified part) in water.

POLYGALACEAE (MILKWORT FAMILY)

239. *Polygala paucifolia* Willd.   FRINGED MILK-WORT, GAY-WINGS, FLOWERING WINTER-GREEN, BIRD-ON-THE-WING

W1912A DJ Abscesses on limb. Make good-sized bunch and steep in 2 qt. warm water. Drink 1 cup. Also, smash leaves and boil well, put in small cotton bag and poultice.

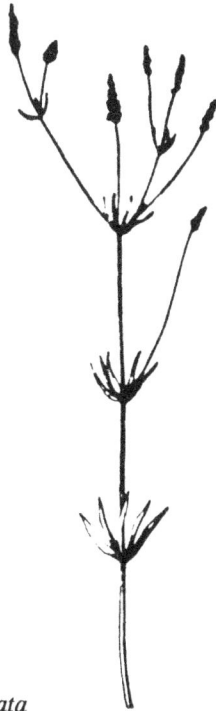

P. verticillata

W1914A DJ See 158, *Pyrola elliptica*, W1914A.

F1933A JS (See 157, *Chimaphila umbellata*) Venereal disease. For babies with syphilitic sores. Put handful of plant in warm water and boil. Use as wash.

F1933B SS Boils. Boil whole plant in 1 qt. water. Use as wash.

240. *Polygala verticillata* L.   WHORLED MILK-WORT

F1938A CHJJ For summer complaint in little babies. Bundle whole plant. Steep it in water. Give the baby ½ teaspoonful.

STAPHYLEACEAE (BLADDERNUT FAMILY)

241. *Staphylea trifolia* L.   BLADDERNUT

W1912A SH For sore faces. Scrape bark down, dry, grind to powder. Put ½ teaspoon powder in ½ qt. water and steep. Rub on with sponge.

W1912B BB See 41, *Sassafras albidum,* W1912A.

W1914A JD For when a woman swells after copulation as result of some bad medicine being administered to influence them [*sic*]. Also from a root having been made to represent a woman that has been pierced with thorns. With: (1) 34, *Pinus strobus;* (2) 180, *Physocarpus opulifolius,* W1914A. Scrape bark off from 3 small switches of each. Wet with warm water and apply as a poultice.

WndA ED To keep children from crying. Scrape bark down from green twigs. Make infusion and wash child with it.

HIPPOCASTANACEAE (HORSE-CHESTNUT FAMILY)

242. *Aesculus hippocastanum* L.   HORSE-CHESTNUT (ESCAPE)

W1912A KD See 375, *Inula helenium,* W1912D.

ACERACEAE (MAPLE FAMILY)

243. *Acer pensylvanicum* L.   STRIPED MAPLE, MOOSEWOOD, GREENSTRIPED MAPLE, WHISTLEWOOD

W1912A EH Laxative. With: (1) 81, *Juglans cinerea,* W1912F; (2) 307, *Fraxinus americana,*

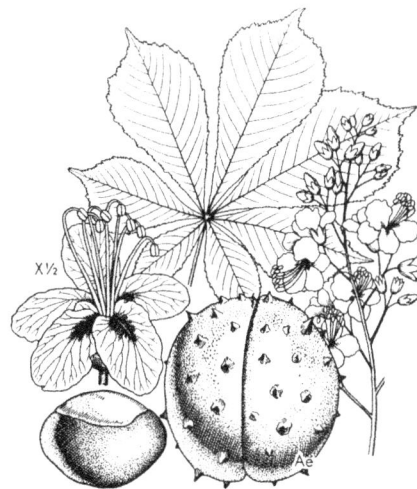

A. hippocastanum

A. pensylvanicum

A. rubrum

W1912C; (3) 308, *Fraxinus nigra,* W1912E; (4) 212, *Dirca palustris,* W1912G. Take equal parts of the barks of each, fill a 5 qt. pail. Put in a large brass kettle with water and boil down to 5 qt. Add 1 qt. of black strap molasses syrup and boil a little again. Take a teaspoonful before retiring and again in the morning if necessary.

F1938A CHJJ Emetic. Handful of bark, scraped upward, bring to boil. Do not boil too strong. Drink all you can in the morning before breakfast.

H1973A ON Paralysis. Boil bundle in water—apply as poultice. Reapply.

244. *Acer rubrum* L.   RED MAPLE, SOFT MAPLE, SCARLET MAPLE, SWAMP MAPLE

W1912A JJJr (when leaves are red in the spring) For sore eyes—white spots (cataracts). Scrape bark toward center of young, fresh 2 in. switch. Put in ½ cup warm water. Let 2 or 3 drops fall in the eye.

F1933A SS See 254, *Aralia nudicaulis,* F1933A.

FndA SG Trapping medicine. Cleans away all the human scent. Muskrats will not even see the trap.

Take short shoots, bend them, and make into a bundle. Tie 2-3 bundles. Put in pot filled with water. Boil. Put traps in the hot solution and boil.

245. *Acer saccharum* Marsh.   SUGAR MAPLE, ROCK MAPLE, HARD MAPLE

W1912A ED See 187, *Rosa acicularis,* W1912A.

F1933A SS See 254, *Aralia nudicaulis,* F1933A.

F1938A CHJJ See 96, *Carpinus caroliniana,* F1938A.

A. saccharum

A. spicatum

246. *Acer spicatum* Lam.   MOUNTAIN MAPLE

W1912A KD See 91, *Alnus incana*, W1912A.

ANACARDIACEAE (SUMAC FAMILY)

247. *Rhus* spp. (*R. glabra* L. and *R. typhina* L.)
SUMAC

W1912A JJ Bubo (lumps that remain after having syphilis). With: (1) 40, *Lindera benzoin*, W1912C; (2) 307, *Fraxinus americana*, W1912D; (3) *tcedⁿ-doⁿ*. Scrape bark upward and add scraped bark of (1) and (2) and an unspecified part and quantity of (3). Mix together, wet the concoction with warm water, and bandage afflicted parts. Bathe with the liquid.

W1912B JJ Smallpox preventative. Scrape bark and put double handful in pail of water. All drink this—but not if you have smallpox.

W1912C LJ Irregular menses. Take roots of 1 small tree, put in 3 qt. water, boil down to about ½. Take ½ teacup 4 times a day before meals. Also a laxative and good for rashes in children's throats. Take young shoot and scrape bark downward for a piece about 7-8 in. long. Add to ½

teacupful of water and steep. Give teaspoonful 2 or 3 times a day or oftener.

W1912D ED Horse's bellyache. Smash 3 groups of the red berries, let stand in 2 qt. water that is warm. Have horse drink 1 qt.

W1912E HJn For consumption of 4-5 yr. standing. Steep seed cluster ½ hr. in 2 gal. water. Drink warm, vomit, use fingers. Drink more right away. Do this 3 times a week.

W1912F KD Falling of the womb. With: (1) 255, *Aralia racemosa*, W1912F; (2) 337, *Lonicera oblongifolia*, W1912A. Take 5 seed branches and add a big bundle of (1) and a double bundle or handful of (2), after stripping leaves. Boil in ½ gal. until 1 qt. remains. Take wineglassful 2 hr. apart. Also for pain when urinating. Scrape bark bottom to top. Heat ½ gal. of scrapings in the oven and tie onto the abdomen warm.

W1912G LJ For worms in children which cause convulsions. Steep root, not as strong as for adults. Give every hour or 2 in severe cases.

W1912H DJ Evacuation of placenta. With: (1) 269, *Apocynum androsaemifolium*, W1912A; (2) 409, *Carex brevior*, W1912A. Take a 3 in. piece of root plus 1 bunch of each of (1) and (2). Add to 4 qt. water, let stand for 5 min. Drink as much as possible. Repeat if necessary.

W1912I JA Coughs, phlegm, gas on stomach. With: (1) 139, *Salix nigra*, W1912A. Place a small bundle of 3 roots with a small bundle of the branches of (1) in 1 gal. water and boil down to ½. The quantity of the ingredients may be increased as desired. Drink as often as you can. If taken warm, it has a quicker effect.

W1912J JJ See 186, *Prunus virginiana*, W1912B.

W1912K KD See 25, *Pteridium aquilinum*, W1912A.

W1914A JJJr For smoking. Leaves dried. Bark also used. Lichens and moss used as a substitute.

F1933A JS Measles, makes children break out when they cannot. Take the one with dark red berries (*Rhus typhina*), cut the ripe fruit, put in warm water like tea. Drink as much as want.
Fever, lumps, or swellings of the breast. Scrape the hairy one (*R. typhina*) up and apply as poultice as hot as possible.
Pains all over intestines and when part of colon protrudes due to internal swelling ("guts stick out back"). Take big bunch of bark, smash it, put in hot water. Tie on as poultice and sit on it hot.

F1933B SS TB. Red tasseled sumach has a red, pear-shaped berry that grows every 7 yr. When it grows big like a tree, split it [the berry] up and put 1 small piece in 1 qt. water. Boil it, drink like water.

F1938A HS Poultice for trouble on breast. Peeled bark on top, steeped with "horns" [berries]. Drank, too.

F1938B CHJJ Poultice for bumps and bruises. Used with apple peelings. Make salve, apply with bandage.
Mumps. Use *Rhus glabra*. No details given.

F1938C CRE/ET Swellings that increase and become red and spread. Boil several roots in just enough water to cover the roots. When the root is done, the bark peels off. Squash the peeling and place like a bandage around the part affected.

F1938D CHJJ See 179, *Pyrus malus*, F1938A.

F1938E CRE/ET Physic for newborn babies. Said there was 2 kinds of sumach and that the tall one was used for medicine. Steep pieces of bark 3-4 in. long in cup of warm water. Give by the teaspoon. Repeat several times a day until all the poison or black substance in waste is removed.

F1938F JG Tobacco. Smoke leaves.
Scarlet fever. Scrape bark up from 3 shoots, add 2 qt. water, boil 5 min. Drink as often as can.
Appetizer for sick person. Preparation not given.
Dislocated joints—elbow or shoulder. Scrape bark off 4 shoots, use as a poultice, keep on until dries, then remoisten with hot water.

R. typhina

F1939A CG Measles. Steep red berries when tops get ripe. Drink.

F1939B SS Mumps. With: (1) 339, *Sambucus canadensis*, F1939C. Take berries from cone and cook handful in 1 qt. water. Gargle. Put [in] bark of (1) and poultice swelling.
Diphtheria. Split the galls, boil a small piece in 1 qt. water, drink.

F1939C PH For cuts. No details given.

F1939D NL Very good for healing wounds. Scrape inner bark of first year growth into fuzzy balls. Put in boiling water and cover for 10-20 min. (the longer it soaks, the better). Apply the soaked bark to cut and bandage and keep bandage well soaked with the juice.

FndA KD Piles. With: (1) 363, *Cirsium discolor*, FndA. Pick roots of well-matured plant. Can use whole plant, but must take care in crushing it. Scrape bottom to top a handful of bark. Boil like potatoes in 1½ qt. water, reducing it to 1 qt. In severe cases, reduce to 1 pt. Take ½ tumbler frequently.

FndB AJm Berries made into tea and gargled for sore throat.

FndC HH Itch. Galls for diphtheria. Decoction of inner bark.

FndD CHJJ Not used as medicine.

FndE JG Mild fevers. Scrape bark up, put 3 shoots in 3 qt. water, boil 5 min. and cool. Drink lukewarm, ad lib.

FndF JJm Chancre in mouth, for gargling, for scarlet fever. Separate cones, take middle and steep it.

H1973A JT Berries for colds, whooping cough. No details offered.

H1973B DC Sugar diabetes—makes you puke. Boil up berries and drink.

248. *Toxicodendron radicans* (L.) Kuntze   POISON IVY, POISON OAK

W1914A JD Said was cured by taking juice of 1 plug of chewing tobacco mixed with the juice from 3 lemons and rubbing it on the blisters. Said it runs 10 days and then takes 10 more days to get better.

H1973A Anyms. For boys and girls that are nervous.

H1973B TL Infections. Had sores on his lips that would not heal. Howard Sky put poison ivy poultice on them and they healed and never bothered him again.

RUTACEAE (RUE FAMILY)

249. *Zanthoxylum americanum* Mill.   PRICKLY ASH, TOOTHACHE TREE

W1912A DJ Toothache or neuralgia. Bark smoked.

W1912B JJ When water stops—a type of gonorrhea. Break up root of 1 plant, put in 1 teacup warm water, and steep. Drink as quickly as possible and keep from urinating as long as possible.

W1912C JJJr To promote miscarriage. Scrape bark towards middle off 3 sticks about 1½ ft. long. Boil in 2 cups water, down to 1 cup. Take all of it. Cramps and worms, but not for children. Same as above.

W1912D SH Pain after confinement. Take handful of bark scraped downward and put in 2 qt. water and steep. Take ½ cup ad lib until pain goes.

W1912E JA Toothache. Pack the bark into the tooth.

W1912F JJ See 194, *Spiraea* sp., W1912B.

W1912G BB Toothache (cf. 72, *Hamamelis virginiana*, W1912C). Put bark into the tooth.

F1938A CHJJ See 380, *Polymnia uvedalia*, F1938B.

H1973A JT Toothache. Put in bad teeth, breaks them up.

H1973B Anyms. To break up the teeth. Chew the bark.

Z. americanum

OXALIDACEAE (OXALIS FAMILY)

250. *Oxalis stricta* L. (*O. europaea* Jord.)   LADY'S-SORREL, TALL YELLOW WOOD SORREL

F1938A CRE/ET Summer complaint characterized by cramps, nausea, fever. Crush 4 plants in teacup of lukewarm water. Drink 1 cup entirely, repeat until relieved.

F1938B CRE/ET Blood medicine. With: (1) 306, *Plantago* sp., F1938D. Boil roots of both in 1 gal. water until it goes down 2 qt. Drink as water.

FndA HRE One of 10 unspecified ingredients in antiwitch medicine. Preparation not given.

H1973A Anyms. To refresh mouth. Cook first, keep in bottle, gargle in morning, use for snuff.

GERANIACEAE (GERANIUM FAMILY)

251. *Geranium maculatum* L.   WILD GERANIUM, SPOTTED GERANIUM, ALUMROOT, PURPLE CRANESBILL

W1912A KG Sore mouth, sores on face, itch. Put 2 roots in 1 qt. water and boil down to ½. For sore mouth, wash mouth 5 times a day. For itch and sores, wash infected parts.

W1912B DJ Venereal disease. With: (1) 217, *Cornus alternifolia*, W1912C; (2) 339, *Sambucus canadensis*, W1912C. Take roots of 2 plants, add barks of (1) and (2) scraped from 1½ ft. lengths. Put in 1 qt. warm water and let stand 15 min. Wash affected parts. Root is also powdered and sprinkled on.

W1912C DJ For navel of baby when does not heal. Also good for blisters. With: (1) 217, *Cornus alternifolia*, W1912D; (2) 339, *Sambucus canadensis*, W1912B. Dry and powder the root. Scrape bark off from ends to center of 4-5 ft. lengths of (1) and (2). Dry and powder barks. Mix. Put a pinch or 2 of mixed powders on navel and bandage.

G. maculatum

W1912D JSl Sore mouth in children caused by being kissed by mother during her menses. Put 3 pieces of root about 2 in. long in 1 pt. water. Boil just a little. Wash mouth often.

W1914A DJ Venereal disease which has the appearance on the skin of being burnt. Also for sore or trench mouth in children. Put small handful of roots in 2 qt. water, boil quite a bit. Wash the skin and drink a little whenever you like. Wet cloth and put it on.

W1914B RS For sore mouth in children or adults. Chew up.

F1933A SS Sore mouth, to clean out innards, to heal severed umbilical cord. Put 2 pieces in quart of water. No further details given. For umbilical cord, chew root and spit on navel.

F1933B CHJJ Sore mouth. No details.

F1933C PS Gonorrhea. Chew the root and spit on the penis, or smash the root and put in slightly warm water and heat. Make a tea by smashing the root and putting it in cold water. Let stand, drink a teaspoonful at a time.

F1939A HJ Sore throat. See preparation above.

F1939B BD Chancre sores. Chew root raw, for adults. For little babies, steep a root size of finger in cup of water.

FndA ED(?) Sore mouth and venereal disease. No preparation given.

FndB JG Heart trouble. Boil 1 root in ½ qt. water. Boil 3 min. Drink as often as possible.
Sores in the mouth. Chew the root.
An emetic. Boil 5 roots in 5 qt. water. Drink a lot and vomit early in the morning before sunrise if possible.
For countering the effects of love medicine. Not specified, but may involve putting it into the victim's tea without his/her knowing it.

FndC SS Sores in the mouth, when mouth is coated and cannot taste anything. Put in 3-4 qt. of water [the roots] until it [the water] gets warm. Wash mouth with it.
Diarrhea. Steep and drink.

FndD HJ Sore throat. Crush roots, put in cupful water, boil a little, and pour hot water over it. It turns reddish color. Gargle with it.

BALSAMINACEAE (TOUCH-ME-NOT FAMILY)

252. *Impatiens capensis* Meerb.   SPOTTED JEWELWEED, TOUCH-ME-NOT, SNAP-WEED

W1912A KD See 396, *Taraxacum officinale,* W1912B.

W1913A ?? When you cannot make water. Take about 6 roots, cutting at first joint. Put in 1 qt. water, steep well. Drink as much as you can. Better if drink all at once.

W1914A DJ For fever. Put 4 or 5 small plants in 1 qt. cool water, rub up a little. Drink all up right away. Fix more if required.

W1914B JJJr For stricture or difficult urination caused by copulation with woman during menses. Put a small handful of plants in 3 qt. water, boil down to 1 qt. Drink quite a bit as hot as possible. Hold water in bladder as long as possible to let medicine work. Trouble sometimes extends to kidneys. Also for dropsy, because stem is watery and grows in wet ground. Put small handful of plants in 2 qt. water, boil a little. Drink as often as you feel like, and also wash body with it.

W1914C JD For sore or raw eyelids. Smash stems and tie on.

FndA JG No use given.

253. *Impatiens pallida* Nutt.   PALE JEWEL-WEED, TOUCH-ME-NOT, SNAPWEED

W1912A MR For women's breast injuries. Mash and poultice.

W1912B DJ Fever. Put 2 whole plants in 2 qt. cold water and let stand for ½ hr. Drink as soon as you can, then soak same plants in more water and repeat.

F1933A CHJJ For poison ivy and mosquito bites. Smash plant, rub on blisters or bite.

F1933B SS To stop suffering while having a baby. Fold up 2 stalks, cook for 10 min. Drink cupful at a time until you have had a gallon.
For poison ivy and mosquito bites, too.

F1933C J & JS To induce childbirth and stop suffering. Wash whole plant (including root) and put it in water. Drink a panful, cease suffering—immediate delivery.

F1938A FJ Poison ivy. May also use 68, *Sanguinaria canadensis,* F1938H. Crush and apply juice to blisters. Used separately.

F1938B CRE/ET For women in difficult labor [256, *Panax quinquefolius,* F1938B, is preferred if it can be found]. Put seed pods in cup of warm water and drink the water.

F1938C AJ To induce birth. Pour warm water in a cup over leaves and stems. Drink like tea. Give just before birth. Do not dilute.

FndA CHJJ Poison ivy. Smash stalks and use juice.

FndB CRE/ET Use not given.

FndC HRE Poison ivy. Crush leaves and rub on blisters. Indicated that juice of stalk is secondary to juice of leaf.

H1973A FK Poison ivy. Pulp put in water.

ARALIACEAE (GINSENG FAMILY)

254. *Aralia nudicaulis* L.   WILD SARSAPARILLA

W1912A MS Blood medicine, upset stomach. Steep small handful of roots in 4 qt. water. Drink any quantity, anytime.

W1912B JJ See 34, *Pinus strobus,* W1912G.

W1912C SH See 26, *Onoclea sensibilis,* W1912B.

W1912D SH See 397, *Tussilago farfara,* W1912A.

W1914A JD Sore eyes. Put 1 small root in 1 cup water, boil. Put liquid in saucer, then put eye under water and open.

F1933A SS Blood purifier. With: (1) 75, *Ulmus rubra,* F1933A; (2) 244, *Acer rubrum,* F1933A; (3) 245, *Acer saccharum,* F1933A; (4) 82, *Juglans nigra,* F1933A; (5) 35, *Tsuga canadensis,* F1933A; (6) 93, *Betula lutea,* F1933A; (7) 34, *Pinus strobus,* F1933B; (8) 84, *Fagus grandifolia,* F1933A; (9) 255, *Aralia racemosa,* F1933A; (10) 332, *Mitchella repens,* F1933D; (11) 157, *Chimaphila umbellata,* F1933C; (12) 192, *Rubus odoratus,* F1933B; (13) 335, *Lonicera canadensis,* F1933A; (14) 306, *Plantago* sp., F1933C; (15) 317, *Verbascum thapsus,* F1933C; (16) niyane^n?daka:?; (17) niyohge^n?gwasá:?; (18) nõ?washo:? djoe^n'gawa:yas. Use a handful of the split roots, and combine with 4 bark pieces from the east side of the tree, 6 in. long of (1)-(8). Add a small handful of the split roots of (9), a small bundle of the whole plant of (10), the whole plant, roots and all of (11), the root only of (12) and (13), 2 entire plants' roots only of (14), an unspecified amount of (15), a bunch of split roots of (16), an unspecified amount of bark from east side of (17), and an unspecified amount of bark of (18). Further details of preparation not given.

F1933B SS Blood medicine. Smash roots and boil in 3 qt. water. Drink as much as you want.

F1933C CHJJ See 21, *Osmunda claytoniana,* F1933A.

F1938A AJm Used in blood medicines, for colds. No details given.

F1939A HJ Blood medicine, fever sores. With: (1) 157, *Chimaphila umbellata,* F1939A. Bundle roots about 3 in. long into bundle size of thumb and forefinger. Use less than a handful of (1). Steep these. Drink a little at a time. Keep drinking.

FndA KD See 30, *Abies balsamea,* FndB.

H1973A Anyms. Cancer and sugar diabetes—as a preventative. Details not offered.

H1973B Anyms. See 99, *Ostrya virginiana,* H1973B.

H1973C Anyms. Poultice for sore throat. With: (1) 68, *Sanguinaria canadensis,* H1973A; (2) 42, *Asarum canadense,* H1973C. Details not offered.

H1973D Anyms. Split skin between the toes. Powder root, put on ½ hr., remove.

255. *Aralia racemosa* L.   SPIKENARD, PETTY-MORREL, LIFE-OF-MAN

W1912A BB Rheumatism. Fill 1 qt. whisky with roots. Drink cupful at night.

W1912B PT Diarrhea. Cut roots in small pieces and place in quart of water with a couple of pieces (of root) 3 in. long, mashed up. Add boiling water, steep, and stir. Drink as much as you can drink.

W1912C DJ For a cut anywhere. Cut root into 2 in. pieces of a root 4 in. long. Remove outside bark. Boil in 1 qt. water. Use pulp as poultice.
Also for cough or threatening consumption. Drink the above liquid.

W1912D SH See 26, *Onoclea sensibilis,* W1912B.

W1912E KD See 25, *Pteridium aquilinum,* W1912A.

W1912F KD See 247, *Rhus* sp., W1912F.

W1914A DJ Whooping cough—"pulling cough." Boil roots of 1 plant in 6 qt. water down to 3 qt. Take a teaspoon 4 times a day.

W1914B JD To get more strength. Put 1 root in 1 qt. or less of cold water. Drink lots.

W1915A JJJr Saw his sister eat this, then she reported passing a whole tapeworm the following day.

F1933A SS See 254, *Aralia nudicaulis,* F1933A.

F1933B CHJJ See 332, *Mitchella repens,* F1933C.

F1933C CHJJ See 225, *Celastrus scandens,* F1933A.

F1933D CHJJ See 315, *Scrophularia lanceolata,* F1933A.

F1938A HRE & SRE Tonic for women. With: (1) 157, *Chimaphila umbellata,* F1938C; (2) 116, *Rumex crispus,* F1938A; (3) *?o:yeⁿ:.* Boil roots in 1 gal. water down to 1 qt. Drink glassful before meals (with a little water).

F1938B CRE/ET See 117, *Rumex obtusifolius,* F1938B.

F1939A NL To promote menstruation when stopped by having a cold. With: (1) 332, *Mitchella repens,* F1939A; (2) 29, *Taxus canadensis,* F1939B. Roots used with vine of (1) and unspecified part and amount of (2). Preparation not detailed.

F1939B HJ Blood medicine or tonic. With: (1) 212, *Dirca palustris,* F1939A; (2) 157, *Chimaphila umbellata,* F1939B; (3) 117, *Rumex obtusifolius,* F1939A. Root of plant with bark of (1) and unspecified part and amount of (2) and (3). Boil down. Take a tablespoon twice. Drink no alcohol.

F1939C CG Blood medicine. No details given.

FndA CHJJ See 198, *Baptisia tinctoria,* FndC.

FndB KD See 34, *Pinus strobus,* FndB.

H1973A Anyms. See 434, *Smilacina racemosa,* H1973B.

256. *Panax quinquefolius* L.   GINSENG, SANG (COMMERCIALLY EXPLOITED)

W1912A HJn Sore eyes in 2-year-old child (ulcerated cornea). Slightly steep 1 small root in cup of water. Bottle it. Wash eyes every hour with clean rag, squeezing drops into eye. Keep up treatment after cure for 2 months.

W1912B MS To stop vomiting from cholera morbus, sick stomach. Grate roots and pour a teacupful of boiling water over ½ teaspoon of powder. Take ½ hr. apart if bad, 1 hr. apart if normal, taking a mouthful or 2 at a time.

W1912C DJ Sores on body anywhere. Steep root of 1 plant in 1 qt. warm water. Make more with the same root—do 3 times. Drink what you like 3 times.

W1912D MS See 56, *Hydrastis canadensis,* W1912A.

W1914A JD For when you vomit gall. Put ½ root in nearly full cup of water. Let stand a while and drink. Can use same root second time, but let stand longer. Use anytime.
Also for bad appetite. Same preparation as above. Use 3 times a day before meals.

F1933A SS Upset stomach, when yellow gall comes up. Boil roots in 1 qt. water. Drink all you can.

F1938A ?? Birth inducement. Give just before birth.

F1938B CRE/ET See 253, *Impatiens pallida*, F1938B.

F1938C AJ See 72, *Hamamelis virginiana*, F1938.

F1939A JJm See 56, *Hydrastis canadensis*, F1939A.

FndA BD Asthma. Smoke pulverized root.

H1973A DC See 434, *Smilacina racemosa*, H1973A.

H1973B Anyms. See 42, *Asarum canadense*, H1973F.

H1973C JT See 116, *Rumex crispus*, H1973D.

H1973D Anyms. Tonic, to liven you up. Preparation not offered.

H1973E TL Used for anything, fainting spells. Usually put in with other medicines. Dry root, smoke it.

257. *Panax trifolius* L.   DWARF GINSENG, GROUND-NUT

W1912A SH Fishing medicine. Slice bulb up and put a little in a cup of water. Put hook and line in this solution overnight. Chew root while fishing, spit on hands and wash in water. Also spit on the bait, keeping some root in mouth.

W1912B KD See 375, *Inula helenium*, W1912D.

W1914A PJ Lacrosse medicine. When a rival is trying to get the ball, the captain shouts "*djitgáye^n?*" and that makes the other lose the ball for sure. A decoction is rubbed on the arms and legs.

W1914B DJ For divining illness. Put in water, drink, and wash with the liquid.

APIACEAE (CARROT FAMILY)

258. *Angelica atropurpurea* L.   ALEXANDERS, PURPLE-STEM ANGELICA

W1912A DJ See 262, *Heracleum lanatum*, W1912B.

W1914A DJ When a person has been exposed and has been about half frozen—limbs or body may swell. Smash 1 root, put in boiling water, and steam the body by sitting over a pot.

W1914B DJ Witchcraft. Find a lump on a tree, take the bark off of it, and make a picture on the inside with the burnt end of the root. Shoot this picture and the baby will be born prematurely and die. This is done as a punishment by evilly disposed persons whose rifles have been tampered [with].

W1914C JD Good for ghosts inside. About sundown you hear a rattling. Caused by improper feast for the dead. Put 1 root in 2 qt. water, smash, and steep a little. Wash body and sprinkle inside house and out, also around places dead person was liable to go.

F1938A CHJJ Fever and chills. Take root in fall, dry, make decoction.

F1939A KD For flu—used in epidemic of 1924. Pulverize root and boil ½ glassful. Drink ½ tumbler.

F1939B HRE Pneumonia. Steep 4 pieces of root in 1-2 qt. water. No further details given.

FndA JG Rheumatism. Smash roots taken any season. Put on piece of white cloth and poultice—can use raw. Also take a sweat bath. Can also boil roots (1 big one). Smash and boil in 2 qt. water for ½ hr. Put foot or leg on board and cover with a thick blanket so that steam will not escape.
Also for female weakness. Wash roots from 1 plant and put in pot. Cut up roots, but do not

smash. Put in 3 qt. water, boil 5 min., cool. Let her drink as hot as can stand as often as thirsty.
Also for colds. Same preparation as for rheumatism, above.

H1973A JT Poison, poultice of root for broken bones.

H1973B EP & BC Said Dr. Pierce used to buy it. Used it as a blood purifier. Used dried root.

259. *Angelica venenosa* (Greenway) Fern. DEADLY ANGELICA, HAIRY ANGELICA

F1938A HRE For sprains and twisted joints, apply poultice.
For suicide, eat roots.

260. *Cicuta maculata* L.   WATER-HEMLOCK, SPOTTED COWBANE, MUSQUASH-ROOT, BEAVER-POISON, POISON HEMLOCK

W? KD For improperly healed bone, sprain, or dislocation. Reduces node that forms on improperly healed bone. Pour boiling water over crushed roots (if used raw, will cause blisters). Cover until cold, then poultice. Do not use on cuts—too strong.

W1912A DJ Twisted joints and sprains. Smash whole plant and add warm water. Make poultice. Tie on for ½ hr. only—otherwise, leg stiffens.

W1912B LJ Lameness, old, running sores. Also for horses. Mash roots, apply as poultice. Leave on 2-3 hr. Will stiffen joint if left on too long.

W1912C PS Suicide. Chew root.

W1914A JD Sore and lame joints. Smash and tie on roots. Will make leg hard as wood if left on too long.

W1914B JJJr Suicide. To handle without washing hands afterwards causes fits. Used in old times to wash floors, thereby preventing disease. For floor wash, put a double handful of roots in 6 qt. water, boil quite a bit. Sprinkle around floor, also outside of house.

F1933A CHJJ Not used as medicine, but heard that Mrs. Snow thinks women used it for suicide.

F1938A HS For bruises, but do not apply to cuts. Also for swollen joints. Mash root and poultice.

F1938B JH Water on the knee. Mash root and apply as poultice.

F1939A NL For sprains and torn muscles, pains in joints. Smash roots and apply as poultice.

F1939B PH For cuts (?) and sprains. Mash roots to fineness of tea dust. Wet dust and apply as poultice. Leave on 2 or 3 nights.

F1939C WP Dislocations. Crush root and make poultice.
Suicide. Woman takes it when husband dies.

261. *Daucus carota* L.   QUEEN-ANNE'S-LACE, WILD CARROT (NATUR.)

F1938A JH Falling womb. Preparation not given.

F1938B HRE Stoppage of urine. Boil roots of 1 plant in 1 gal. water. Drink like water, but do not drink anything else.

F1938C CRE/ET Disorder of blood, for men—pimples, pale, no appetite. Boil 1 root in 1 gal. water. Drink like water throughout the day.

262. *Heracleum lanatum* Michx.   COW-PARSNIP, MASTERWORT

W1912A JJ For chancre or lump on penis. With: (1) 264, *Pastinaca sativa*, W1912A. Finely cut roots of both and add 2 tablespoons to 1 qt. water and boil. Wash parts. In severe cases, smash roots and poultice.

W1912B DJ Rheumatism, headache caused by cold. With: (1) 258, *Angelica atropurpurea*, W1912A. Steep 1 good-sized plant of each in 8 qt. water. Hold limb over the steam, covering with a quilt. When you sweat, the bad will come out.

W1914A DJ Hunting medicine used to clean rifle. Longhouse preachers (at present as well as in past) say it should not be used. If a pregnant woman handles a rifle, the hunter will not be able to hit anything afterwards. Put 1 root in 8 qt. water, boil a lot, wash rifle.

F1933A KD When you get a bruise on back of stomach. With: (1) 351, *Anaphalis margaritacea*, F1933A, or 371, *Gnaphalium uliginosum*, F1933A; (2) 375, *Inula helenium*, F1933C. Take double handful of roots along with handful of flowers of (1) and a handful of roots of (2). Put in 1 gal. water, boil down to ½. Cool. Take wineglass 2 hr. apart.

263. *Osmorhiza claytonii* (Michx.) Clarke   SWEET CICELY

W1912A ED Fishing medicine. Chew roots and spit on bait.

W1912B BB For chewing.

W1912C BT Fishing medicine. Chew root and spit on bait.

W1912D HJn Diarrhea. Steep 3 2-in. pieces 5 min. in 1 cup water. Take 1 teaspoonful for boy or girl.

W1914A DJ Hunting medicine, such as for muskrat. With: (1) *u?náʰsāā?*. Place the roots of a single plant of each in a 2 or 3 oz. bottle of spirits (alcohol). Wet a straw in it and place it in the trap. It will attract the muskrat from a distance.

W1914B RS Intermittent fever. For children, put 1 root in 1½ qt. water that is warm, cover, and let stand. When the liquid turns a little blue, it is ready. Drink a little at a time. For adults, put 2 roots in 3 qt. water, smash roots, and boil to make stronger. Boil down to 1 qt.

W1914C JD See 355, *Aster macrophyllus*, W1914A.

F1933A CHJJ See 91, *Alnus incana*, F1933A.

F1938A HS Tonic and physic. Collect roots in fall when tops have fallen away. Steep.

F1938B JH Physic. Make tea of roots.

FndA JG Fever and headache. Steep root.
Antilove medicine. Chew roots for immediate effects.
Seed medicine in spring. Put 3 small roots and whole plant in 1 qt. water. Boil 3 min., cool, put in pail and put corn, watermelon, and maple sugar in to sweeten the water. Will sweeten melons when they grow.

FndB JG Blood purifier, also good for venereal disease. With: (1) 304, *Prunella vulgaris*, FndB; (2) 42, *Asarum canadense*, FndD; (3) 332, *Mitchella repens*, FndA; (4) 27, *Polystichum acrostichoides*, FndB; (5) 353, *Anthemis cotula*, FndB; (6) 348, *Achillea millefolium*, FndB; (7) 81, *Juglans cinerea*, FndD; (8) 157, *Chimaphila umbellata*, FndA; (9) 153, *Gaultheria procumbens*, FndB; (10) 173, *Dalibarda repens*, FndA; (11) 53, *Coptis trifolia*, FndB. Take root and whole plant of (1), (4), (8), (9), and bark of shoots from (7), and vine and roots of (3), plus unspecified parts and amounts of (2), (5), (6), (10), and (11). Dry ingredients and make into a powder. Put powder in 3 qt. water. Boil. Take a cupful as often as you can.

264. *Pastinaca sativa* L.   WILD PARSNIP (NATUR.)

W1912A JJ See 262, *Heracleum lanatum*, W1912A.

265. *Sanicula marilandica* L.   BLACK SNAKE-ROOT, SANICLE

W1914A JJJr For sore navel in little children. Put the roots of 2 plants in 3 qt. water, boil quite a bit. Give a little with teaspoon or finger every time they will take it. Wash body, commencing with head.
Also for adults. Put 3 plants in 4 qt., boil a little as for tea. Drink whenever you can.

W1914B JJJr Dropsy. Put a small handful of plants in 2 qt. water. Boil a little. Wash body and drink as much as you want. It is used because its stem is watery and the plant grows in wet ground.

W1914C JD See 355, *Aster macrophyllus*, W1914A.

W1914D DJ See 300, *Mentha canadensis,* W1914A.

266. *Sium suave* Walt.   WATER-PARSNIP

W1912A KD For pain in broken limb. Cut and smash roots. Take double handful and put in just enough water to make it thick. Stir until cools. Place between cheesecloth and spread on limb.

W1912B JJJr Sprained or out of joint limbs. Fry the turnips, peel off skin, tie on with cloth for 10 min. Apply again next day or so.

W1912C KD See 57, *Ranunculus abortivus,* W1912D.

GENTIANACEAE (GENTIAN FAMILY)

267. *Gentiana andrewsii* Griseb. (might be *G. clausa* Raf.)   CLOSED GENTIAN, BOTTLE GENTIAN

W1912A JJJr Kills pain, headache, sore eyes. Steep roots of 3 plants in cup of water. Spray on with mouth or squeeze drops in eyes.

W1912B JSl Pain in the side. With: (1) 193, *Spiraea alba,* W1912A. Mash and powder dried roots. Put tablespoon of the 2 roots together in 1 qt. water and boil down ½. Take tablespoon 3 times a day before meals.

W1914A JD For lonesomeness (melancholy and craziness). With: (1) 46, *Actaea spicata,* W1914B.

G. andrewsii

Put 1 root of each in ½ qt. water. Steep. Drink a very little. Wash head with medicine, also drink some.

F1938A CHJJ Liver medicine. With: 8 other unspecified plants. Preparation not given.

F1938B CRE/ET Antiwitch charm. Hang piece of dried root in house.

FndA SS Chills. Cut up 4 roots and put in 1 pt. water until just warm. Do not let boil. Drink when cool.
Also for muscular soreness. Pound root and poultice.

FndB SS Fever, headache, cure for witchcraft brought about by jealousy. Dry and scrape roots of both the male and the female of this plant. Put on stove in 1 pt. water. Cover it and steep, keeping it covered. Drink little by little.

268. *Gentianella quinquefolia* (L.) Small   STIFF GENTIAN, AGUE-WEED, GALL-OF-THE-EARTH

F1933A CHJJ For cramps and when worms come up. With: (1) 269, *Apocynum androsaemifolium,* F1933A. Use same amounts of roots of both. Smash with mortar and pestle and boil in 2 qt. water. Strain and drink any amount, but for only 2 days. Otherwise, it will spoil. Make new batch if needed.

F1938A SS Stomachache, worms in adults. Steep whole dried plant—including flowers.

FndA SS Diarrhea or sore chest. Make a bunch of 1 or 2 whole plants and put in ½ qt. water until warm. Take off and drink. Drink hot for sore chest.

FndB HRE Worms. Bellyache is sign of worms. No preparation given.

APOCYNACEAE (DOGBANE FAMILY)

269. *Apocynum androsaemifolium* L.   SPREADING DOGBANE, PINK DOGBANE

G. quinquefolia

W1912A DJ See 247, *Rhus* sp., W1912H.

W1913A PJ Milk used for warts.

F1933A CHJJ See 268, *Gentianella quinquefolia*, F1933A.

F1938A HRE Bugs in horses (horse bots). Pound roots and boil them. Put water in horse's feed.

F1939A CG Said old folks used it, but not used much anymore. Use not specified.

F1939B HJ Liver medicine. Boil root. No further details.

FndA CHJJ Use not specified. Use root.

FndB SS Did not know use.

FndC DJn Did not know use.

270. *Apocynum cannabinum* L.   INDIAN HEMP, DOGBANE

W1912A ED Diarrhea in children having no mother. Smash root and steep for 1 hr. in 1 qt. water. Wash children first thing in morning and again at bedtime for 3 days.

F1933A JS Blood medicine. With: other unspecified roots. Preparation not given.

F1938A AJ Spring or summer emetic, to clear up yellow eyes. Cut root in pieces. Put tablespoon in 1 qt. boiling water. Set aside and drink 2 teaspoons at bedtime and ½ cupful before breakfast. Before supper, take whole cup. At night, take rest. Next morning, take laxative.

FndA CHJJ Blood purifier, laxative, biliousness. Use root. No further details.

H1973A Anyms. For babies. Scrape stem, boil, give 2 drops to baby.

ASCLEPIADACEAE (MILKWEED FAMILY)

271. *Asclepias exaltata* L.   POKE MILKWEED

W1912A DJ To give wrestler strength. Preparation not given.

FndA CHJJ To make string, to sew moccasins. Take bunch of peels from outside of stalk. Boil with wood ashes until brown, take to creek to wash them, then dry and take 3 fibers and roll under palm of hand like making rope on thigh as shoemaker does.

272. *Asclepias incarnata* L.   SWAMP MILKWEED

W1912A JJJr Stricture by having intercourse with menstruating woman. With: (1) 273, *Asclepias syriaca*, W1912A. Take root of 1 plant, wash, boil down ½ in 5 qt. water. Put in jar, not tin pail. Drink 1 cup 6 times before noon, beginning before breakfast. Hold urine as long as possible. Wet end of penis with a little of the liquid.

A. incarnata

W1912B DJ See 67, *Podophyllum peltatum,* W1912G.

W1914A ?? Remedy against bad witchcraft medicines by a woman who has had a child. The newborn's navel is also tied with bark of the stem. Boil the root of 1 plant in 2 qt. water. Take whole amount at once—only way to cure. This medicine is taken after child is born. If taken before, a smaller dose (about 1/3) is used. The same plant may have been used to cause trouble to the woman who is about to have a child.

W1914B DJ Gives strength. Make a tea and wash body—use root.

W1914C JD To heal navel in babies said to be caused by too much gall. Put 1 piece of root 3 in. long in 1 qt. cold water. Take running water from a creek, dipping with the stream. Wet white cloth and tie on. Also give a tablespoon to drink pretty often.

W1914D RS Tooth extraction. Remove part of the stem in the fall when dry, rub up, and make into a cord (similar to that used for sewing moccasins). This is very tough and is tied around children's first teeth. When this type of cord is used, children will never have rotten teeth.

W1918A JJJr Diuretic. No details given.

W1918B JJJr See 224, *Cornus sericea,* W1918A.

FndA SG To make burden strap.

FndB DJ Burden straps.

FndC JG (Note: May be a misidentification of *Eupatorium maculatum,* 368.) Lame back and kidneys—too much or too little urine. One whole plant bundled in 1 qt. water—including root. Boil 3-5 min. Let cool a little. Drink hot as often as she likes. For either sex.

FndD NL No use given.

273. *Asclepias syriaca* L.   COMMON MILKWEED

W1912A JJJr See 272, *Asclepias incarnata,* W1912A.

W1913A PJ Milk used for warts.

W1914A DJ Warts. Apply milk.
Stomach medicine. In making greens, drink the water that remains.

W1914B DJ Warts. Stick a pin or needle in the wart, then apply milky juice.

W1915A JJJr Warts. Apply milky juice to wart.

W1915B JE Warts. Scratch wart until it bleeds, then add milky juice.

W1915C DJ Warts. Tie one of your own hairs around it (the wart) and apply milky juice.

F1933A CHJJ Warts. With: (1) punk water. Squeeze out milk and apply to wart that has been cut open. Wash with (1).

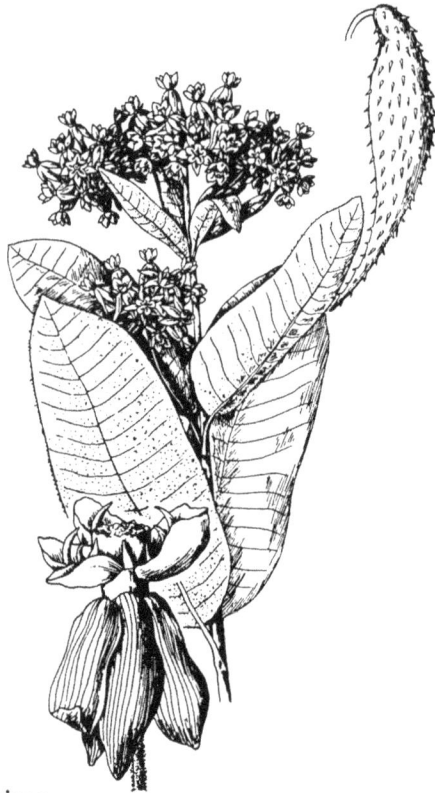

A. syriaca

A. tuberosa

F1938A CRE/ET (some question about identification) See 97, *Corylus americana*, F1938A.

F1939A JJm See 157, *Chimaphila umbellata*, F1939C.

FndA ?? Bee sting. Apply white sap.
Food. Tops cooked as greens.

H1973A Anyms. Cuts. Put milk on wound, bandage.

H1974A ON Warts. Apply milk.

274. *Asclepias tuberosa* L.   BUTTERFLY-WEED, PLEURISY-ROOT, ORANGE MILKWEED, TUBER-ROOT, CHIGGER-FLOWER, INDIAN-PAINTBRUSH

W1914A DJ For lifting or running. Smash roots of 1 plant, put in 1 qt. warm water. Wash arms,

shoulders, and body for lifting. For running, put on legs, dampen or wash running shoes, then step in tracks of leader in running—this will give him cramps.

SOLANACEAE (NIGHTSHADE FAMILY)

275. *Datura stramonium* L.   JIMSONWEED, STRAMONIUM, COMMON THORN-APPLE, JAMESTOWN WEED (PURPLE THORN-APPLE)

F1939A KD No use, or no use given.

276. *Nicotiana rustica* L.   INDIAN-TOBACCO, WILD TOBACCO (RARE INTROD.)

W1912A JJJr To cure insanity caused by masturbation. Boil medium-sized pinch in ½ qt. water, down ½. Patient should drink it when he awakens. Also, burn tobacco.

D. stramonium

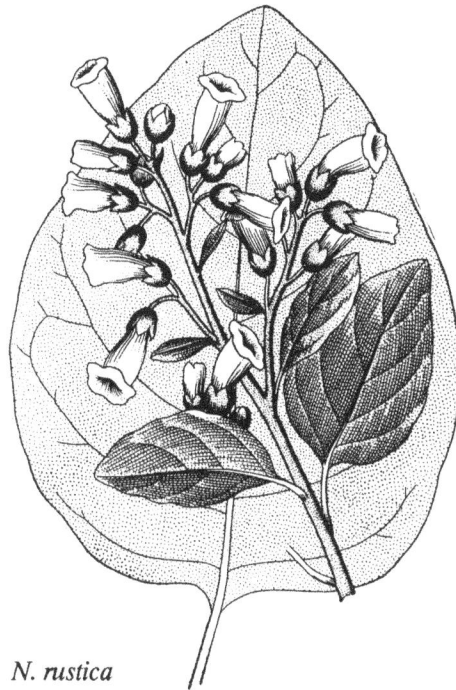

N. rustica

W1913A PJ For all sorts of insect bites. Chew a little and apply spit to the place.

W1914A JE Cure for consumption caused by swallowing a leg of a "thousand legs," or by being bitten by one. Preparation not given.

W1914B DJ For smoking—a good medicine. Mixed with 224, *Cornus sericea,* W1914B.

W1914C DJ See 312, *Mimulus ringens,* W1914A.

W1915A JE Cure for a particular type of spider that lives inside and will leave a mark on your flesh where they run. If scratched, will go over body. Wash with native tobacco.

W1918A JJJr See 224, *Cornus sericea,* W1918A.

277. *Physalis heterophylla* Nees   CLAMMY GROUND-CHERRY

W1914A DJ Scalds and burns, venereal disease externally but not internally, bad stomach—rotten stuff vomited. With: (1) 392, *Solidago squarrosa,*

P. heterophylla

P. pubescens

W1914A. Dry and powder leaves and roots of both. Wet with warm water, wash with it, and let dry with powder material on.

278. *Physalis pubescens* L.   GROUND-CHERRY, STRAWBERRY TOMATO, DWARF CAPE GOOSEBERRY, HUSK-TOMATO (RARE)

W1912A BB Use not given.

279. *Solanum dulcamara* L.   CLIMBING OR TRAILING NIGHTSHADE, BITTERSWEET (NATUR.)

W1912A DJ See 388, *Solidago flexicaulis*, W1912A.

280. *Solanum nigrum* L.   BLACK or DEADLY NIGHTSHADE, POISON-BERRY (NATUR.)

H1973A Anyms. For injured person who has relapse—gets him well again. No further details offered.

281. *Solanum tuberosum* L.   POTATO, WHITE POTATO, IRISH POTATO (RARE ESCAPE)

F1938A AJ Eye inflammation. Scrape potato and apply under bandage. Use leaf as bandage.

CONVOLVULACEAE (MORNING-GLORY FAMILY)

282. *Ipomoea pandurata* (L.) Meyer   WILD POTATO-VINE, MAN-OF-THE-EARTH, MAN-ROOT (RARE)

W1914A DJ Has much magical potency. Not to be touched by children. If you touch it and strike someone, you will kill them. It gives you more strength and was formerly used by hunters, allowing them to carry 2 deer with ease.

W1914B JD Bewitchment. With: (1) *uwadásee*; (2) *ukdéha?djigéʰsā* (cf. 25, *Pteridium aquilinum*, W1914A). Had small nails in with (1). These 2 ingredients are placed together in a small coffin with the root shaped like a person of (2) — the nails being used to pierce where it is desired to affect the person. The latter will die in 10 days.

F1933A CHJJ Pains in the abdomen. Scrape root upward, boil, and drink. Also for the initial stages of TB. Scrape roots and boil. Drink all you can. (Note: Root taken standing erect for good medicine, horizontal for witchcraft.)

F1933B FJ Cough medicine. Decoction of root.

F1933C JS Medicine for all kinds of disease. Dig root in March. Dry and store it. Cut off 2 in. and grind it, and put in 1 gal. hot water and cook for 20 min. Cool and drink all you can.

F1933D SS TB and coughs when you spit up blood. With: (1) 375, *Inula helenium*, F1933D. Cut up 4 in. root lengthwise and mix with (1). No further details given.

F1938A AJ Headache and upset stomach. Scrape to powder. Put in hot water about a pinch to a teacup of warm water. Cool. Take a cup 3 times a day. For adults and children.

F1938B CHJJ See 301, *Mentha spicata*, F1938C.

F1939A HJ & CG Good for liver. No further details.

H. canadense

H. virginianum

FndA DJn See 185, *Prunus serotina,* FndB.

HYDROPHYLLACEAE (WATERLEAF FAMILY)

283. *Hydrophyllum canadense* L.   CANADA
WATERLEAF

W1912A ?? Antidote to poisons (bloodsuckers,
snakes) that have been put in someone's food and
cause consumption. With: (1) *skaswagádiʰ* (cf. 288,
Mertensia virginica, W1912A). Smash 1 root of
each plant and steep in 6 qt. water. Take ½ qt.
every ½ hr.

284. *Hydrophyllum virginianum* L.   VIRGINIA
WATERLEAF, SHAWNEE-SALAD, INDIAN-
SALAD, JOHN'S-CABBAGE.

W1912A PT Cracked lips and sores on the mouth.
Chew root or make into a tea. For tea, take 3
roots for a cup. Dry and powder them and put a
teaspoonful to a cup of boiling water. Wash
mouth.

BORAGINACEAE (BORAGE FAMILY)

285. *Cynoglossum officinale* L.   HOUND'S-
TONGUE (NATUR.)

C. officinale

E. vulgare

W1912A DJ Consumption with bad hemorrhage. With: (1) 313, *Pedicularis canadensis,* W1912B. Steep 2 whole plants in 3 qt. water. Drink a bellyful, make more.

W1912B DJ See 365, *Erigeron philadelphicus,* W1912A.

W1914A DJ Cancer on the legs—"rotten legs." Take a single plant just before the flowers come out and boil it a little in 2 qt. water. Wash the place and poultice with mashed plants.
Also for internal venereal disease. Put a root in 2 qt. water and boil down ¼. Wash the sores and take a cupful twice a day.

286. *Echium vulgare* L.   BLUE-WEED, BLUE-DEVIL, VIPER'S-BUGLOSS (NATUR.)

W1914A DJ See 231, *Chamaesyce glyptosperma,* W1914A.

287. *Hackelia virginiana* (L.) Johnst.   STICKSEED, BEGGAR-LICE

H1973A Anyms. Put around potatoes to keep bugs off.

288. *Mertensia virginica* (L.) Pers. ex Link   VIRGINIA BLUEBELLS, VIRGINIA COWSLIP, ROANOKE-BELLS

W1912A ?? See 283, *Hydrophyllum canadense,* W1912A.

H. virginiana

W1914A DJ For venereal disease (when nose is red, it is inside). Put 1 handful of roots in 4 qt. water. Boil a little. Take as much and as often as they want.

289. *Myosotis* sp. FORGET-ME-NOT

W1914A DJ For cows when bearing a calf and the womb comes out too. With: (1) 62, *Ranunculus* sp., W1914A. Put 1 plant of each in a pail full of water. Boil quite a bit. Put it on corn husks and give twice a day, evening and morning. Sprinkle a little salt on husks, too.

290. F1939A HJ No identification made. May be in the Boraginaceae. It is specifically stated that the plant in question is not wild indigo or *Baptisia* [which contains a blue pigment]. For kidneys. Can chew and swallow spit and within 2 hr. if you urinate it will look bluish in color if you watch it.

VERBENACEAE (VERBENA FAMILY)

291. *Verbena hastata* L.   BLUE VERVAIN, BLUE VERBENA, SIMPLER'S-JOY

W1914A JJJr Will make an obnoxious son- or daughter-in-law leave. Mash leaves and let stand in cold water. Give to person you want to leave.

F1933A KD See 56, *Hydrastis canadensis,* F1933A.

FndA CHJJ Stomach cramps and/or worms. With: (1) 53, *Coptis trifolia,* FndC. Take roots anytime, pound up, put in 1 qt. water. Boil until 2/3 left, cool. Drink anytime.

FndB JG & SG (Note: This plant may have been confused with *Aster* sp.) Summer complaint. Boil 1 whole plant in 1 qt. water for 3 min. Take ½ cup 3 times a day until diarrhea stops.

292. *Verbena* sp.   VERBENA, VERVAIN

W1914A DJ Colds. Put 2 small roots in 8 qt. water. Boil down to 1 qt. Take anytime.

V. hastata

LAMIACEAE (MINT FAMILY)

293. *Agastache nepetoides* (L.) Kuntze   YELLOW GIANT-HYSSOP

W1912A DJ See 329, *Galium aparine,* W1912A.

294. *Collinsonia canadensis* L.   RICHWEED, STONEROOT

W1912A Ed Gives strength to babies bathed in it. Smash handful of root in 1½ qt. warm water. Wash child in morning and at night. Make it strong because it is a hard root.

W1912B HJn For listlessness in boys and girls. Smash root and steep 10 min. in quart of water. Let cool. Take tablespoon 4 times a day.

W1912C BB Did not know.

C. canadensis

W1912D ED See 27, *Polystichum acrostichoides*, W1912C.

F1933A CHJJ Rheumatism. With: (1) 51, *Cimicifuga racemosa*, F1933A; (2) 65, *Caulophyllum thalictroides*, F1933A; (3) 24, *Adiantum pedatum*, F1933A; (4) 434, *Smilacina racemosa*, F1933A. Collect roots of all and dry them. Smash them before using or boil them green and cool. Drink all you can. For foot bath, put water in pan first, then roots. Then bathe feet and legs.

F1938A AJ Diarrhea with blood. With: (1) 317, *Verbascum thapsus*, F1938E. Take a piece of root about 1 in. long, add 1 6-in. leaf of (1). Put in cup of boiling water ¼ full. Let cool. Take root and leaf out. Lie down and drink.

F1938B CRE/ET Diarrhea. With: (1) 317, *Verbascum thapsus*, F1938D. Take root size of thumbnail along with leaf of (1). Put in 1½ qt. water and boil

for ½ hr. Cool. Take about ½ cup. Do not take too much, will cause constipation.

F1939A DJn Kidney trouble. Take 1 lb. of roots and boil way down for 2 hr. until syrup. Sweeten with maple syrup. Take ½ tablespoon 3 times a day before meals for 1 week, then 3/4 teaspoon for another week, then tablespoon.

FndA SS Headache. Put powdered leaves on forehead—enough to cover it.

H1973A DC For anything. Heard that some used it for heart trouble. Boil roots.

H1974A ON Blood medicine—for boils. Roots used. No further details offered.

295. *Galeopsis tetrahit* L. HEMP-NETTLE (NATUR.)

W1912A JJJr A cure for a man who has been bewitched by a woman who kisses him after chewing 54, *Hepatica nobilis* var. *acuta*, W1912B. Steep roots of 3 plants in 5 qt. water. Take enough to make you vomit 3 times. Vomit into a pot. Go to where there is a fork in the path and bury vomit. When it dries up, man will recover.

296. *Hedeoma pulegioides* (L.) Pers. MOCK-PENNYROYAL

W1912A ED Headache. Steep small bunch in 3 qt. water. Drink 2 cups like a tea at mealtime.

FndA SG Tea with winter meals. Gather in summer, hang to dry.

FndB JJm Fever, chills—when you go out in winter and come home chilly. Steep whole plant like tea.

297. *Lycopus americanus* Muhl. ex Bart. WATER-HOREHOUND, BUGLEWEED

W1912A DJ See 167, *Tiarella cordifolia*, W1912B.

W1914A DJ Poison—placed in corn soup to kill an obnoxious person. Cured by *gwaʰsudt* (not identified), 2 plants put in ½ qt. cold water. Drink it all, follow with warm water, vomit.

W1914B DJ See 339, *Sambucus canadensis,* W1914A.

298. *Lycopus virginicus* L. WATER-HOREHOUND, BUGLEWEED

W1912A DJ If leaves or roots are eaten, person will be poisoned. Limbs or body will swell. Used with: (1) 434, *Smilacina racemosa,* W1912A. One whole plant mixed with about 3 roots of (1). For a remedy for this, use 1 medium-sized, steeped branch of 233, *Rhamnus alnifolia,* W1912B, and 161, *Ribes americanum,* W1912A, in warm water. Drink a little at a time. Bark used as a poultice for swelling.

299. *Mentha aquatica* x *spicata* = *M.* x *piperita* L. PEPPERMINT, BERGAMOT MINT (NATUR.)

W1912A JW A tea from this will throw off witchcraft, a plant of a certain kind having been thrown at you to cause it.

F1933A CHJJ Colds, fevers. Make infusion of whole plant. Gather in summer, dry it.

F1938A CHJJ See 301, *Mentha spicata,* F1933C.

300. *Mentha canadensis* L. FIELD MINT, AMERICAN WILD MINT

W1914A DJ Cure for poison caused by 144, *Cardamine bulbosa,* W1914A. With: (1) 265, *Sanicula marilandica,* W1914D. Put 1 small cluster of each plant in 2 qt. water and boil quite a bit. Take quite a lot to make you vomit.

W1914B JD See 215, *Oenothera biennis,* W1914A.

301. *Mentha spicata* L. SPEARMINT, CURLY MINT (NATUR.)

W1912A MS Emetic for children—for bad stomach. Make tea of it.

M. aquatica

M. spicata

W1912B MS See 42, *Asarum canadense,* W1912F.

F1938A CRE/ET Headache. Crush in cold water until green color comes out. Soak cloth in this and apply to forehead.

F1938B CHJJ See 97, *Corylus americana,* F1938B.

F1938C CHJJ Ingredient in Little Water Medicine. With: (1) 282, *Ipomoea pandurata,* F1938B; (2) 367, *Eupatorium dubium,* F1938A; (3) 299, *Mentha spicata,* F1938A; (4) squash; (5) 166, *Saxifraga pensylvanica,* F1938A; (6) 213, *Circaea lutetiana,* F1938A; (7) 453, *Platanthera psycodes,* F1938A; (8) 432, *Medeola virginiana,* F1938A; (9) 323, *Lobelia cardinalis,* F1938C; (10) 176, *Geum canadense,* F1938A; (11) 148, *Cardamine* sp., F1938A; (12) 1, *Peziza succosa,* F1938A; (13) an unspecified "vine." Scrape root and grind all in a stone mortar except (11) and (12). Drink or place on any injured part. Use bird bill as measure.

F1938D CHJJ Headache. Use whole plant.
Hay fever. Use leaf.
Food. Leaves and stem used.

F1939A CG Colds. Steep whole plant. Drink a little at a time.

FndA HJ Headache. Snuff fresh tops.

302. *Monarda* sp.　BERGAMOT, HORSEMINT

W1912A MS For softening of the brain. Make a tea. Use as an emetic early in the morning for 3 mornings. Wait 4 days, repeat. Also bathe head and face.

W1912B ED Headache and constipation in little children. Smash handful of roots and steep in 1 qt. water, hot. Take ½ cup 3 times a day before meals.

W1912C MR General lassitude. Make into a tea. Drink when you feel like it.

W1912D KD See 123, *Sarracenia purpurea,* W1912A.

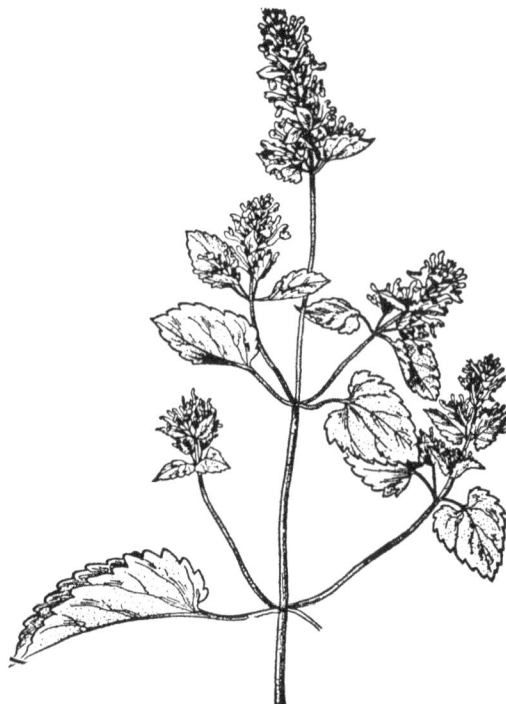

N. cataria

F1933A CHJJ See 54, *Hepatica nobilis* var. *acuta,* F1933A.

F1939A JG Fever, also for sickness caused by dead people. Bundle whole plants with roots in 1 qt. water. Boil 3 min., let cool.

FndA CHJJ Used in Little Water Medicine.

303. *Nepeta cataria* L.　CATNIP, CATMINT (NATUR.)

W1912A BB Diarrhea in children caused by mother's adultery. With: (1) 348, *Achillea millefolium,* W1912G. Take 2 stems of each and boil them to a strong tea in 2 qt. water. Take a cupful twice a day before breakfast and at bedtime.

W1912B LJ For children when they have a bellyache due to cold. Steep and make into a tea. Do not sweeten.

W1914A DJ Vomiting from fever or unknown cause. Put 1 cluster of plants in 1 qt. water, bruising the plants. Drink as much as often as you like. Also for trouble caused by eating gravy, butter, or other rich foods. Same preparation as above, but add warm water.

W1914B JJJr For children when peevish and you do not know what is the matter. Also good for headache. Take ½ dozen tips of plants and a ½ cupful of water. Boil quite a bit. Give a little in a teaspoon or put some on woman's nipples. Wet a little with cold water and tie on aching part.

F1938A AJ Cough, cold, sore throat, chills. Put 1 whole stem in teacup and pour warm water over it. Set aside a few minutes. Drink about ¼ of it 4 times.

F1956A IL Fever in babies. No preparation given.

FndA JS Laxative and astringent in diarrhea, fever in babies. Use whole plant. No further details.

FndB SG & JG Headache, fever, summer complaint. Rub a bunch of whole plants on forehead as a poultice for fever. Also rub and put in ½ qt. cold water—water will turn green. Drink.

H1973A DC For restless babies who cannot sleep. Snip tops, soak in boiling water, drain. Give 2 tablespoons with sugar.

304. *Prunella vulgaris* L.   SELF-HEAL, HEAL-ALL (NATUR.)

?? ?? Colds, coughs, diarrhea. Boil 3 qt. water with 1 plant. No further details.

W1912A MR Stomach cramps. Make small bundle of whole plants, steep to a good strength in 2 qt. water. Take cup at a time when you want.

W1912B KD For when baby cries too much. With: (1) 311, *Linaria vulgaris*, W1912B. Use small bunch along with 2 or 3 pinches of the flowers of (1). Put in ½ glass of boiling water. Let cool. Take 1 teaspoon often.

Also for sickness caused by grieving. With: (1) 323, *Lobelia cardinalis*, W1912F. Put whole plant with 1 raceme of (1) in cup and pour over boiling water. Cool and drain. Take as much as possible at once.

W1912C ED For heaves, puffing of short breath. Steep a root in 1 qt. water. Take ½ teaspoonful 3 times a day.

W1912D DJ To cure consumption caused by drinking a decoction of the root of 43, *Nuphar lutea*, W1912C. With: (1) 414, *Scirpus tabernaemontanii*, W1912A. Preparation not given.

W1914A DJ For sore legs on kneecap when joint is stiff. With: (1) 313, *Pedicularis canadensis*, W1914B. Put 2 plants of each in 8 qt. water, boil down to 1 qt. Take 1/3 of it 3 times a day. Cover legs with blanket and sit over 8 qt. as it steams down to one. Repeat for 4 different days.

W1914B JD See 215, *Oenothera biennis*, W1914A.

F1933A CHJJ Upset stomach, vomiting, diarrhea—when you feel loose inside. With: (1) 68, *Sanguinaria canadensis*, F1933C. Take 2-3 whole plants along with 2-3 roots of (1). Steep both in 1 qt. water. Drink as much as you like.

F1933B HJm Diarrhea. With: 348, *Achillea millefolium*, F1933A. This plant was said to be used as an alternate medicine for 168, *Agrimonia gryposepala*, F1933B. No preparation was given.

F1933C CHJJ See 232, *Ceanothus americanus*, F1933B.

F1938A HRE Emetic. Boil whole plant. Drink all you can.

F1938B CRE/ET Diarrhea and fever. Boil 3 tablespoons in about 2 qt. water—use whole plant. Boil enough to turn green.

F1938C ?? To strengthen womb. Boil about ¼ lb. of the entire plant in a gallon of water. Cook 20 min. Cool. Drink like water.

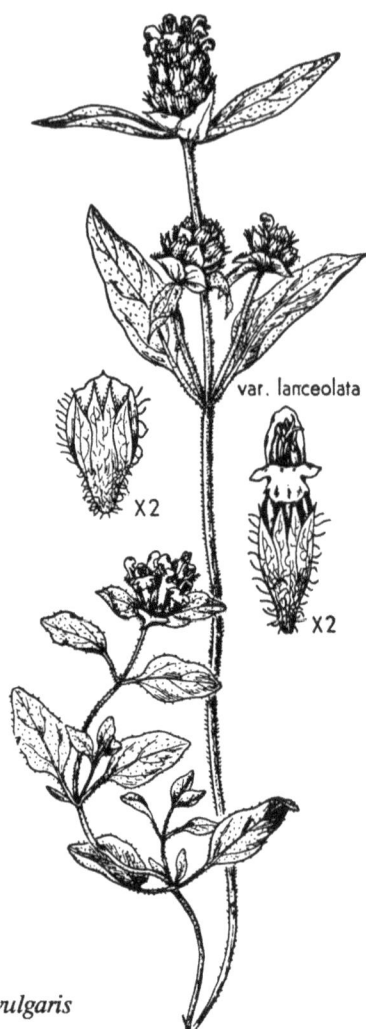

P. vulgaris

F1939A CG Biliousness. Bunch up a plant or 2 and steep it. Not much needed.

FndA JG Fever that comes every other day, short breath, vomiting, and diarrhea. Take all of young plant except the flower. Use 1 qt. of water for each plant. Boil for 2 min., no longer (it acts differently when boiled longer). Drink lukewarm as often as can. Boil longer (5 min. longer) for diarrhea.

FndB JG See 263, *Osmorhiza claytonii,* FndB.

H1973A JT Sugar diabetes. No preparation offered.

H1973B DC For anything. Best for backache. Also for lady's womb—prevents operation. Stew up and drink.

H1973C Anyms. One of highest medicines. Good for anything. Sister of catnip (303, *Nepeta cataria*). Cook 2 handfuls of whole plant in 2 cups water. Boil, drink.

305. *Scutellaria lateriflora* L.   MAD-DOG SKULL-CAP

W1912A SH Smallpox preventative—keeps throat clean. Dry root and take 1 tablespoon of powder and steep in 1 qt. water. Drink a wineglassful 3 times a day.

PLANTAGINACEAE (PLANTAIN FAMILY)

306. *Plantago* spp. (*P. major* L. and *P. rugelii* Dcne.) PLANTAINS

W1912A BB Burns, sores. Tie leaf on burn or sore.

W1912B MR Fever. Make a moderately strong tea from the root. Take ½ cup every ½ hr.

W1912C ED Cuts and bruises. Smash whole plant and apply as poultice. Keep it wet.

W1912D KG Sore joints, sprains. Put roots of 4 plants in ½ qt. water and boil down ½. Wet cloth in the liquid and bind limb or joint. Keep wetting.

W1912E JSI Bleeding or cut. Heat leaves and tie on cut or wound.

W1912F KD See 353, *Anthemis cotula,* W1912C.

W1914A DJ See 406, *Juncus bufonius,* W1914A.

F1933A SS Good for bladder—causes urine to flow fast—an annual female medicine. Root is boiled in 1 gal. water. Drain. Take 1 cup, 5 times a day.

F1933B J & JS Cuts and wounds. Tie on whole leaf.
Spider bites. Grind up leaf and apply.

For blood and pains anywhere. Take 2 heaping handfuls of roots, boil in 2 gal. water. Drink anytime.

F1933C SS See 254, *Aralia nudicaulis,* F1933A.

F1938A AJ Red one (*Plantago rugelii*) for when a woman overworks and has what whites call "nervous breakdown." Wash roots and put in 1 gal. water and boil good for about 40 min. Set aside until ready for use. Drink like water for about 2 weeks. No beverages but water.

F1938B JH Erysipelas. With: (1) 317, *Verbascum thapsus,* F1938C. Apply poultice of leaves.

F1938C CRE/ET See 405, *Symplocarpus foetidus,* F1938B.

F1938D CRE/ET See 250, *Oxalis stricta,* F1938B.

F1939A CG Red one (*Plantago rugelii*) good for pains in chest and stomach. Boil 1 whole plant in kettle of water, keep drinking.

F1939B HJ For drawing sores.

F1939C NL Bruises, swellings, pains. Take a lot of leaves and wilt them. Spread them as hot as patient can bear on the bruise and bandage.

F1956A SJ Use the one with the red stem (*Plantago rugelii*) heart trouble. No preparation given.

FndA ?? One with red base (*Plantago rugelii*) good for blood medicine.

FndB CHJJ Good for lady when cannot have baby. Also for blood. Boil roots of 2 plants.

FndC HS For poultice when face swells with erysipelas. Smash leaves and apply.

H1973A Anyms. For sores anywhere. Poultice with leaves.

H1973B DC Poultice for old sores.

H1973C EP & BC Arthritis. Put in shoe, also poultice with leaf.

H1973D FS Rheumatism. No details offered.

H1973E TL Bruises. No details offered.

H1974A RB Infected cuts. Poultice with leaf.

OLEACEAE (OLIVE FAMILY)

307. *Fraxinus americana* L.   WHITE ASH, AMERICAN ASH

?? ?? Physic. Decoction of bark from 1 ft. saplings.

W1912A PJ Snake bites. Find root where it sticks up like a snake's head. Kick (do not cut or scrape) off bark. Make poultice with warm water. May drink decoction.

W1912B ED See 11, *Lycopodium digitatum,* W1912A.

W1912C EH See 243, *Acer pensylvanicum,* W1912A.

F. americana

W1912D JJ See 247, *Rhus* sp., W1912A.

W1912E ED See 12, *Lycopodium lucidulum,* W1912A.

W1914A DJ When horse is bound up. With: (1) 308, *Fraxinus nigra,* W1914A. Put 2 double hand-fuls of each in 6 qt. water. Boil down to 1 qt. Mix with feed.

W1915A JE Hunting medicine for deer. Chew bark and vomit to clean out insides.

F1938A CHJJ Physic. Scrape bark down from 1 ft. sapling. Boil, drink a cupful at once.

F1939A PH Cramps in stomach. With: (1) 120, *Tilia americana,* F1939A (an optional ingredient). Scrape bark downward from young shoots. Boil 2 bunches (diameter of circle formed by thumb and forefinger) in 2 qt. water down to ½. Drink like water.

308. *Fraxinus nigra* Marsh.   BLACK ASH

W1912A HJn Painful urination from going with a prostitute. In fall, take ash of a year's growth, cut-ting 2 sticks and scraping bark upwards. Steep 1 hr. in 1 gal. water. Drink a quart warm before meals. Rest. Avoid salt and meats.

W1912B ED Rheumatism. With: (1) 425, *Asparagus officinalis,* W1912B; (2) 71, *Platanus occidentalis,* W1912B. Cut handful of bark from east side and combine with a handful of roots of (1) and the downward cut bark of (2). Steep in 8 qt. water. Put feet in this liquid as hot as can stand it. Put quilt over you. Repeat 3 times a day for 1 hr. at a time. Do for 3 days, stop 3 days, repeat.

W1912C SH Stricture from going with a woman when she has her changes. With: (1) 236, *Parthenocissus quinquefolia,* W1912A; (2) 229, *Euonymus* sp., W1912A. Take a patch of bark from the roots on the south side of the tree, about 4-5 in. long. Add a 6 in. length of (1) cut up into 1 in. pieces. Add a handful of blue vines (2). Put in

F. nigra

2 qt. water, boil down to 1 qt. Take cupful, then ½ cup, 5-6 times a day.

W1912D ED See 11, *Lycopodium digitatum,* W1912A.

W1912E EH See 243, *Acer pensylvanicum,* W1912A.

W1912F SH See 24, *Adiantum pedatum,* W1912G.

W1912G RS See 7, *Parmelia conspersa,* W1914A.

W1914A DJ See 307, *Fraxinus americana,* W1914A.

F1933A KD See 56, *Hydrastis canadensis,* F1933A.

309. *Syringa vulgaris* L.   LILAC, PURPLE OR WHITE LILAC (ESCAPE)

W1914A DJ Sore mouth caused by smoking some-one else's pipe or kissing a girl with menses. Also in children. Chew leaves or bark.

C. glabra

L. vulgaris

SCROPHULARIACEAE (FIGWORT FAMILY)

310. *Chelone glabra* L.   TURTLE-HEADS, BAL-MONY, SNAKE-HEADS

W1912A ED Antiwitchcraft. Smash 1 root and place in 1 cup water. Drink at bedtime.

W1912B KD See 169, *Agrimonia* sp., W1912A.

311. *Linaria vulgaris* Mill.   BUTTER-AND-EGGS, WILD SNAPDRAGON, COMMON TOAD-FLAX (NATUR.)

W1912A DJ See 48, *Anemone virginiana*, W1912A.

W1912B KD See 304, *Prunella vulgaris*, W1912B.

H1973A Anyms. Diarrhea. Put handful of leaves in cup with cold water. Stir, drink.

312. *Mimulus ringens* L.   COMMON MONKEY-FLOWER, PURPLE OR ALLEGHENY MONKEYFLOWER

W1912A KD See 57, *Ranunculus abortivus*, W1912D.

W1914A DJ To counteract the poison of poison sumach (*Rhus vernix*). With: (1) 236, *Parthenocissus quinquefolia*, W1914A; (2) 276, *Nicotiana rustica*, W1914C. Put a small sprig or 2 of it with small sprig or 2 of (1) and very little of (2). Boil it a little and wash sores.

313. *Pedicularis canadensis* L.   LOUSEWORT, WOOD-BETONY

W1912A SH Heart trouble. Mash 1 root of 1 plant, put in 1 qt. warm water, let stand for few hours. Take a little at a time all day until well.

W1912B DJ See 285, *Cynoglossum officinale*, W1912A.

M. ringens

P. canadensis

W1914A RS For bellyache caused by eating food prepared by woman during menses. Put 2 plants in 4 qt. water, boil down 1 qt. Drink often a sufficient quantity to make you vomit.

W1914B DJ See 304, *Prunella vulgaris,* W1914A.

314. *Penstemon hirsutus* (L.) Willd.
PENSTEMON, BEARD-TONGUE

W1912A DJ See 101, *Chenopodium album,* W1912A.

P. hirsutus

W1914A DJ Cures the effects of love medicines and reverts it back to the initiator. Also may be used to promote affection in a wife. With: (1) 48, *Anemone virginiana,* W1914B. Smash each plant, put in 1 qt. water, boil a little. Used as an emetic to counteract love medicine. Drink a quart of warm water after taking. Do this 4 times.

315. *Scrophularia lanceolata* Pursh HARE-FIGWORT

W1912A JJJr Sunburn, sunstroke, frostbite. Tie leaves on the place, top side next to the skin. Will take out soreness.

W1912B MS See 332, *Mitchella repens,* W1912E.

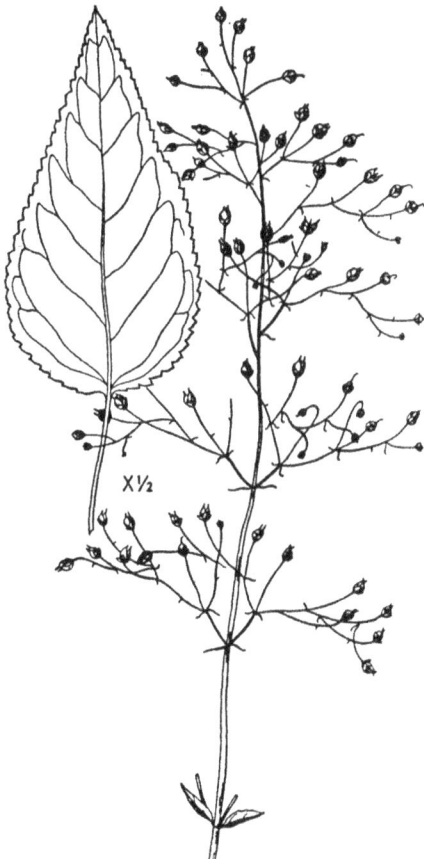

S. marilandica

F1933A CHJJ Dropsy. With: (1) 255, *Aralia racemosa,* F1933D; (2) 166, *Saxifraga pensylvanica,* F1933A; (3) 229, *Euonymus* sp., F1933A. Did not know preparation.

F1938A ET Prevents cramps and colds after birth. Make tea of root, drink like tea.

F1938B AJ Hemorrhage after childbirth. Used with honey as it comes from the comb if bleeding does not stop. Boil 1 root in 1 qt. water for about 15 min. Take ¼ cup at a time until hemorrhage stops.

F1939A DB Blood purifier. With 1 or 2 unspecified ingredients. Preparation not given.

FndA CHJJ For woman's blood after childbirth. Boil nearly a handful of the roots and bring nearly to a boil in 3 qt. water. Drink lukewarm when thirsty.

316. *Scrophularia marilandica* L. CARPENTER'S-SQUARE, FIGWORT

W1912A MS Female weakness due to irregular menses. Put a couple of roots in 3 qt. water and steep. Drink as much as you can, anytime.

317. *Verbascum thapsus* L. MULLEIN, MULE'S-EAR, FLANNEL-PLANT, DEVIL'S-TOBACCO (NATUR.)

W1912A MS Catarrh, asthma. Dry leaves and smoke, exhaling through nose.

W1912B LJ To regulate bowels of baby. Take the inside part of the first rosette of leaves. Put 3 in small cup water, steep. For babies, give teaspoon every little while.
Also for toothache. Heat leaf with hot water, add vinegar, put on face.

W1912C DJ Swellings or abscesses. Cut a piece of the leaf the size of the abscess (if too big, abscess will open to that size). Apply when the abscess is about to open.

W1912D ADy Bad hiccoughs. Dry leaves on stove. Break it up, smoke it in a pipe, swallow the smoke.

W1912E KD See 110, *Polygonum aviculare,* W1912A.

W1914A DJ For lump or swelling anywhere that results from an injury. Boil 1 plant in water, make a mash of the warm leaves, and tie them on as poultice. Drink the concoction and wash the swelling as well.

W1914B RS Earache. Heat the leaves up with sweet milk and apply as a poultice. Remoisten.

F1933A BB Baby chafe. Leaves applied with the upper side towards chafed place.
Consumption. Dry leaf and smoke.

F1933B JS Swellings and mumps. Smash big bunch of leaves and apply as poultice.

F1933C SS See 254, *Aralia nudicaulis,* F1933A.

F1938A CHJJ Swellings on face. Heat leaf and apply directly on swelling.

F1938B HRE Consumption. Boil the roots.

F1938C JH See 306, *Plantago* sp., F1938B.

F1938D CRE/ET See 294, *Collinsonia canadensis,* F1938B.

F1938E AJ See 294, *Collinsonia canadensis,* F1938A.

FndA NL Sores. Leaves as poultice.

FndB SG Rash on babies. Put whole plant in pot, add water, let boil. Drink.

H1973A JT Piles. Use leaves as toilet paper.

H1973B DC Put leaf in shoe for feet that smell, poultice for sores, leaves boiled for fever.

H1973C Anyms. Put in shoe for softness.

V. thapsus

318. *Veronica officinalis* L.   SPEEDWELL, GYPSY-WEED (NATUR.)

W1912A DJ (cf. 227, *Euonymus europaeus,* W1912C) For cow drying up. Steep 1 plant in 2 qt. water. Feed cow corn husks soaked in the liquid. Wash udder with remaining liquid.

W1914A DJ Will neutralize witchcraft intended to spoil your hunting. Used as an emetic and a wash for rifles. Place small handful of the plants in 6 qt. water and boil down to 5 qt. Drink part to make vomit, clean rifle with balance.

V. officinalis

319. *Veronicastrum   virginicum   (L.)   Farw.* CULVER'S-ROOT

?? ?? Rheumatism. Preparation not given.

F1933A J & JS Biliousness—gallstones will be softened so that they can be passed. Smash root and use a little in warm water. Steep it like tea. Drink.

F1933B CHJJ Fevers and too much gall. Boil a small part of the root in 2-3 qt. water. Drink anytime like water.

F1934A CRE/ET For too much gall, fever. Take root of 3 fingers length and steep it in lukewarm water. When water turns green, remove it. Do not make it too strong. Drink like water.
Also an antiwitch medicine.

F1939A HJ Said some fear it is poison, but not if it is used properly.

F1939B CG For overdoses—makes you physic. Also for a bad heart. Put 2 tablespoons of root and boil up about 2 qt. water. Boil down to ¼ teacup and put in a tablespoonful of Epsom salts. Take 3 mornings.

FndA SS Coughs, physic. Split 1 root and put in 1 qt. water until lukewarm. Drink little by little when you think you are going to cough.
Also an emetic. Make 3 qt. of medicine for 1 morning. Do 3 mornings, 3 times each morning.

FndB HRE Any kind of sickness, fever, witch medicine. Take tendrils from roots, put in 2-3 qt. hot water and steep. Drink like water for fever.

FndC CHJJ Said Howard Jimmerson used it for chills and diarrhea. Roots steeped until water turns yellow.

FndD KD When person is warm in summer and takes a cold drink and it leaves a lump in the chest. Smash root and stalk in 1 pt. water. Do not boil.

OROBANCHACEAE (BROOME-RAPE FAMILY)

320. *Epifagus virginiana* (L.) Bartr.   BEECH-DROPS, CANCER-ROOT

W1914A JD For diarrhea caused by eating food prepared by woman during menses. Put 5 or 6 small plants in ½ qt. water, steep to make tea. Drink ½ of what is made at a time, anytime.

CAMPANULACEAE (BLUEBELL FAMILY)

321. *Campanula americana* L.   TALL BELLFLOWER

W1912A ED Whooping cough. Smash 3 roots and let stand a ½ hr. in a teacup of hot water. Take 3 tablespoons before meals.

322. *Campanula aparinoides* Pursh   BLUE MARSH BELLFLOWER

W1912A DJ Induces childbirth for young women giving birth for the first time. Boil 8-9 stems in

8 qt. water until 6 remain. Drink right away. Take after the birth as well.

323. *Lobelia cardinalis* L. CARDINAL-FLOWER, INDIAN-PINK

W1912A DJ For gathering (abscess) in the breast. Steep 3 plants in 1 qt. water. Drink ½ of it, sit down, rest. Drink some warm water, then drink remainder, leaving some to wash breast with. Poultice with fresh plants.

W1912B DJ Stricture caused by connections with a woman when she has her changes. With: (1) 435, *Smilacina stellata*, W1912A. Steep 2 whole plants of each in 6 qt. water. Drink ½ of this.

W1912C KD Chancre and fever sores. With: (1) 362, *Cichorium intybus*, W1912A. Take a small handful of each root and boil them together in a pint of water until well done. Then take out the roots and pound them. Wash the sores with the liquid and apply the roots in a thin rag as a poultice, moistening them occasionally with the liquid. Use poultice for a day, then make a new one. Bury the old poultice in the ground, removing the disease from you.

W1912D ED Love medicine. Smash the roots and add enough water to wash the face. This attracts the opposite sex.

W1912E KD See 57, *Ranunculus abortivus*, W1912C.

W1912F KD See 304, *Prunella vulgaris*, W1912B.

W1912G KD Cramps. Pulverize root, plant, and blossom. Add boiling (not lukewarm) water, let cool. Drink down. First dose should be moderate. If take too much at start, makes you throw up. Small quantity very frequently. If prepared with lukewarm water, it will cause further cramps. Water must be very hot.

W1913A PJ For pain or trouble caused by witchcraft, also to ward off witchcraft. Make a poultice of root, apply to spot, and bone, insect, etc. will be drawn out. Take internally to ward off.

W1914A DJ When a woman's breast cakes, cracks, and the glands on the neck and under the arms swell. Smash plant and boil a little in 2 qt. water. Take ½ cup once a day and wash sores. Mashed plants applied as poultice.
Also an emetic for when someone has given you something bad to eat (candy, tea, etc.) and your stomach is sore and you do not know what is the matter. Put plant in 6 qt. water that is warm. Drink a lot, vomit, repeat.

W1914B JD Love medicine—will make you seem best looking in a crowd. Bundle up 2 plants so that bundle is 4-5 in. thick. Put in 5 qt. water, boil quite a bit. Wash face, neck, and hands. Can save and use again 3 times.

W1915A DJ Bad stomach caused by consumption. With: (1) 197, *Amphicarpea bracteata*, W1915A. Put 2 whole plants of both in 1 qt. water and boil quite a bit. Take as much as you want as often as you like.

F1938A HRE Fever. No preparation given.

F1938B JH No use recorded.

F1938C CHJJ See 301, *Mentha spicata*, F1939C.

FndA SS Love and basket medicine. Steep stalk and flower in teacup of water. Wash face in entire solution through 1 day. The other person will do whatever you want. Also, sprinkle on baskets.

H1973A Anyms. For stomach or anything. Will strengthen all other medicines. This plant should be kept outside. Must burn tobacco to keep it inside.

H1973B Anyms. Add to all medicines to give strength. One of highest plants.

324. *Lobelia inflata* L. INDIAN-TOBACCO

W1912A MS For counteracting sickness produced by witchcraft. Bundle 1 whole plant, put in teacup, adding boiling water. Cover, let stand until morning. Take it all in morning before sunrise. Have clear, warm water ready and use this as an emetic.

Take more water, repeat 3 times. Do this for 3 mornings. Take medicine at night, i.e., when it is dark.

W1914A DJ A cure for the tobacco or whisky habit, or a love medicine that has been mixed with tobacco. Mash 3 whole plants, put in ½ qt. water. Drink whole quantity, follow with warm water, vomit. Take for 3 mornings when stomach is empty.

W1914B JJJr Venereal disease. Smash 4 roots in 1 qt. warm water, put in a sealer and let stand for a short time. Not fit to drink. Wash sores and let dry on—will heal up. May also dry up root, pulverize, and apply to sores. This plant some times disappears when you go to find it again.
Also for ulcers or sores on legs. Same preparation as above.
Also for abscess at side of neck. Leaves are smashed and applied as a poultice.

F1939A JG Antilove medicine, antiwitch medicine, a divining agent, emetic and physic, not for sore eyes—too strong. Put whole plant in cold water overnight, the smallest plant you get can do this. Do 3 times [no further details given].

F1939B HJ Emetic. Put root in water, boil a little. Take ½ teacupful or less.

325. *Lobelia kalmii* L.   KALM'S LOBELIA, BROOK LOBELIA

W1912A DJ Earache and abscess—when the ear gathers and breaks inside. Cut off 4 in. of top of young plant, mash, steep in hot water. Dip in cloth and squeeze into ear.

W1912B DJ To cure the effect of love medicine made from 327, *Lobelia spicata*, and 326, *Lobelia siphilitica*. Mash 1 fair-sized plant in a cup of warm water. Drink before breakfast and vomit.

326. *Lobelia siphilitica* L.   GREAT LOBELIA, BLUE CARDINAL-FLOWER

W1912A ED Antibewitchment. Smash up the plant and put it in a teacup of hot water. Drink it at bedtime.

H1973A ON Cough medicine. Gargle, do not drink. No further details offered.

327. *Lobelia spicata* Lam.   PALE-SPIKED LOBELIA

W1912A DJ Love medicine. No further details.

W1912B BB Good for old, bad blood. Also for breaking out under the jaws and at the side of the neck. Take a couple of stalks and a quart of water, boil until dark. Use as a lotion, washing the sores at morning, noon, and night.

F1933A SS & JS Emetic for lovelorn. Steep in lukewarm water, ½ qt. Drink 1 teacupful and then drink lots of water—throw up. It will make you forget the one who ran away.

328. *Lobelia* sp.   LOBELIA

FndA JG Divining sickness and its cure. A person who is suffering sleeps on it to learn the cause of his suffering. Put it under your pillow at night and dream about your sickness during the night to discover what is wrong with you. Only the sick person does this and the next morning he will know what medicine to use.

H1973A Anyms. Love medicine. Cannot be witched if you carry it with you.

RUBIACEAE (MADDER FAMILY)

329. *Galium aparine* L.   BEDSTRAW, CLEAVERS, GOOSEGRASS (NATIVE & NATUR.)

W1912A DJ Poison ivy and itch. With: (1) 49, *Aquilegia canadensis*, W1912B; (2) 293, *Agastache nepetoides*, W1912A; (3) 365, *Erigeron philadelphicus*, W1912B. Steep 1 whole plant of each in 2 qt. water that is warm. Use as a wash.

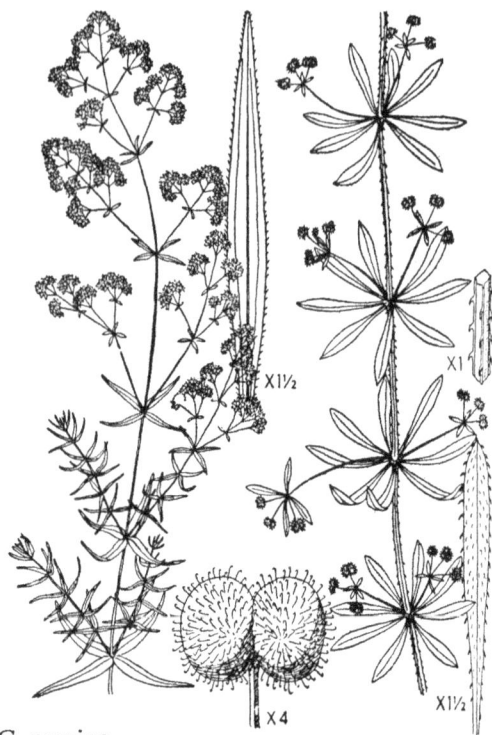

G. aparine

330. *Galium triflorum* Michx.   SWEET-SCENTED BEDSTRAW

F1933A CHJJ See 332, *Mitchella repens*, F1933A.

F1933B CHJJ See 332, *Mitchella repens*, F1933C.

F1973A Anyms. For dreams. Take hair, chew, swallow.

FndA CHJJ Backache in babies. Whole plant is applied as poultice.

331. *Galium* sp.   WILD LICORICE, BEDSTRAW

W1912A KG Venereal disease (gonorrhea). Put 4 plants in 2 qt. warm water, wash the parts with this.

W1912B ED (cf. 200, *Desmodium glutinosum*) Basket or peddler's medicine. Smash root, put handful in cupful of cold water. Sprinkle solution on baskets.

G. triflorum

F1938A CHJJ See 42, *Asarum canadense*, F1938A.

H1973A ON Basket medicine. Steep in warm water, put on baskets.

H1974A Anyms. Blindness. With: (1) 53, *Coptis trifolia*, H1974A. Preparation not offered.

332. *Mitchella repens* L.   PARTRIDGE-BERRY, TWIN-BERRY, RUNNING-BOX, TWO-EYED BERRY, MITCHELLA

W1912A DJ Typhoid-like fever caused by odor from killed snake. You become crazy. With: (1)

M. repens

202, *Melilotus alba,* W1912A. Steep 1 plant of each a little in 1 qt. water. Take ½ cup at a time.

W1912B SH Inward fever thought to be caused by stomach trouble, revealed by thirst. With: (1) *ga^hderahóndji*; (2) 27, *Polystichum acrostichoides,* W1912F. Make a small bunch and add vine and roots of 1 plant of (1). Put in ½ qt. water, steep. Take a little at a time all day. If little ones have this fever, it turns into convulsions. Then pound stripped leaves of (2), add hot water, and apply as poultice to the small of the back at night.

W1912C JSl For cross babies when they will not suckle. They twist as though something were wrong with their backs. Put small handful of vine in 2/3 cup of water and boil a little. Take ½ teaspoonful 4 times a day and 4 times through the night if necessary.

W1912D BB When baby is through sucking, give him some in a teaspoon. Wash the baby with the liquid. The mother also drinks ½ qt. in the morning. Steep a handful well in 4 qt. water.

W1912E MS For "whites" or leukorrhea (sick womb). With: (1) 359, *Aster* sp., W1912A; (2) 315, *Scrophularia lanceolata,* W1912B. Two handfuls of leaves and roots. Boil in 5 qt. water, down to 4. Drink like water, anytime.

W1912F ED Cuts. Smash and apply as a poultice to stop bleeding.

W1914A DJ For babies when their abdomens swell. Put small bunch of vines in 1 qt. water and boil quite a bit. Give teaspoon anytime. Mashed vine can be used as poultice—wet occasionally.

W1914B RS Rash on babies. Put small handful of plants in ½ cup boiling water, cover, let stand. Let the baby suck the liquid off of fingers.
Also for pregnant woman who has pain in sides caused by eating food prepared by woman with menses. With: (1) 54, *Hepatica nobilis* var. *acuta,* W1914B. Put a double handful of the plant along with the whole plant of (1) in 4 qt. water and boil well. Drink a cup every time she wishes it. Should lie down. The trouble referred to often causes abortion.

F1933A CHJJ Swollen testicles or ruptures. With: (1) 330, *Galium triflorum,* F1933A. Boil in water and drink, then smash up the plant, mix it with (1), and apply as poultice.

F1933B SS Good for the kidneys, will take the chill out of the blood, good for bladder stricture. Boil 1 big bunch in 1 gal. water. No further details.

F1933C CHJJ Love medicine. With: (1) 330, *Galium.*

F1933D SS See 254, *Aralia nudicaulis,* F1933A.

F1938A CRE/ET Physic for newborn babies with stomachache. Boil roots of 2 plants in a cup of water. Pour off liquid and cool. Give a teaspoon at a time. Acts like a physic.

F1938B CHJJ See 380, *Polymnia uvedalia,* F1938B.

F1939A NL See 255, *Aralia racemosa,* F1939A.

F1940A CHJJ See 152, *Epigaea repens,* F1940A.

FndA JG See 263, *Osmorhiza claytonii,* FndB.

FndB SS See 340, *Triosteum aurantiacum,* FndB.

H1973A DC Pregnant mothers use it so that children will not get rickets. Also for gas on the stomach. Boil whole plant in 1 qt. water. Steep for 10 min.

H1973B EP BC Fever. No details offered.

H1973C FS Fever. Make hot poultice, put on chest.

CAPRIFOLIACEAE (HONEYSUCKLE FAMILY)

333. *Diervilla lonicera* Mill.   BUSH HONEY-SUCKLE

W1912A DJ For little children when the mother has connections with another man. This spoils the child. Medicine is good for the blood. Boil either the whole plant or 1 root in 6 qt. water down to 5 qt. Baby takes ½ teaspoonful, mother takes 1 teaspoonful at a time, often.

W1912B PT See 21, *Osmunda claytoniana,* W1912A.

W1912C KD See 25, *Pteridium aquilinum,* W1912A.

F1939A KD No use given.

334. *Linnaea borealis* L. ssp. *americana* (Forbes) Hultén   TWINFLOWER, LINNAEA

W1914A DJ Cramps, fever in children, or for when they cry. Boil 2 small sprigs in ½ cup water. Give teaspoon anytime. Also for intermittent fever in adults. Boil small bunch of plants in 1 qt. water pretty well. Give ½ teaspoon anytime.

335. *Lonicera canadensis* Bartr.   FLY HONEY-SUCKLE

W1912A DJ For when a child is cross and cries all night. Seems mad and has no pain anywhere. Also for homesickness. Scrape bark (not too much so as to make too strong for a baby) into 4 qt. water and steep. For adults, put 1 medium-sized branch in 6 qt. water and steep. For baby, give 1 small teaspoon and wash whole body with liquid. Adult, take ½ qt. at a time.

W1912B ED Syphilis chancres. Take the young growth, a good handful, and boil it in 8 qt. water

L. canadensis

until 6 remain. Drink as much as frequently as desired, taking a lot each time.

F1933A SS See 254, *Aralia nudicaulis,* F1933A.

336. *Lonicera dioica* L. var. *glaucescens* (Rydb.) Butters   WILD OR MOUNTAIN HONEY-SUCKLE

W1912A PT See 21, *Osmunda claytoniana,* W1912A.

W1914A DJ To make an unfaithful wife stay home. A love medicine, also counteracts love medicine. Put 1 whole vine in 8 qt. water, boil down to 1 qt. Put along with tea or other foods. Both parties must take it. Also taken as an emetic to throw off effects of love medicine.

W1914B JJJr For sickness in little child caused by being picked up by a pregnant woman. Baby becomes feverish and sick. Put 1 small plant in 1 qt. water, boil quite a bit. Give to baby on finger, to larger children in spoon, quite often. Wash body also.
Also good for a pregnant woman to take when she feels sore inside, legs, or elsewhere. This is a sign that the baby is sick in the same place. Put a

L. dioica

L. oblongifolia

double handful of the mature plant in 5 qt. water, boil down nearly ½. Drink nearly a cupful often. Makes baby quiet.

F1933A CHJJ See 214, *Epilobium angustifolium,* F1933A.

337. *Lonicera oblongifolia* (Goldie) Hook.   SWAMP FLY HONEYSUCKLE

W1912A KD See 247, *Rhus* sp., W1912F.

W1914A DJ When you are lonesome and are not well enough to go anywhere, but are restless. Scrape bark of vine upward off of about 6 small branches. Put in 6 qt. warm water. Let stand until strong enough. Drink a quart, a little at a time for 6 days. Wash with remainder 6 times. This medicine is often prescribed by old women fortune tellers.

338. *Lonicera* sp.   HONEYSUCKLE

W1912A JJ See 186, *Prunus virginiana,* W1912B.

W1912B RS See 96, *Carpinus caroliniana,* W1912A.

339. *Sambucus canadensis* L.   BLACK ELDER-BERRY, COMMON ELDER

W1912A HJn Measles. Scrape the bark upward from a couple of young shoots or switches. Steep in 2 qt. water for ½ hr. Drink a teacup every day.

W1912B DJ See 251, *Geranium maculatum,* W1912C.

W1912C DJ See 251, *Geranium maculatum,* W1912B.

W1912D KD See 122, *Malva neglecta,* W1912A.

W1913A PJ Headache. Bark scraped and made into poultice with hot water.

W1914A DJ Laxative for child when passage is hard. With: (1) 297, *Lycopus americanus,* W1914B. Preparation unclear. Put a sprig of water [water horehound?] in 1 qt. water and boil quite a bit. Dip a little on the fingers and let the child suck. Wash child as well and it will stop crying.

W1914B DJ Laxative (when costive). Scrape bark downward off of 2 small branches. Put in 2 qt. water for a man. Take a bellyful. For a child, just give a spoonful. Also wash body, commencing with the head.

S. canadensis

F1938A CHJJ Physic for tiny babies. Blossoms steeped in warm water. Give baby a little at a time.

F1938B CHJJ Physic. Scrape bark down, boil, drink a cupful at once.

F1938C AJ Baby physic. Make tea of bark. Give to baby 3-4 times until all black is removed. Black considered poison.

F1938D AJ Wash baby after birth. Steep bark in hot water. Do not wash eyes, ears, or nose with anything special—just mother's milk.

F1939A PH Poultice for cuts. Scrape the bark from the tip to the bottom of young shoots. Take off the outer bark and save the inner bark. Dry and pulverize. Then wet the dust and poultice it on the cut.

F1939B CG Heart disease. Take the young green shoots, earlier than July. Scrape out the pith, about 2 handfuls. Put in 3 qt. of barely lukewarm water. Let stand about 3 hr. Do not steep it. Keep drinking every little while.

F1939C SS See 247, *Rhus* sp., F1939B.

FndA CG For when children are bloated in the bowels. Scrape the bark into fuzzy balls. Poultice this on the abdomen. It will not take long to make them physic. Do not give it internally.

FndB JG For kidneys, bowels. Scrape bark downward off 3 small shoots. Boil in 3 qt. water for 3 min. Let cool. Drink ad lib. as much as possible.

FndC JG Gonorrhea. Boil about a handful of pith in a quart of water. Water should turn milky. Drink as often as can.

FndD AJ Spring emetic. Take the bark from 5 sticks, 1 yr. old. Do this in the spring. Scrape up, put bark in basin, pour warm water, let stand about 2 min. Drink the whole thing. Sit still about 4 min., and have about 2 gal. of lukewarm water ready. Drink all of this that you can. The yellow bile will come up. Fill belly with warm water 3 times before breakfast. Will make you feel like a young man.

H1973A JT Stomach troubles. No preparation offered.

H1974A DC To cut down on gall. Scrape bark upward from about 4 30-in. sticks. Squeeze juice out of bark by rubbing in hands. Put in 2 teaspoons water, put in ½ cup, drink, wait 1 hr. after, then give lukewarm water by the quart after becoming nauseous. They will vomit up the gall. Use again only once.

340. *Triosteum aurantiacum* Bickn.   WILD COFFEE, HORSE-GENTIAN

W1912A DJ Irregular or profuse menses. Take piece of root 1½ in. long. Put in 6 qt. water and steep. Take quite a bit at a time.

W1912B JJJr No use given.

W1912C Mrs. PS Used to fatten puny babies. Scrape about a teaspoon of the fresh root. Put in a cupful of warm water. Give as much as they want to drink—give according to age. Good for adults, too.

W1912D KD Bad cold, the throat gets cold and dry. Cut up the roots fine and put them in hot water, 1 bunch to a teacupful. Let stand until cool. Take a teaspoonful into the mouth and do not swallow for 10 min., then swallow. Do 1 hr. apart, repeat if bad.

W1914A DJ Excessive menstruation caused by flying squirrel eating corn soup or other food. Put 2 pieces of root from 1 plant in 4 qt. water. Boil quite a bit. Take a little at a time until person gets better.

W1914B JD Cure for witchcraft. Smash the roots of 4 plants and apply as a poultice. If the trouble is in the stomach, take a root of 1 plant and 1 qt. water. Boil a little. Take a good drink for 4 or more times a day.

FndA SS Venereal disease, diuretic. Use root. No further details.

FndB SS For pain when urinating. With: (1) 157, *Chimaphila umbellata,* FndC; (2) *gayowāke^{n}s*; (3) 332, *Mitchella repens,* FndB; (4) 218, *Cornus amomum,* FndA; (5) 81, *Juglans cinerea,* FndE. Take split roots in a bundle, boil in 3 qt. water for 10 min. Drink 4 glasses a day before meals and bed. Will work in 2 days. Then take a laxative made of (1)-(5). Unspecified parts and amounts of these ingredients are put in 5 qt. water and boiled down to 1 pt. of syrup. Add teaspoon of Epsom salts and take 3/4 cup early in morning.

T. aurantiacum

V. acerifolium

H1973A DC Heard it was good for pneumonia—makes you sweat. Use root. No further details offered.

341. *Viburnum acerifolium* L.   MAPLE-LEAF VIBURNUM, ARROWWOOD, DOCK-MACKIE, PURPLE-LEAF VIBURNUM

W1912A ED To kill witchcraft manifested by a pain that moves about. It will take out ants, worms, etc., which cause pain. Also, charcoals, water beetles, crickets, broom splints, hair, etc. Scrape bark from young twigs downward. Warm ½ cup of whisky and add 2 handfuls of the bark. Let stand until you think it is ready to use. Apply to pain as a poultice. The doctor also takes the liquid in his mouth before the poultice is applied and sprays where the pain is felt. Patient also takes a mouthful of the liquid every 3 or 4 hr.

W1912B HJn Will suppress excessive menses. Bunch up 3 trees (?) and steep in a gallon water. Drink lukewarm a cupful every hour. Fix again, if necessary.

W1912C MR For stricture and painful urination in men. Take a good bunch and steep it in a quart water. Drink a quart a day if possible.

342. *Viburnum lantanoides* Michx.   HOBBLE-BUSH, WITCH-HOBBLE, MOOSEWOOD, TANGLEWOOD

W1912A PT For sore chest and lost breath. Boil a big handful down to 2 fingers in a gallon water. Drain. Drink as much as you can.

W1912B JJ See 227, *Euonymus europaea,* W1912B.

FndA CRE Love medicine, but not sure.

FndB CHJJ See 26, *Onoclea sensibilis,* FndA.

343. *Viburnum lentago* L.   SHEEPBERRY, SWEETBERRY, NANNY-BERRY, BLACK HAW, COWBERRY, WILD RAISIN, WILD TEA

W1912A PT For spitting blood all the time. Cut the root into 3 in. lengths and split them. Take a handful and add 1 pt. water. Boil a little. Drink as much as possible. The more you drink, the quicker you will get better.

W1912B JJ See 194, *Spiraea* sp., W1912B.

FndA CHJJ No use given.

FndB ?? No use given.

344. *Viburnum recognitum* Fern.   ARROWWOOD

W1912A JJJr Prevents conception. Take 2 good handfuls of fresh green twigs. Boil in 5 qt. water, down to ½. Take when woman has her changes, as soon as possible or medicine will spoil. Once is enough. Menses will continue, but the blood is better and the woman grows fat and feels well afterwards.

W1914A DJ See 222, *Cornus foemina,* W1914B.

345. *Viburnum trilobum* Marsh.   CRANBERRY-BUSH, CRANBERRY-TREE, HIGHBUSH CRANBERRY, GROUSEBERRY, SQUAW-BUSH, PIMBINA

W1912A DJ Fallen womb after birth and facilitates passage of afterbirth. Put 1 handful of fresh branches and twigs in 4 qt. water and boil down to 3 qt. Drink a cupful often.

W1912B HJn Bad blood and fever (perhaps typhoid and gall in the blood). Scrape bark upward from a 5 in. stick. Steep it very little in a cup of water. Squeeze the bark out. Drink ½ cup before breakfast. Rest 10 min., drink bellyful of warm water, and vomit. Do for 2 days.

W1912C BB Laxative. Scrape bark down and steep in ½ cup water. Put in bottle, use as required. Very strong. The effects of this medicine are stopped by a medicine made of 88, *Quercus macrocarpa,* W1912B.

W1912D KD See 34, *Pinus strobus,* W1912B.

W1914A DJ Blood purifier for anybody. Take 2 6-in. lengths that are 3/4 in. in diameter. Put in 6 qt. water and boil quite a bit. Drink all you can.

F1938A CHJJ Fever, for baby. Boil root in water. Take a teaspoon at a time.

F1938B CRE/ET See 97, *Corylus americana,* F1938A.

F1938C CRE/ET See 68, *Sanguinaria canadensis,* F1938A.

F1938D AJ See 68, *Sanguinaria canadensis,* F1938E.

F1938E CRE & AJ See 186, *Prunus virginiana,* F1938A.

346. *Viburnum* sp.   VIBURNUM

?? ?? Contraceptive. Drink 3 qt. People should not use it in preaching at Longhouse or you will never have a child again.

DIPSACACEAE (TEASEL FAMILY)

347. *Dipsacus fullonum* L. (*D. sylvestris* Huds.) COMMON TEASEL, FULLER'S TEASEL, WILD TEASEL (NATUR.)

W1912A JJJr Used as a brush for the hair. Also good for acne or "worms in the face." The water that collects at the base of the connate leaves is used to wash the face.

FndA SS (Note: This plant is described as "growing in swamp, no flower, green color, sharp needles." cf. 421, *Sparganium eurycarpum*, W1914A) To kill someone you dislike. Plant is poison. Scrape root, make into a powder. Put a pinch of it in coffee or tea. A pinch or less lasts 1 or 2 days.

Also, to take another's land. Grind dried root and put a pinch in another's tea. They die in 2 days. Then, take over their land. They get diarrhea and do not stop until they die.

Also, to kill a cute baby you're jealous of. Old folks used to dip finger in this and put it on the lips of a cute baby that they jealous of. Then, when the baby died, they would enjoy the grief of the parents and eat ghost bread at a 10 day feast. Now this is known to happen when the baby is half white.

ASTERACEAE (ASTER FAMILY)

348. *Achillea millefolium* L.   COMMON YARROW, MILFOIL (NATUR.)

W1912A ED Headache. Handful of roots put in 4 qt. water. Steep. Drink as much as often as want.

W1912B HJn For babies with any kind of sickness or fever. Steep 3-4 leaves in ½ teacup of water for 1 min. or less. Give the teacup every little while.

W1912C SH Headache. With: (1) 354, *Arctium* sp., W1912G. Smash leaves, put on leaf of (1), tie on head as poultice.

W1912D KD Diarrhea with blood. With: (1) 174, *Fragaria* sp., W1912A; (2) 191, *Rubus occidentalis*, W1912A; (3) 190, *Rubus idaeus*, W1912B; (4) 116, *Rumex crispus*, W1912E; (5) an unspecified rose *yagodûⁿdákta*. Pound up large single handful, bundle up to about 7 in. long. Add 1 handful of roots of (1)-(5). Put in ½ gal. water and boil to a quart. Take ½ wineglass 2 hr. apart.

W1912E KD See 353, *Anthemis cotula*, W1912B.

W1912F KD See 353, *Anthemis cotula*, W1912C.

W1912G BB See 303, *Nepeta cataria*, W1912A.

W1914A DJ Neuralgia. Chew and hold juice in mouth. If bad, make a poultice of leaves and tie on sore spot.
Also good for when "sore through joints." Make tea and drink.

F1933A HJm See 304, *Prunella vulgaris*, F1933B.

F1938A AJ When people fall and lie unconscious but limp, but do not have fits and spit at mouth. Take 3-4, break up. Put ½ handful in teacup. Add ½ cup cold water. When turns green, remove plants and sprinkle the body with water. Then rub hard. Put 2 drops in his mouth.

F1938B CRE & AJ Convulsions in babies. Boil 1 entire plant in 1 pt. water until water turns green. Feed 1 teaspoon to baby, rub rest over body.

F1938C CRE/ET Worms in children. Leaves put in cup, warm water poured over. When water turns green, remove leaves and drink spoonful.

F1938D CRE/ET Headache and worms in children. Gather and dry leaves. Pour boiling water over leaves.

F1938E HRE Gonorrhea. Take the tops, boil in 2-3 qt. water. Wash affected parts.

F1939A KD Emetic for sunstroke accompanied by diarrhea and fever. Mash 3 whole plants, steep them in boiling water. Let cool. Drink ½ wineglassful frequently.

FndA CHJJ Summer complaint. See 442, *Sisyrinchium angustifolium*, FndA.

FndB JG See 263, *Osmorhiza claytonii*, FndB.

349. *Ambrosia artemisiifolia* L.   RAGWEED, HOGWEED, BITTERWEED

W1914A DJ See 350, *Ambrosia trifida*, W1914A.

H1973A DC Cramps from picking berries. Boil up, make tea.

H1973B Anyms. Cook root for stroke. Take 4-5 cups a day.

350. *Ambrosia trifida* L.   GIANT RAGWEED, GREAT RAGWEED, BUFFALO-WEED

W1914A DJ Bad diarrhea with bleeding. With: (1) 349, *Ambrosia artemisiifolia*, W1914A. Put 4 plants of each in 2 qt. water. Boil quite a bit. Drink all you can. Make more from same plants, but boil more.

W1914B JJJr Blood medicine. No details given.

351. *Anaphalis margaritacea* (L.) Benth. & Hook. f. ex Clarke   PEARLY-EVERLASTING

W1912A MR Sore eyes. Steep plants like tea and wash eyes.

F1933A CHJJ Diarrhea, dysentery. Use roots and stalks. No further details.

F1933B KD (possibly 371, *Gnaphalium uliginosum*, F1933A) See 262, *Heracleum lanatum*, F1933A.

352. *Antennaria plantaginifolia* (L.) Richards. EVERLASTING, PUSSY'S-TOES, LADY'S TOBACCO

W1912A DJ Toothache. Boil bunch in teacup down to ½. Wash out mouth, keeping in mouth as long as possible. May also chew the plant or fill cavity with its pulp.

W1912B SH Whites (leukorrhea). Steep the roots of 4 plants in 1 qt. water. Drink ad lib.

W1914A DJ For toothache caused by rats or mice gnawing food that is prepared and eaten. Put 1 plant in 3/4 cup water. Boil a little. Rinse mouth and hold in mouth. Do often. Also chew leaves and put in hollow tooth.

353. *Anthemis cotula* L.   MAYWEED, STINKING CHAMOMILE, STINKWEED, DOGWEED (NATUR.)

W1912A JW Toothache. Chew root. Will stop the toothache and will break up decayed tooth.

W1912B KD To cure vomiting. With: (1) 348, *Achillea millefolium*, W1912E. Take ½ single handful without roots, combine with good-sized single handful of whole plant of (1). Add 1 pt. cold water and let stand 4 hr. Drink a wineglassful about an hour apart.

W1912C KD Summer disease—vomiting, diarrhea of white, watery fluid accompanied by cramps. With: (1) 97, *Corylus americana*, W1912B; (2) 74, *Ulmus americana*, W1912C; (3) 306, *Plantago* sp., W1912F; (4) 348, *Achillea millefolium*, W1912F. Take ½ handful, omitting leaves, add ½ handful of whole plant of (4), ½ handful of bark of (2), ½ handful of root of (3), and a small handful of the roots, branches, and leaves mixed of (1). Boil all in 1 qt. water down to ½ pt. Take ½ wineglass 3 hr. apart.

W1914A DJ When child breaks out on body with red spots—caused by when he catches cold and you cannot tell what is the matter. Put 2 handfuls of plant in 2 qt. water and boil pretty well. Drink any amount, anytime—the more, the quicker the cure.

W1914B JJJr When short of wind—too much gall. Used to clean stomach. With: (1) 34, *Pinus strobus*, W1914D. Put 1 small handful in 4 qt. water and boil very little. Drink as often as you can (not necessary to vomit, unless you feel like it).

W1914C JD When not feeling well—throw up and do not know what makes it. Also for ptomaine poisoning. Bundle up 1 stalk, put in ½ qt. cold water, let stand for 5 min. Take ½ when bad. Take other ½ later on, if necessary.

F1933A J & JS Emetic for spring fever, sedative. Break up 1 whole, dried plant. Put stems and roots of 3-4 plants in 1 qt. water. Drink cold to vomit and settle stomach.

F1938A FJ Biliousness—throwing up bile. Roll whole plant, put in 1 qt. water. Let boil down once or twice. Drink as water when thirsty.

FndA SG Summer complaint, cramps in stomach of adults or children. Use whole plant. Put in 3 qt. water. Let boil 5 min. Let cool. Drink any amount, anytime. Do not eat meat or potatoes—only fish. Be still.

FndB SG See 263, *Osmorhiza claytonii*, FndB.

354. *Arctium* spp. [*A. lappa* L. and *A. minus* (Hill) Bernh.]   BURDOCK

W1912A DJ Sore head. Tie on with handkerchief.

W1912B DJ Black magic. With: (1) 445, *Smilax hispida*, W1912B; (2) 187, *Rosa acicularis*, W1912B; (3) 214, *Epilobium angustifolium*, W1912D. Put roots with whole plants of (1) and (2) in 1 qt. water and boil. First make a doll of *Arctium* and stick it with a pointed stick where you wish to affect the person you wish to injure. For counteracting, take (3), fixed the same way (see 214, *Epilobium angustifolium*, W1912B).

W1912C SH Bee sting. Mash leaf and apply as poultice.

W1912D SH For big pimples on face and neck. With: (1) 157, *Chimaphila umbellata*, W1912C. Dry 1 lb. of the root and 1 lb. of (1). Mix in 2 qt. water, steep, drink ad lib.

W1912E KD Boils. See 363, *Cirsium discolor*, W1912B.

W1912F ED See 11, *Lycopodium digitatum*, W1912A.

W1912G SH See 348, *Achillea millefolium*, W1912C.

W1913A PJ Headache. Leaves tied on head.

W1914A DJ In case of sorcery where a doll is made of root and stuck with thorns. Put 1 root in 4 qt. water, boil quite a bit. Drink a small amount after. Vomit to remain in stomach (?). A leaf is tied wherever there is pain.

F1933A JS See 185, *Prunus serotina*, F1933A.

F1938A CRE/ET Earache, bruises, fever, backache. For bruises, make leaves limp with heat, apply. For earache, pour ½ cup boiling water over piece of leaf, drop ½ teaspoon liquid into ear, plug ear with leaf. For backache, heat leaf and apply.

F1938B AJ For pains anywhere, strained back. Smash 3 medium-sized leaves, then warm them. Put on cloth, keep it warm. Leave on all night. Remove next day and wash area with warm water and soda.

F1938C HS Blood medicine. No preparation given.

F1939A CG Rheumatism. Apply leaves directly to back and chest. Will draw poison out and make sweat.

FndA DJn See 185, *Prunus serotina*, FndB.

H1973A DC Makes urinate a lot. Boil up root and drink.

H1973B TL Cuts. Pound with stone, poultice cut.

H1973C ON Sore legs in horses, sore muscles. Root used. No further details offered.

H1974A ON Wash legs of race horses. Warm liquid sponged on, wipe with towel, wrap with towel.

H1974B DC Kidneys. Stew roots for 10 min., cool, drink.

355. *Aster macrophyllus* L.   BIGLEAF ASTER

W1914A JD Venereal disease. With: (1) 68, *Sanguinaria canadensis*, W1914B; (2) 61, *Ranunculus recurvatus*, W1914A; (3) 265, *Sanicula marilandica*, W1914C; (4) 263, *Osmorhiza claytonii*, W1914C. Put 4 roots of each plant in 5 qt. water. Boil down to 1 qt. Drink ½ qt. at bedtime, then ½ qt. the next morning. Do not use meat or salt for 4 days. Will loosen bowels. Once is enough.

F1933A CHJJ Goes with blood medicine. Use roots. No further details.

356. *Aster novae-angliae* L.   NEW ENGLAND
ASTER

W1913A JJJr Love medicine. Same as for 359,
Aster sp., W1913A.

W1914A DJ For different kinds of fevers, and
cures nearly all kinds of disease (fevers). Put roots
and leaves of 2 plants in 6 qt. water, boil quite a bit
down to 1 qt. Drink pretty often when you are bad.
Not very strong. Therefore, drink as often as you
can.

FndA ?? See 393, *Solidago* sp., FndA.

H1973A DC Her mother used the root for something, but she did not know for what.

H1973B Anyms. For weak skin—when anything
irritates. Sensitive women use it as a douche. Boil,
let cool.

357. *Aster prenanthoides* Muhl.   ZIG-ZAG OR
CROOKSTEM ASTER

F1938A CHJJ Fever in babies. Boil root in water.
Teaspoonful at a time is given to baby.

FndA SRE Kidneys, colds, or when cold settles in
kidneys. With: (1) 369, *Eupatorium perfoliatum*,
FndE. Boil 2 roots in gallon of water, do not let
boil down. Cool, drink ad lib. Use (1) when there
is a cold. Boil roots in 2 qt. water. Take hot, lie
down, cover with blanket.

358. *Aster puniceus* L.   PURPLE-STEMMED
ASTER

W1912A MS Fever, colds, typhoid, pneumonia.
Three-four roots steeped in 2 qt. water. Take
warm if patient is chilly, or cool if fevered. Take
any amount, anytime.

W1912B ED Consumption. Steep handful of roots
in gallon water. Drink any amount, anytime. The
more, the better.

359. *Aster* sp.   ASTER

W1912A MS See 332, *Mitchella repens*, W1912E.

W1912B ED See 195, *Waldsteinia fragarioides*,
W1912A.

W1913A ?? Love medicine. The leaves of the aster
where the leaf of one plant leans over and inserts
point in cordate base of another are used. The first
represents a man, the second a woman. The roots
are taken after Indian tobacco is used beside
another plant of the same kind. The roots are
kept. The girl's or woman's name is mentioned
when tobacco is sprinkled. The woman then will
come and pass 1 night with him. This is what the
name signifies.

H1973A JT Hunting medicine. No details offered.

H1973B Anyms. Put in cotton bag, pin to shirt for
good luck.

360. *Balsamita major* Desf. *(Chrysanthemum balsamita* L.)   COSTMARY, MINT-GERANIUM
(RARE ESCAPE)

W1912A KG Earache. Smash up the leaf. Add 1
tablespoon of warm water to 1 leaf. Keep in a
saucer, well covered. Wet a cloth and drop in the
ear as much as preferable. Plug up ear with cloth
or something.

361. *Bellis perennis* L.   ENGLISH DAISY, LAWN-
DAISY (NATUR.)

W1914A DJ When your stomach is bad and you
throw up yellow stuff. Put 2 clusters of the plant in
4 qt. water and boil quite a bit. Take as much and
as often as you like.

362. *Cichorium intybus* L.   CHICORY, BLUE-
SAILORS, CORNFLOWER, SUCCORY,
WITLOOF (NATUR.)

W1912A KD See 323, *Lobelia cardinalis*, W1912C.

F1933A J & SG Not used as medicine.

363. *Cirsium discolor* (Muhl. ex Willd.) Spreng. FIELD-THISTLE

W1912A BB Flowers in full bloom chewed by old men when they had no tobacco. Boys use it occasionally.

W1912B KD Boils. With: (1) 354, *Arctium* sp., W1912E. Take 7 roots and 1 big handful of the roots of (1). Add 1 qt. water, boil down to 1 pt. Take ½ wineglassful 1 hr. apart. Follow with salts (67, *Podophyllum peltatum,* W1912I, used anciently instead of salts).

F1938A CHJJ See 440, *Veratrum viride,* F1938B.

FndA KD See 247, *Rhus* sp., FndA.

364. *Cirsium vulgare* (Savi) Tenore   BULL-THISTLE, COMMON THISTLE (NATUR.)

H1973A JT Bleeding piles and cancer. No details offered.

365. *Erigeron philadelphicus* L.   FLEABANE, DAISY

W1912A DJ Running sores. With: (1) 285, *Cynoglossum officinale,* W1912B. Mash 1 plant of each, steep in 2 qt. water. Use as a wash and poultice for both (i.e., running sores and dropsy) afflictions.

W1912B DJ See 329, *Galium aparine,* W1912A.

H1973A Anyms. Good for opening lungs. Cook whole plant, then drink.

366. *Erigeron pulchellus* Michx.   ROBIN'S-PLANTAIN, POOR-ROBIN'S-PLANTAIN

W1914A DJ For quick consumption. Put 5 young plants without flowers in 10 qt. water, boil down to 1 qt. Take a cupful or so as often as you like.

W1914B JJJr For cough or cold that is tight, i.e., has no expectoration. Put 2 large roots in 2½ qt. water, boil. Drink as much as you can (does not cause vomiting).

367. *Eupatorium dubium* Willd. (or *E. fistulosum* Barratt ex Hook.)   JOE-PYE-WEED, PURPLE BONESET, TRUMPET-WEED

F1938A CHJJ See 301, *Mentha spicata,* F1938C.

368. *Eupatorium maculatum* L. and *E. purpureum* L. SPOTTED JOE-PYE-WEED AND SWEET JOE-PYE-WEED OR GREEN-STEMMED JOE-PYE-WEED

W1912A ED Heals soreness of womb and abdomen after childbirth. Also good for kidneys. Steep well 3 roots in 4 qt. water. Drink a lot anytime until soreness goes away.

W1912B JJ Gonorrhea. Split hollow root, leaving rootlets on. Boil handful in pail of water. Drink all you can.

W1912C KD See 433, *Polygonatum pubescens,* W1912A.

W1914A DJ See 132, *Populus alba,* W1914A.

F1933A CHJJ See 232, *Ceanothus americanus,* F1933B.

F1938A CHJJ See 198, *Baptisia tinctoria,* F1938A.

F1939A KD Consumption—when the stomach starts to dry. Root dried and ground to powder. Decoction of handful of powder in ½ gal. water is boiled down to ½ qt. Drink like water.

FndA FSt Kidneys. Steep roots of 1 plant in 1 gal. water for 15 min. Cool, drink like water.

FndB CHJJ See 198, *Baptisia tinctoria,* FndC.

H1973A DC Chills and fever with colds. Cook roots. Give hot if chills, cold if fever.

369. *Eupatorium perfoliatum* L.   THOROUGH-
WORT, BONESET

W1912A KD Laxative. With: (1) 396, *Taraxacum
officinale,* W1912G; (2) 196, *Cassia hebecarpa,*
W1912A; (3) Epsom salts. Take a big handful of
flowers, same with (1). Add ¼ lb. of leaves of (2).
Put in ½ gal. of soft water and boil down to 1 qt.
Drain. Add 2 lb. of sugar. Some times add a pint
of gin to make it keep. Drink a wineglassful at
bedtime and before breakfast.

W1912B ED Syphilis with chancres. Put root in
cup of water and smash it. Let it stand ½ hr., and
wash sores with it. Then warm leaves on stove and
tie on dry. Do for 3 days, stop for 3 days, repeat
until cured.

W1912C MR Headache. Crush the plant and tie
on forehead.

W1912D LJ For stricture caused by going with a
girl during the menstrual period. Put 1 bunch of
roots in 3 qt. of water. Boil down to ½. Take all
you can drink.

W1912E PT Pains in the stomach and on the left
side (?). Steep roots of 3 plants in teacup of boiling water. Cool, drain. Drink all as soon as can.

W1912F KD Broken bones. With: (1) 120, *Tilia
americana,* W1912F; (2) 34, *Pinus strobus,*
W1912K; (3) 122, *Malva neglecta,* W1912C. Pound
a gallon of young leaves, add cold water to make a
mash. Place on piece of cotton and wrap cold
around limb. Leave on for 24 hr., then renew.
Continue for 3 weeks. Splints are made from bark
of young white pine (2), the inner side of the bark
being placed against the limb, but over the boneset
poultice. Bind tightly. If poultice gets hot, skin will
blister, but a poultice of basswood (1) leaves will
correct this. If bruised badly, add a poultice of (3),
cold.

W1912G DJ Sorcery. Put in enemy's liquor flask
and he will drink until he kills himself. Cut root in
short lengths and put in liquor flask. See 47,
Anemone canadensis, W1912A for a cure for this.

W1913A JJJr Fever. Roots used. No further
details.

W1914A DJ Used in divination to learn if a man
and woman will come together or separate, or to
test faithfulness of wife. Two pieces of root are
taken from different plants. They will curl one toward the other, showing that the couple will untie.
Name roots for wife and suspected lover.
Also for stopping the liquor habit. Smash 2 roots
in 4 qt. water, boil down ½. Take a cup, 3 times a
day.

F1933A CHJJ Typhoid. Smash plants with roots
and boil. Drink.

F1933B CRE/ET Pleurisy. No preparation given.

F1939A HJ Fever. Steep tops, when in flower, and
drink.

F1939B JG Fever, ague. No preparation given.

FndA DJn Colds. Steep it. No further details.

FndB NL Fever in person or horse. Steep whole
plant. No further details.

FndC CHJJ Pneumonia. Decoction of roots.

FndD CG Medicine against onset of cold. Steep
up lower part of stem with leaves taken toward
fall. Not the roots. No further details.

FndE SRE See 357, *Aster prenanthoides,* FndA.

H1973A Anyms. Fever, sore shoulders, broken
bones, piles. No preparation offered.

H1974A DC Fevers. Cook 2 roots in a quart of
water. Drink lukewarm.

370. *Eupatorium rugosum* Houtt.   WHITE
SNAKEROOT, WHITE SANICLE

W1914A DJ If a woman has menses slightly and a
man goes with her then, she may put a pin or
needle in her mouth. This will cause difficulty in

urination and the penis becomes "rotten" inside. Plant should be pulled while person is still walking. Put stems of plant in 8 qt. water, boil down to 1 qt. Take a good bellyful right away.

F1933A CHJJ Sweat bath. Boil roots in gallon water. Drink a little too. Put patient in bed and keep him there until well cooled.

F1938A CHJJ Good for anything.

F1938B ET Inflamed womb. Make tea of roots and drink like tea.

FndA CHJJ Physic. Boil 1 whole plant including roots in 2 qt. water until it goes down ½. Take at night. Drink. Lie down.

FndB SS When venereal disease gets in the blood, this medicine separates the blood from the disease. Take root of 2 plants, boil for 5 min. in 3 qt. water.
Also for fallen womb. Put 2 roots in 3 qt. water, bring to boil. Drink 4 or 5 glasses a day.

H1973A Anyms. Horse medicine—to make stop sweating. Cooked up. No further details offered.

371. *Gnaphalium uliginosum* L. LOW CUDWEED OR EVERLASTING

F1933A KD (possibly 351, *Anaphalis margaritacea*, F1933B) Asthma. Steep. No further details.
Also for bruises. With: (1) 262, *Heracleum lanatum*, F1933A; (2) 375, *Inula helenium*, F1933A. No preparation given.

372. *Helianthus strumosus* L. WOOD-SUNFLOWER

W1914A JD For worms in young or old. Kills worms and makes others come or pass out. Put the root of 2 plants in 1 qt. water. Boil quite a little. Take all you can.

373. *Hieracium pilosella* L. MOUSE-EAR HAWK-WEED (NATUR.)

W1912A ED Diarrhea. Steep small bunch in quart of water. Drink glass at a time an hour apart.

W1912B JW Toothache. Chew root.
Diarrhea. Steep and drink.

374. *Hieracium* sp. HAWKWEED

W1912A JJ Consumption. With: wood ashes. Put sunflower (*Helianthus*) into the ashes (2 qt.). Place on the side of the stove, stir, cook, sift again, pound in the pounder, sift again, pound again. Put 2 qt. of this into a pot holding 10 qt. water, stir. The oil will rise, skim it off and put into another dish. Let oil boil slowly. Put in bottle. Take 1 teaspoon in morning after meals.

H1974A ON For sore close to bone (e.g., ankle). Put roots in water and put on sore.

375. *Inula helenium* L. ELECAMPANE (NATUR.)

W1912A BB For sores and cuts from a rusty nail that will not heal. Mash and apply as poultice.

W1912B KD Gas on stomach. Dry and pound root to flour. Put 3 tablespoons flour in a bottle, add 1 pt. cool water, cork up. Wrap bottle in paper and bury bottle slanted in ground for 5 days. Use on sixth day. Do not drain. Take 1 tablespoon in glass early in morning and before meals.

W1912C SH Heaves in horses. Dry roots, powder up. Put tablespoon of the powder with the feed twice daily, morning and night.

W1912D KD Pains in the chest that make you hold your breath. With: (1) 242, *Aesculus hippocastanum*, W1912A; (2) 257, *Panax trifolius*, W1912B; (3) 44, *Nuphar* sp., W1912A; (4) 146, *Cardamine diphylla*, W1912D. Take 1 thimbleful of root cut not closer than 3 in. from the main root stock. Add the meat of (1), powder the roots of (2), (3), and (4). No further details.

W1912E ED See 37, *Juniperus virginiana*, W1912A.

W1912F KD See 433, *Polygonatum pubescens*, W1912A.

W1912G MS See 42, *Asarum canadense*, W1912F.

W1914A DJ For when a girl leaks rotten. With: (1) 22, *Osmunda regalis*, W1914A. Women in such a condition cannot raise young ones. Dry 1 root, put in 2 qt. water, add 1 root of (1), and boil quite a bit. Take ½ at a time, anytime.

F1933A CHJJ Cough. Boil roots. Sometimes scrape off the roots and swallow them dry. Drink all you can.

F1933B CHJJ Cough. With: (1) 42, *Asarum canadense*, F1933B; (2) 68, *Sanguinaria canadensis*, F1933D; (3) 72, *Hamamelis virginiana*, F1933B. Put together and boil. Drink like tea.

F1933C KD See 262, *Heracleum lanatum*, F1933A.

F1933D MS See 282, *Ipomoea pandurata*, F1933D.

F1938A CRE/ET For fever, especially in TB when it first starts. Also used for heart trouble. Roots are dried, then boiled in water, and drunk.

F1938B CHJJ Good medicine for anything, especially colds and coughs.

F1938C JC Asthma. Pinch root, steep in pint of water. Drink as needed.

F1939A CHJJ Root used. No further details.

F1939B JG Consumption. Two preparations. First preparation: Use just leaves. One leaf in ½ qt. water. Boil 3 min. Drink when go to bed—1 tablespoon. Second preparation: Use the root in wintertime. Use ½ large root in ½ qt. water. Scrape root that has been dried. Take 1 tablespoon at bedtime.

F1939C JJm Stroke. Take 2 roots about finger long and split them. Put in 4 qt. water and boil down to 3 qt. Drink.

F1939D HJ For healing cough. Said it was frequently confused with 129, *Epilobium angustifolium*. No further details.

FndA SS For cough or heaves in people or horses. With: (1) 42, *Asarum canadense*, FndE. Take small quantities of both roots sliced. Boil in 3 qt. water. Drink ad lib.

H1973A Anyms. Rheumatism. Make poultice of leaves, then put lard over it. Also for asthma in children. Leaves are dried. No further details offered.

H1973B JT Cleans out intestines. With: other unspecified ingredients. Cut root, roast until brown. No further details offered.

H1973C Anyms. Arthritis. Cook root and all, put on where sore.

H1974A Anyms. Rheumatism—swellings in joints. Poultice.

376. *Lactuca canadensis* L.   WILD LETTUCE

W1912A SH See 396, *Taraxacum officinale*, W1912F.

F1938A CRE & AJ Severe bleeding from cut. Root smashed and applied as poultice.

377. *Lactuca tatarica* (L.) Meyer ssp. *pulchella* (Pursh) Stebbins   BLUE LETTUCE (RARE INTROD.)

H1973A Anyms. Piles—goes into pores and closes them like a bandage. Pound, apply as poultice.

378. *Onopordum acanthium* L.   SCOTCH THISTLE, COTTON-THISTLE (NATUR.)

W1912A JJJr For witchcraft poison, and for counterwitchcraft as an emetic. Make decoction of 3 roots. No further details.

379. *Polymnia canadensis* L.   SMALL-FLOWERED LEAF-CUP

H1973A FS Thinks it was smoked once—just to smoke. Maybe toothache.

380. *Polymnia uvedalia* (L.) L. BEAR'S-FOOT, YELLOW OR LARGE-FLOWERED LEAF-CUP (RARE)

W1914A DJ Against ghost that disturbs you while you are asleep—such as when the quilt is pulled off and you do not know what did it. Burn end of dried root, smell the smoke, and rub a little of the black on face under eyes. You can then sleep all right.

F1933A HRE To frighten ghosts away—ghosts are a cause of sickness. Burn root, let smoke fill room. Rub soot on face. Put an "x" on cheek.

F1938A JH Fever. Tea from root. Break up root and steep a teaspoonful in hot water.

F1938B CHJJ Pains in back, vomiting—kidney trouble. With: (1) 249, *Zanthoxylum americanum*, F1938A; (2) 332, *Mitchella repens*, F1938B; (3) *oyeⁿ:ō:do*, cf. 166, *Saxifraga pensylvanica;* (4) 161, *Ribes americanum*, F1938A; (5) *oda?enō:?k.* Take root about size of finger, add a little bark from the root of (1), the whole, crushed stalk of (2), the root of (3), peel the root of (4), and bark from the root of (5). Put in quart of water. Take ½ teacup 3 times a day.

F1940A CHJJ Fumigant for ghosts, returning souls, etc., nightmares. Dried root. No details given.

381. *Prenanthes alba* L. WHITE LETTUCE, RATTLESNAKE-ROOT

W1914A ?? When you feel weak and almost paralyzed—will give you life. Smash 1 root and put in ½ cup water. Cover, let stand for 5 min. Wet body with it.

F1933A SS Dog or rattlesnake bites. Pound up roots and put on wound—put juice on root.

F1933B HRE Said Jonas Snow breaks the root, holds it toward the snake, and then captures it.

Break root until juice exudes.

F1938A CRE/ET Remedy for bites of animals—prevents blood poison. Squash the root and put on bite smoothly. Apply until bite heals.

FndA CHJJ Prevents and cures rattlesnake bites. Use roots. No details given.

382. *Prenanthes altissima* L. RATTLESNAKE-ROOT, WHITE LETTUCE, TALL LETTUCE

F1933A SS Rattlesnake bites in the same way that 381, *Prenanthes alba,* is used.

F1939A CG Against rattlesnake bites. Pound many roots and poultice bite. Steep some in warm water and drink also.

F1939B KD Use not given.

383. *Prenanthes trifoliolata* (Cass.) Fern. GALL-OF-THE-EARTH

W1912A DJ Sore eyes. Boil roots of 1 plant in ½ teacup water and steep. Squeeze a few drops into the eye. Dampen cloth with the liquid and tie over the eyes. Use all.

W1912B SH Deer hunting medicine—must make no pounding or whistling noises for 12 days (?). With: (1) an unspecified plant. Crush 1 root of 1 plant into ½ qt. water from a running stream or spring. Let stand overnight, wash rifle with this solution. Pulverize, dry root of (1), and place in small pot containing hot coals. Hold rifle over smoke that rises. Smoke these leaves in a pipe for 12 mornings before hunting and also at the place where the deer are. Also sprinkle some of the dried root on a log near your hunting post. Stand quiet and the deer will come to you. One leaf of this plant is called the "buck," the other (the hastate) is called the "doe."

W1914A DJ Love medicine. Break bulbous root open, rub on hand, face, chew some.
Also a hunting medicine. Make a tea from 2 roots in 2 qt. water. Wash body and gun, put inside moccasins, also soak bullets.

F1933A CHJJ For rattlesnake bite. Use roots.

F1938A CHJJ For bristles on toe or foot. Smash root and apply to affliction.

F1938B JC Skin swelling. Mash up entire plant, apply to skin with bandage.

FndA KD Said it was a mysterious root. If the stalk is cut first to get at the root, the root will disappear. When she sees it, she is surprised. She makes a big circle around it and digs up the whole plant with a spade. It is suggested that if cattle eat it, it harms the milk in some way.

384. *Rudbeckia hirta* L.   BLACK-EYED-SUSAN, YELLOW DAISY

W1914A JD For worms in children. Put 2 roots in 1 cup, steep to a tea. Give a tablespoon at a time. Must not take anything greasy or salty at this time.

H1973A JT For the heart. Boil (root?). Take a spoonful at a time.

385. *Senecio aureus* L.   GOLDEN RAGWORT OR GROUNDSEL, SQUAW-WEED

W1912A MS Fever in children. Steep 2 small or 1 large rosette in 1 pt. water. Drink small amount warm whenever thirsty.

W1912B DJ Good for kidneys and the blood. Boil 3 plants in 4 qt. water until 2 qt. remain, then add 3 qt. water and boil down ½. No further details.

W1912C JSl Broken bones. Put 5 plants in 6 qt. water. Boil down to not quite ½. Drink cupful 3 times a day. Lie down, do not move. Splints not used, bandage tied firmly enough, and patient remains inactive.

FndA ?? Diaphoretic. Use root. No further details.

386. *Silphium perfoliatum* L. (probably *Polymnia* sp.) CUP-PLANT, INDIAN CUP (RARE INTROD.)

F1939A JG Goes against sickness caused by the dead. Burn the root, mark cheek of sick person. Do same for child who cries at night.

FndA SS Paralysis. Slice root, bundle to size of thumb and forefinger. Boil in 4 qt. water. Wash face with 1 qt., use other 3 qt. as an emetic.
Also for preventing seeing ghosts. Burn the root. Put soot on finger and mark child's cheek. Will prevent child from seeing ghosts at night.

387. *Solidago canadensis* L.   CANADA GOLDEN-ROD, COMMON GOLDENROD

W1912A BB When have no appetite in morning because there is too much gall—an emetic. Goldenrod removes the gall because the flower is yellow. Steep handful in 4-5 qt. water until ½ left. Add cool water. Drink 3 qt. or until stomach hurts, then vomit using fingers. Then take more. Do this 3 days before breakfast. This medicine is used in the spring.

W1912B KD For baby who starts suddenly, especially in sleep. With: (1) 424, *Allium* sp., W1912A. Pound tubers of (1), add ½ teacup of hot water. Give a little at a time.

W1912C MS See 56, *Hydrastis canadensis*, W1912A.

W1914A DJ (referring to the goldenrod gall) Gambling medicine for peach stone game, snowsnake, etc. Take 6 galls and mix them with some kind of red paint (?), and put on the palm of the hand and on the peach stones, snowsnakes, etc.

F1939A ?? No use recorded.

388. *Solidago flexicaulis* L.   ZIG-ZAG GOLDEN-ROD, BROAD-LEAF GOLDENROD

W1912A DJ Biliousness—when excrement is yellow. With: (1) 279, *Solanum dulcamara*, W1912A. Boil 1 of each plant in 3 qt. water, down to ½. Take cup before breakfast and ad lib.

389. *Solidago juncea* Ait.   EARLY GOLDENROD

F1933A SS It will stop vomiting yellow. Good for fever, also. Put 1 root in 2 qt. boiling water. Drink anytime.

F1938A HS For jaundice—yellow skin, dizziness, eyeballs turn yellow, nausea. Put a near handful in gallon until scum forms on top. Drink as much as stomach can hold.

F1938B AJ When you throw up and cannot stop, biliousness. Pour boiling water over 1 flower in basin. Drink that, then drink ½ cupful for a while, then ½ if it does not stop—keep drinking.

F1938C CHJJ Emetic. Take 3 or 4 blossoms. Boil in water. Drink all you can.

F1938D HRE Emetic—when yellow gall comes up into mouth, the stomach is upset. Boil bunch of flower tops in gallon water. Drink until it comes up. Will clean your stomach up.

F1938E CRE/ET Upset stomach due to liver disorder. Pour hot water on flower—1 qt. water to 2 flowers. When liquid turns green, cool. Drink like water during day—the more, the better.

FndA HJim Emetic. Flower tops used. No further details.

FndB SS Emetic. Flower used. Put 4 bunches of flowers in 4 qt. water. Bring to boil, let cool, and strain. Drink repeatedly until full. Do this 3 mornings before eating. Face east before vomiting, vomit into an old pail, and tie a piece of cloth around forehead or else you will get a headache. Throw the vomit in the river in the direction it is flowing.

390. *Solidago nemoralis* Ait.   ROUGH GOLDEN-ROD, GRAY GOLDENROD, OLD-FIELD GOLDENROD, LOW GOLDENROD

W1912A Ed Good for the kidneys. Boil handful of roots in 6 qt. of water until 4 qt. remain. Take a glass 3 times a day before meals.

391. *Solidago rugosa* Mill.   TALL HAIRY GOLD-ENROD, BUTTERWEED

F1938A AJ Dizziness, weakness, sunstroke. Remove top, boil 3-4 flowers and leaves in 1 gal. water for 5 min. Strain. Drink like water.

FndA AJm Biliousness and liver medicine. Whole plant used. No further details.

392. *Solidago squarrosa* Muhl.   RAGGED OR STOUT GOLDENROD

W1914A DJ See 277, *Physalis heterophylla*, W1914A.

393. *Solidago* sp.   GOLDENROD

W1912A KD See 91, *Alnus incana*, W1912A.

FndA ?? See 356, *Aster novae-angliae*, FndA.

394. *Sonchus asper* (L.) Hill   SPINY SOW-THISTLE (NATUR.)

W1912A KD See 42, *Asarum canadense*, W1912G.

395. *Tanacetum vulgare* L.   TANSY, GOLDEN-BUTTONS (NATUR.)

W1914A JJJr For pains all over the body caused by too much gall. Cut tops of plants, a double handful. Smash leaves, wet with cold water, and apply to aching part.

F1938A AJ Headache, cold, anything. Crush tops, put handful in cloth, put cold water on. Apply to forehead. For cold, put small piece up nose. Said was dangerous to drink.

FndA NL Headache. Bandage the leaves around the head.

FndB J & SG Bruises, cuts, bone decay, headache. Preparation not given.

396. *Taraxacum officinale* Weber ex Wiggers
COMMON DANDELION (NATUR.)

W1912A RS Love medicine. Root is taken when a forked root is found with another root between the fork—resembling male anatomy. Name of person wished to attract is spoken when plant is taken. Root is then thrown behind him or her. When this is done, the other will be sure to follow.

W1912B KD Liver spots. With: (1) 207, *Trifolium repens,* W1912A; (2) 253, *Impatiens* sp., W1912A; (3) rhubarb. Take the root, add a large handful of (1), and ½ handful of whole plant of (2). Put in 1 pt. hot water. Let cool. Drain. Take 1 glass 4 hr. apart. Then take a leaf of (3), cut off stem, and pound it and wring out the juice while taking medicine above. Wash face with warm, soft water after washing with juice of (3) in morning, but not at night.

W1912C LJ Emetic for throwing off yellow gall since flower is yellow. Steep root, make strong. Take when necessary, ½ small cup.

W1912D DJ Swollen testicles brought about by local trauma. Mash flowers, tie on as poultice—keep wet.

W1912E JJJr Worms in the teeth that cause decay. Chew flower stem.

W1912F SH Kidney trouble—pains in back, dark circles, and puffiness around eyes in morning. With: (1) 376, *Lactuca canadensis,* W1912A; (2) 212, *Dirca palustris,* W1912H. Take roots of 3 plants, add ½ of the bark of a small bush of (2). Put in 1½ qt. water and steep for 2 min. separately. Put 1 root of (1) into 1 qt. cold water after mashing it. Take wineglass of first mixture 5-6 times a day, starting early in morning and finishing before retiring. Take second mixture in between doses of first mixture.

W1912G KD See 369, *Eupatorium perfoliatum,* W1912A.

W1914A DJ For when testicles get caught in a fence rail. Boil 3 plants in ½ qt. water and wash affected parts. Apply the mashed plants as a poultice.

W1914B JJJr Antiwitch medicine. When you have been bewitched in some way (e.g., something shot into body, or otherwise). Get a forking root, good sized. Use tobacco in fixing (with invocation) calling on the witch to cure the sick person. One of forking roots is called by the name of the witch and the other for the sick person. This will cause the sickness to be thrown back on the witch. The root is placed in 2 qt. water and boiled down to 1. Whole body is washed with the liquid. Save the liquid, take it to a creek, sprinkle tobacco in the creek, throw out the liquid and say, "you had better go to the end of the land." This carries the disease away. The dandelion is chosen because flowers are visible at night.
Also a love medicine when the roots are twined together. Put roots in 1 qt. water, boil quite a bit. Wash on face and fingers. Liquid is saved and tried again, if not successful at first. See 175, *Geum aleppicum,* W1914B.

F1933A JS See 185, *Prunus serotina,* F1933A.

F1939A CG Dropsy. Gather a big bundle, steep. Drink every little while.

FndA KD For when lungs swell and you get a pain between shoulder blades. Comes from neglected colds, rheumatism, tonsil infections. With: (1) 43, *Nuphar lutea,* FndB. Take whole plant, add pinch of dried whole plant (with root) of (1). Pour 1 qt. boiling water over these and set on stove for 20 min. Set aside to cool. Drink cold every 2 hr., ad lib.

FndB AJm Food. Leaves eaten.

H1973A DC Anemia. Boil up and drink.

397. *Tussilago farfara* L.   COLTSFOOT (NATUR.)

W1912A SH Consumption cough medicine. With: (1) 19, *Botrychium virginianum,* W1912A; (2) 68,

Sanguinaria canadensis, W1912I; (3) 254, *Aralia nudicaulis,* W1912D. Put 1 root of each (a small root of *Sanguinaria canadensis*) in a $.50 liquor flask with the liquor. Let stand 3 days. Take a teaspoon 4 times a day before meals and at bedtime.

398. *Xanthium strumarium* L.   COMMON COCKLEBUR, CLOTBUR

H1973A Anyms. Witching medicine—if you want to marry a girl and others do not want you to. No details offered.

ALISMATACEAE (WATER-PLANTAIN FAMILY)

399. *Alisma plantago-aquatica* L.   WATER-PLANTAIN

W1912A ED Consumption. Steep handful of roots in 3 qt. water. Drink as much as you like.

W1912B KD Liniment for runners. With: (1) 70, *Dicentra cucullaria,* W1912A. Handful of leaf steeped like tea in 1 pt. water. Rub solution on body. Strengthens veins. Eat root of (1).

W1913A JJJr Lame back or kidneys. Use plant with one side of root split.

W1914A DJ Consumption. Boil down 1 plant or 3 bunches of the roots in 3 qt. water until ½ remains. Take 1 cup, 4 times a day. (Note: Compare names for 306, *Plantago* sp.)

F1938A AD Womb trouble. Put in hot water. No further details.

F1938B HS Use not given.

400. *Sagittaria latifolia* Willd.   WAPATO, DUCK-POTATO, ARROWHEAD LILY

W1912A KD For boils around the abdomen of children caused by licking at an old wooden sugar spoon. Also for constipation. With: (1) 402, *Acorus calamus,* W1912C. Put 3 roots with 3 teaspoons of (1) that has been dried and scraped in a pint of water. Boil 20 min. Cool, drink ½ wineglassful 2 hr. apart.

F1938A CHJJ See 96, *Carpinus caroliniana,* F1938A.

FndA CHJJ No use given.

FndB CG Rheumatism. Steep up.

FndC FS Corn medicine, when starting to plant corn. Take the root and steep it, boil for 10 min. Let cool. Put corn in oven right before planting.

POTAMOGETONACEAE (PONDWEED FAMILY)

401. *Potamogeton* sp.   PONDWEED

W1912A ED For catching muskrats. Smash and boil the roots and leaves. Put in kettle with a little warm water. Put in trap.

W1912B DJ Soreness all over in men from being witched. With: (1) 421, *Sparganium eurycarpum,* W1912A; (2) 9, *Rhytidiadelphus triquetrus,* W1912A. Mash each plant and tie on as poultice, where seat of soreness is (e.g., where muscle twitches).

ARACEAE (ARUM FAMILY)

402. *Acorus americanus* (Raf.) Raf. (*A. calamus* L.)   SWEETFLAG, FLAGROOT

W1912A JSl Toothache. Dry roots, cut up like tobacco. Put in pipe and smoke, sucking smoke into hollow tooth.

W1912B RS Toothache. Make a piece of the root the size of the hole, pack it in. This will stop aching. In about 6 months, tooth will all break up.

W1912C KD See 400, *Sagittaria latifolia,* W1912A.

W1912D KD See 437, *Trillium* sp., W1912B.

W1912E KD See 57, *Ranunculus abortivus,* W1912D.

H1973A JT Headache, sore throat, earache. One drop of boiled liquid in ear for earache, use root with just a little water. Gargle with water for sore throat.

H1973B DC See 434, *Smilacina racemosa,* H1973A.

H1973C TL Colds, sore throats from colds or singing too much. No preparation given.

403. *Arisaema triphyllum* (L.) Schott ex Schott & Endl.   JACK-IN-THE-PULPIT, INDIAN-TURNIP

?? KD To induce pregnancy in mare. When horse is sick and listless. Dry it and make it into a flour by grating with a nutmeg grater. Mix with animal's food (bran), wet a little, and feed. This will bring the mare around right. If overdose occurs, mare will not bear more foals.

W1912A KG Diarrhea in children. Take a turnip 2 in. in diameter and boil in 1½ qt. water until 1 qt. remains. Take 1 teaspoon about 6 times a day.

W1912B DJ Cramps. With: (1) 68, *Sanguinaria canadensis,* W1912H. Take a piece of the turnip about the size of a small walnut, add root of (1) about 1½ in. long. Both are dried and mixed to powder. Add powder to 1 qt. water, steep a little. Drink ½ of what is made anytime, then the remainder.

W1912C SH For sore on face. No details given.

W1912D MJW For pains. Root scraped and drunk.

W1912E SH See 75, *Ulmus rubra,* W1912A.

W1914A JJJr For baby when it lies still all day and night as if thinking about something . . . is lonesome, not satisfied with home (they know as much as old people). Smash 2 roots, put in 1 qt. warm water. Do not let stand long. Wash whole body.

W1914B JD For nonconception caused by cold blood. Smash 1 root and put in 1 qt. cold water. Drink in 3 drinks, anytime.

W1914C DJ See 40, *Lindera benzoin,* W1914A.

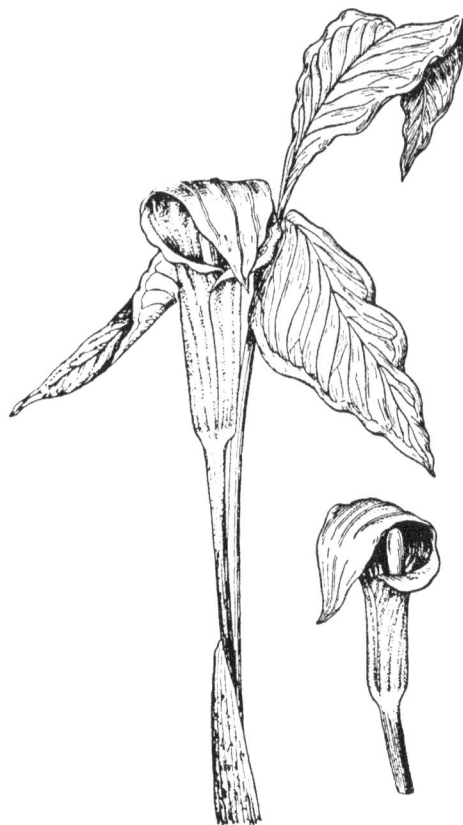

A. triphyllum

F1938A JH Bruises. Mash plant and make poultice. Apply hot.

F1938B JC Bronchial colds. Chop root and put in whisky. Drink occasionally.

F1938C CHJJ See 437, *Trillium* sp., F1938A.

F1938D CHJJ See 440, *Veratrum viride,* F1938C.

F1938E HS Bruises and lameness. Make a hot poultice.

H1973A JT To make you sneeze and for sore eyes. For sore eyes, boil and put eyes over steam.

H1973B DC Mother would splice around root and rub on her face, but never knew why.

H1973C ON Poultice. Use root.

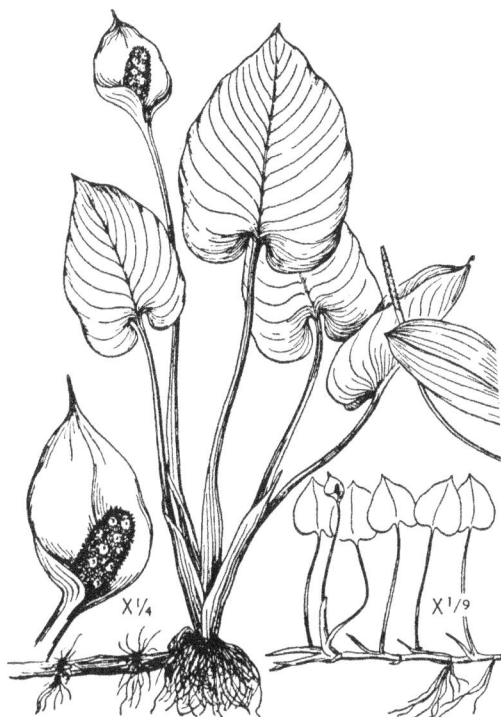

C. palustris

H1974A Anyms. Sore joints. Dry, powder, mix with rubbing alcohol.

404. *Calla palustris* L. WATER ARUM, WILD CALLA

W1914A DJ Snakebite. With: 414, *Scirpus tabernaemontanii*. Take roots and lower parts of stems, and the whole plant of *Scirpus tabernaemontanii*, 1 plant of each. Smash up, put in 2 qt. water, boil down ½. Wash bite, put mashed plants on as poultice, wetting it occasionally.

405. *Symplocarpus foetidus* (L.) Salisb. ex Nutt. SKUNK-CABBAGE

W1914A DJ If bitten in a fight or by a dog, this will cause the biter's teeth to fall out. Crush leaves and put on bite.

W1914B JD Consumption. Dry 1 root, powder it. Put 1 teaspoon of powder in 2 in. warm water in a tumbler. Let stand a little while. Drink this dose 3 times a day.

W1914C JD To cure strong smell under your arm. Wash with liquid [as prepared in W1914B, above] and place wet cloth on. Put 2 roots in ½ qt. water.

F1938A CRE/ET Displacement of womb. Put ½ cup crushed stalks in 2 qt. water. Boil until the water turns yellow. Use 1 qt. as a douche (used a sumac stem in olden days as a syringe).

F1938B CRE/ET Falling of the womb. With: (1) 306, *Plantago* sp., F1938C; (2) *deyaidjagagowa* 'something that comes out of the earth'. Put a handful of leaf with 2 roots of (1) and a handful of upper parts with seeds of (2). Boil in 1 gal. water until water turns green. Drink abundantly like water. Take until all symptoms go away, even a short while afterwards. Takes about 2 months.

F1938C CHJJ Rheumatism. With: (1) 51, *Cimicifuga racemosa*, F1938C; (2) 444, *Smilax herbacea*, F1938B; (3) 29, *Taxus baccata*, F1938A; (4) *deyoyondahgo* 'berries on either side'. Take a handful of the roots of all except (3), from which the bark is scraped upward. Put in 2 qt. water and boil down to 1 qt. Put pail between legs, cover with blanket, sweat, and rub down.

F1938D CHJJ See 444, *Smilax herbacea*, F1938A.

H1973A JT Pass seed over female genitals to bring about childbirth.

H1973B DC Worms in children. No details given.

H1973C RB Bad wounds. No details given.

H1974A DC Worms.

JUNCACEAE (RUSH FAMILY)

406. *Juncus bufonius* L. TOAD-RUSH

W1914A DJ To give strength to runners and other athletes. An emetic. With: (1) 306, *Plantago* sp., W1914A. Take small handful or bunch, add 1 plant of (1). Smash and put in 6 qt. water and boil quite a bit. Drink a lot to induce vomiting in the early morning. Also, wash entire body. Repeat procedure 3 times.

FndA SG & JG Spring emetic and emetic for ball players.

407. *Juncus tenuis* Willd.   YARD-RUSH, PATH-RUSH, SLENDER RUSH

W1914A JJJr Running and ball playing medicine. Put 1 bunch of plants in 5 qt. water, boil. Drink a good bellyful, vomit. Wash body with remainder. Do this for 3 days before a match. Should run a little each time, using whole strength.

F1933A JC Runner's emetic. No details given.

F1956A SJ For colt that has had too much feed. Steep plant, mix in with milk.
Lacrosse medicine. Steep, drink, and vomit. Also bathe in it.

408. *Luzula* sp.   WOODRUSH

W1912A DJ See 67, *Podophyllum peltatum*, W1912G.

CYPERACEAE (SEDGE FAMILY)

409. *Carex brevior* (Dewey) Mackz. ex Lunell SEDGE

W1912A DJ See 247, *Rhus* sp., W1912H.

410. *Carex oligosperma* Michx.   SEDGE

W1912A DJ See 116, *Rumex crispus,* W1912D.

411. *Carex platyphylla* Carey   BROAD-LEAF SEDGE

W1912A DJ See 17, *Equisetum fluviatile,* W1912A.

W1914A DJ Snowsnake medicine, a witch medicine. Boil 3 plants pretty well in a quart of water. Wash the snowsnake with it. This will make the snake beat. A few drops sprinkled on the track from a bottle will make the opponent's snakes stick or go off the track. Another snake following

the one treated will stick.

412. *Carex prasina* Wahl.   SEDGE

W1914A DJ When stomach is bad from an unknown cause. Said to have been brought to a hunter by his dog in the fall when hunting. Dog eats it as medicine. Put a small handful of it in 4 qt. water. Boil down a little. For an emetic, take lots of it. Otherwise, just ½ cup when dry.

413. *Carex vulpinoidea* Michx.   SEDGE

W1914A DJ See 434, *Smilacina racemosa,* W1914A.

414. *Scirpus tabernaemontanii* Gmel. (*S. validus* Vahl)   SOFT-STEM BULRUSH, TULE, GREAT OR SOUTHERN BULRUSH

W1912A DJ See 304, *Prunella vulgaris,* W1912D.

W1914A DJ See 404, *Calla palustris,* W1914A.

POACEAE (GRASS FAMILY)

415. *Agropyron repens* (L.) Beauv. QUACKGRASS, WITCH-GRASS, QUICK-GRASS (NATUR.)

?? KD Worm remedy. No further details.

416. *Bromus ciliatus* L.   FRINGED BROME GRASS

FndA SG Corn planting medicine. Use whole plant. No further specifics given (cf. preparation of 418, *Elymus canadensis,* F1938A).

417. *Cinna arundinacea* L.   STOUT WOODREED, WOOD REEDGRASS

F1933A CHJJ Sugar diabetes. Man gets thinner and thinner and his skin gets yellow. With: (1) 426, *Clintonia borealis,* F1933A; (2) 232, *Ceanothus americanus,* F1933C. Take all 3 plants whole, smash them, boil to get juice out of them. Drink.

418. *Elymus canadensis* L.   CANADA WILD-RYE, STOUT NODDING WILD RYE

W1912A ED For stricture. With: (1) 156, *Vaccinium* sp., W1912A. Boil a handful of each in 6 qt. water, down to 4 qt. Take 1 glass 5 times a day.

F1933A CHJJ Kidneys. With: (1) 223, *Cornus rugosa*, F1933A; (2) 431, *Maianthemum canadense*, F1933A; (3) 157, *Chimaphila umbellata*, F1933E. Take roots of 5-6 plants of each, put in water, and boil down to ½ qt. Cool. Drink when thirsty, the more, the better. Drink in 2 days, but drink no water. Eat anything.

F1938A CHJJ Corn medicine. Take roots in fall, boil together. Cool. Soak corn in this solution before planting.

FndA FS White corn medicine. No preparation given, but included with 400, *Sagittaria latifolia*.

419. *Festuca obtusa* Biehl.   NODDING FESCUE GRASS (NATUR.)

W1914A DJ Heart disease. Make decoction of root by smashing it and putting ½ a bunch in 2 qt. water and boiling it. Take 4 tablespoons a day.

W1914B DJ Corn medicine. With: (1) 420, *Phragmites australis*, W1914A. Preparation not given.

420. *Phragmites australis* (Cav.) Trin. ex Steud. COMMON REED, REEDGRASS (NATUR. & NATIVE)

W1914A DJ Corn medicine. See 419, *Festuca obtusa*, W1914B.

SPARGANIACEAE (BUR-REED FAMILY)

421. *Sparganium eurycarpum* Engelm. ex Gray GIANT BUR-REED

W1912A DJ See 401, *Potamogeton* sp., W1912B.

W1914A ?? See 20, *Osmunda cinnamomea*, W1914A.

TYPHACEAE (CAT-TAIL FAMILY)

422. *Typha latifolia* L.   COMMON CAT-TAIL, BROAD-LEAF CAT-TAIL

W1912A BB Gonorrhea in women. The woman chews the root.

W1912B SH Sprains. Smash root, mix with hot water. Apply as hot poultice.

W1913A ?? To stop cuts on horses or men from bleeding. Smash roots, put in warm water, let stand ½ hr. Wash cut with the liquid.

FndA KD Cyst of breast. Preparation not given. Also for yellow fever. Patient sleeps on a mattress made of it and improves.

LILIACEAE (LILY FAMILY)

423. *Allium tricoccum* Ait.   WILD LEEK, RAMP

FndA SS Worms in children. Boil, fry, eat with salt. Prevents worms from rising through child's throat.

H1973A TL Spring tonic, cleans you out. Boil up and drink.

424. *Allium* sp.   WILD ONION

W1912A KD See 387, *Solidago canadensis*, W1912B.

425. *Asparagus officinalis* L.   ASPARAGUS (NATUR. & ESCAPING)

W1912A ED See 82, *Juglans nigra*, W1912B.

W1912B ED See 308, *Fraxinus nigra*, W1912B.

426. *Clintonia borealis* (Ait.) Raf.   WOODLILY, BLUEBEADS, CORN-LILY, CLINTON'S LILY, YELLOW CLINTONIA

F1933A CHJJ See 417, *Cinna arundinacea*, F1933A.

H1973A JT For the heart. Boil and drink.

427. *Clintonia umbellulata* (Michx.) Morong SPECKLED OR WHITE WOODLILY, WHITE CLINTONIA

F1933A SS Basket medicine. Makes people buy baskets. Boil whole plant in water.

F1938A CHJJ Chills. Steep entire plant in water. Take while warm.

428. *Erythronium americanum* Ker   YELLOW ADDER'S-TONGUE, TROUTLILY, DOG-TOOTH VIOLET, FAWN-LILY

W1912A DJ Prevents conception. With: (1) 103, *Claytonia virginica*, W1912B. Young girls may eat these plants, but not the roots. They will never have children.

F1940A CHJJ Poultice for swellings. Others used, but not sure of what they are. Smash, make poultice.

FndA SS Did not know.

FndB FSt For removing slivers. Smash roots, make poultice.

429. *Hemerocallis fulva* (L.) L.   ORANGE DAY-LILY, TAWNY DAY-LILY, CORN-LILY (ESCAPE)

FndA SS Did not know.

430. *Lilium philadelphicum* L.   WOODLILY, ORANGE-RED LILY

W1912A DJ Divination of love, to discover unfaithfulness of wife or way of determining love between a young man or woman. Cut 2 lengths of root and put them side by side. Name one the wife; the other, the suspect. As they dry in the sun, they may twist together. If so, suspicion confirmed.

W1914A JD Used by a woman in a case where husband is chasing another woman. Put 2 roots in 5 qt. water. Boil, drink for an emetic, and wash body. Husband will then think that she is the one.

W1914B DJ See 446, *Coeloglossum viride*, W1914A.

FndA SS Did not know.

431. *Maianthemum canadense* Desf.   FALSE LILY-OF-THE-VALLEY, TWO-LEAF SOLOMON'S-SEAL, WILD LILY-OF-THE-VALLEY

W1914A DJ Use not specified. With: (1) 105, *Agrostemma githago*, W1914A. Small handful mixed with 2 stems of (1) in 8 qt. water. Boil down to about 1 qt.

F1933A CHJJ See 418, *Elymus canadensis*, F1933A.

432. *Medeola virginiana* L.   INDIAN CUCUMBER-ROOT

F1933A CHJJ Makes fish bite. Chew root, spit on hook.

F1938A CHJJ See 301, *Mentha spicata*, F1938C.

FndA KD Convulsions in babies. Dry berries and leaves, crush, put pinch according to baby's size in lukewarm water. Leave until soaked. Give teaspoon according to size.

FndB SS Did not know.

433. *Polygonatum pubescens* (Willd.) Pursh SOLOMON'S-SEAL, DOWNY SOLOMON'S-SEAL

W1912A KD Gas on the stomach. With: (1) 68, *Sanguinaria canadensis*, W1912J; (2) 375, *Inula helenium*, W1912F; (3) 368, *Eupatorium*

maculatum, W1912C. Take ½ handful cut into 3 in. lengths, and do the same with (2). Add ½ handful of (1) and 3 roots of (3). Boil in ½ gal. water down to 1 qt. Cool and drain. Take 1 wineglassful 3 hr. apart.

W1912B DJ Snow blindness. Smash about 3 or 4 in. of the root. Let stand a little (in water). Soak cloth and squeeze into eyes.

W1912C DJ Fishing medicine. With: (1) 434, *Smilacina racemosa,* W1912B. Smash 2 in. lengths of roots of each in 1 cup water. Soak or pass fishing line through this liquid. Leaving in over night is best. Put smashed roots in with the worms. Will get a fish on each hook, every cast.

W1914A DJ Sore eyes. Put roots of 1 plant in ½ teacup warm water. Then root is simply smashed (see W1912B, above).

434. *Smilacina racemosa* (L.) Desf.   FALSE SOLOMON'S-SEAL OR SPIKENARD

W1912A DJ See 298, *Lycopus virginicus,* W1912A.

W1912B DJ See 433, *Polygonatum pubescens,* W1912B.

W1914A DJ Rooster fighting medicine. With: (1) 413, *Carex vulpinoidea,* W1914A. Put cluster of roots from 2 plants in 2 qt. water. Boil down about ½. Put under rooster's wing and other rooster will put his head under rival's wing. When this is done, the opponent rooster will be blinded. Does not affect fowl it is put on. Wash off after fight.

F1933A CHJJ See 294, *Collinsonia canadensis,* F1933A.

FndA FSt Spoiled snakebites. Bandage with root.

H1973A DC Blood remedy and tapeworms. With: (1) 53, *Coptis trifolia,* H1973A; (2) 68, *Sanguinaria canadensis,* H1973C; (3) 402, *Acorus calamus,* H1973B; (4) 256, *Panax quinquefolia,* H1973A; (5) 196, *Cassia marilandica,* H1973A; (6) 41, *Sassafras albidum,* H1973A; (7) 153, *Gaultheria procumbens,* H1973B; (8) 133, *Populus balsamifera,* H1973A;

plus 3 other unspecified ingredients. Take roots of all and the tree buds of (8) and mix with whisky. Let stand 1 week. Take a tablespoon each morning.

H1973B Anyms. For when woman has miscarriage. With: (1) 255, *Aralia racemosa,* H1973A. Boil roots in water.

H1973C Anyms. For witching. With: (1) 42, *Asarum canadense,* H1973E; and a second, unspecified plant. No details given.

435. *Smilacina stellata* (L.) Desf.   STARFLOWER

W1912A DJ See 323, *Lobelia cardinalis,* W1912B.

436. *Streptopus roseus* Michx.   ROSE MANDARIN, TWISTED-STALK

W1912A KG Fallen womb. Put the roots of 2 plants in 1 qt. lukewarm water. Drink 1 teaspoon of the liquid and sit down in the remaining liquid after warming it.

437. *Trillium* sp.   TRILLIUM

W1912A DJ Rheumatism. With: (1) fish worms. Take bulbs, dry, smash up to powder. Fill pickle jar ½ full of (1). Cover up and place in sun. Pour off the oil. Rub oil on joints, then put on powder.

W1912B KD To detect bewitchment. With: (1) 43, *Nuphar lutea,* W1912E; (2) 127, *Viola sagittata,* W1912A; (3) 402, *Acorus calamus,* W1912; (4) 42, *Asarum canadense,* W1912K. No specifics given. If person spits out medicine, raves and uses foul language, person is a witch. If not, person may be used to protect oneself against bewitchment.

W1912C KD Stiff muscles. With: (1) 49, *Aquilegia canadensis,* W1912C; (2) 57, *Ranunculus abortivus,* W1912F; (3) 55, *Hepatica nobilis* var. *acuta,* W1912C. Take a handful of root, add handful of root of (1), handful of whole plant of (2), and 3 whole plants of (3). Put in quart water, boil down to ½. Leave roots until cool. Drink ½ wineglass 2 hr. apart.

W1914A RS For the itch. Wash the irritated parts with its milky liquid.

W1914B JD For fishing. Smash 1 root, put in small cup, add a little water. Put fish line in and leave all night.

F1933A CHJJ See 440, *Veratrum viride*, F1933B.

F1938A CHJJ Headache, catarrh. With: (1) 403, *Arisaema triphyllum*, F1938C; (2) 189, *Rubus allegheniensis*, F1938C. Snuff it up.

F1939A CG Food for woman in the womb. Boil handful of the root. Keep drinking.

FndA FSt Did not use, said was poisonous.

FndB SS For the blood. Did not know preparation.

FndC KD Chapped hands, suntan oil. Crush plant and rub on.

H1973A Anyms. For luck and protection of teeth. Dry root and carry it with you.

438. *Uvularia perfoliata* L.   STRAWBELL, BELL-WORT, PERFOLIATE BELLWORT

W1912A MR Cough medicine for children. Make tea of root. Give ½ teacup when thirsty.

W1914A DJ Sore eyes of any degree. Smash roots of 1 plant and steep in ½ cup warm water. Wash eyes.

W1914B JD Sore eyes. With: (1) $a^h se^n$ $niyunra^h dont$; (2) $gane^{n'} d\bar{o}?ta?$. Preparation not given.

F1933A CHJJ For setting bones. Smash whole plant, put in medium hot water, soak. Then apply plant to fracture and bind with white cloth. Then drink solution that remains.

F1933B JS Fractures. Smash roots, bind to fracture. Pound and boil root (1 handful) in 1 gal.

water. Drain, drink when necessary.

F1938A HRE & SRE Broken bones. Smash the lateral roots and apply as poultice.

F1938B CRE/ET Broken bones. Squash roots to a paste. Place directly over the set bone. Dampen if it dries. Do not use a splint.

F1938C JC Bone injury. Make a tea of the root. Apply plant to region injured.

439. *Uvularia sessilifolia* L.   WILD-OATS, SESSILE-LEAVED BELLWORT

F1933A CHJJ No use given.

F1933B SS Sets bones. Also a blood purifier. Pound root, poultice. Also put roots in ½ teacup water, lukewarm. Strain and drink. Apply roots as poultice.

F1938A HRE & SRE Set bones. See 438, *Uvularia perfoliata*, F1938A, above.

440. *Veratrum viride* Ait.   FALSE OR WHITE HELLEBORE, INDIAN-POKE

F1933A JS Head cold, TB. Dry root, scrape, and snuff powder.

F1933B CHJJ Catarrh, headaches, colds. With: (1) 437, *Trillium* sp., F1933A. Dry roots, scrape, snuff powder. Roots gathered in fall.

F1938A CHJJ Snuff. Dry root.

F1938B CHJJ To break open boils. With: (1) 363, *Cirsium arvense*, F1938A; (2) 33, *Pinus rigida*, F1938B. Mash and mix roots. Mix in roots of (1), steep bark of (2). Mix all together, apply as poultice.

F1938C CHJJ Snuff for headache, catarrh. With: (1) 403, *Arisaema triphyllum*, F1938D; (2) 189, *Rubus allegheniensis*, F1938D. Grind up roots, do same for (1). Grind up the bark of (2). Snuff dried powder.

F1939A CG Colds. Dry root, snuff.

FndA SS Head cold. Dry, scrape, snuff.

FndB FSt Catarrh. Dry root, scrape, snuff.

IRIDACEAE (IRIS FAMILY)

441. *Iris versicolor* L.   BLUE FLAG, WILD IRIS, POISON FLAG

W1912A MS To induce pregnancy, paralysis. Gather root at night. Pound in cloth, put cloth in teacup water. Squeeze juice out. Drink early in the morning. Remain in room 2 days. Take at menses. Patient must not be seen or she will die.

W1914A JD Physic. Take a piece the length of the second joint of forefinger, put in ½ cupful water. Let stand for a while and drink.

F1938A CHJJ See 97, *Corylus americana,* F1938B.

442. *Sisyrinchium angustifolium* Mill. (*S. graminoides* Bickn.)   STOUT BLUE-EYED GRASS

F1938A CRE/ET Constipation. Boil ½ cupful of the roots and stalks together in 1½ qt. water until 1 qt. remains. Take a cupful before meals in the morning.

FndA CHJJ Summer complaint. With: (1) 348, *Achillea millefolium,* FndA. Preparation not given.

443. *Sisyrinchium montanum* Greene   BLUE-EYED GRASS

W1912A ?? Fevers such as malaria and scarlet fever, but not typhoid. Put 3 plants in 2 qt. water. Boil down ½. Take ½ teacup 3 times per day before meals.

FndA SS Physic for old people. Did not give preparation. Feared it was poison.

SMILACACEAE (GREENBRIER FAMILY)

444. *Smilax herbacea* L.   JACOB'S-LADDER, CARRION-FLOWER

W1912A KD See 114, *Polygonum punctatum,* W1912A.

F1938A CRE/ET Rheumatism. With: (1) 51, *Cimicifuga racemosa,* F1938D; (2) 405, *Symplocarpus foetidus,* F1938D; (3) 32, *Picea* sp., F1938A; (4) *deyondagwenta*; (5) *deyagodago.* Use a handful of each. Boil 1-2 qt. water until some water evaporates. Cover affected parts and take steam bath.

F1938B CHJJ See 405, *Symplocarpus foetidus,* F1938C.

FndA HRE Stomach troubles in elderly. Boil roots of 2-3 plants in 8 qt. water about 10 min. Drink like water.

H1973A Anyms. Deodorant. Cut root, put in oven, brown. Make powder.

445. *Smilax hispida* Muhl. ex Torr.   BRISTLY GREENBRIER, HAGBRIER

W1912A MS Eye poultice. Scrape bark, apply between 2 cloths to the eyes as a poultice to draw out foreign substances.

W1912B DJ See 187, *Rosa acicularis,* W1912B.

W1914A DJ See 102, *Amaranthus retroflexus,* W1914A.

ORCHIDACEAE (ORCHID FAMILY)

446. *Coeloglossum viride* (L.) Hartm.   LONG-BRACTED ORCHID, LARGE-BRACTED ORCHID, FROG ORCHID

W1914A DJ To bring away placenta after child-birth. With: (1) 430, *Lilium philadelphicum,* W1914B. Put the whole plant along with the root of (1) [in water]. Boil a little. Take 3 tablespoons 3 times per day.

447. *Corallorhiza maculata* (Raf.) Raf.   SPOTTED CORALROOT

W1912A ED For heaves in horses. Put a handful of whole plant in 2 qt. water that is hot, steep a little. Add to horse's grain for 3 days, stop for 3 days, repeat.

W1912B DJ To bring money or to get your wife back. Also a hunting medicine. For money, keep root in pocket. For wife, put a little of the root in ½ cup water, cover up, and call the name of the woman. After 15 min., open up and put it on clothing of the person (the woman). She will return in 2-3 days. Fix the same way for hunting. Use tobacco when picking and tell animal's name. Wash guns and cloths with it.

W1913A PJ Basket medicine or for anything you want to sell. Decoction of root is sprinkled over items. If prospective buyer picks one of these articles up, he will not be able to let go.

W1914A DJ For selling skins or baskets, and for bringing women. Smash small handful of roots, put in cupful water. Sprinkle on articles to be sold. Wet cob with this, first using tobacco and telling it what you want. Sprinkle on bed where woman used to lie. Place corn cob with one end in the fire, saying "come home right away." Let the cob burn slowly, push it in a little at a time. When the cob finishes burning the woman will surely feel it.

W1914B RS Cure for love medicine—you will find out who gave the love medicine. Also for selling baskets. With: (1) *diyusaiitgona.* Smash up a small quantity of the roots of both, tie in a cloth, let stand a little in warm water (½ cup). Drink and wait for 5 min., then drink a lot of warm water to vomit. For selling things, sprinkle the decoction of the articles for sale.

F1933A SS Antiwitch medicine, blood medicine, love and basket medicine. Pound roots, put in lukewarm water.

F1933B CHJJ See 214, *Epilobium angustifolium,* F1933B.

F1938A CHJJ Basket medicine. Put 3-4 roots in a teacup of water. Sprinkle on whatever is to be sold.

448. *Cypripedium acaule* Ait.   PINK LADY-SLIPPER, MOCCASIN FLOWER

W1912A DJ For pains all over the body and skin, caused by bad blood. Boil roots of 1 plant in 3 qt. water down to ½. Take 1 cup every hour.

W1912B DJ (cf. 431, *Maianthemum canadense*) Cure bite of mad dog. Smash leaves, wet, tie on as poultice.

449. *Cypripedium arietinum* R. Br.   RAM'S-HEAD LADYSLIPPER (RARE)

W1914A DJ Intestinal trouble with inflation and pains (lumps). Put 2 whole plants in 3 qt. water. Boil quite a bit. Also smash plants and poultice with them. Drink ½ cup anytime you like.

450. *Cypripedium calceolus* L. var. *parviflorum* (Salisb.) Fern.   SMALL YELLOW LADYSLIPPER

W1912A KD See 35, *Tsuga canadensis,* W1912C.

F1939A HJ Nervousness, lack of energy. Steep root like tea. Drink ½ cup at a time when thirsty.

FndA CHJJ General lassitude. Boil root, drink ad lib.

FndB FSt Blood medicine, for when the blood is bad from scrofula. With: (1) 116, *Rumex crispus,* FndB. Boil both roots in 4 qt. water. Let cool. Drink 1 teacup 3 times per day.

H1973A DC Nerve medicine.

451. *Platanthera grandiflora* (Bigel.) Lindl.   LARGE PURPLE FRINGED ORCHID

FndA HRE & SRE To frighten away ghosts. Dry roots, smash 1 root, put in 2 qt. water. Drink like water.

452. *Platanthera orbiculata* (Pursh) Lindl.
ROUND-LEAVED ORCHID, PAD-LEAF

FndA FSt Scrofula, swollen neck glands. Bandage sores with leaves.

FndB L & AK Cuts. Make poultice of leaf.

453. *Platanthera psycodes* (L.) Lindl.   SMALL PURPLE FRINGED ORCHID

W1912A KD Cramps in young children. Pound a single root and a pinch of the flowers. Put in teacup of hot water. Let cool. Take 1 teaspoon every 1½ hr.

F1938A CHJJ See 301, *Mentha spicata,* F1938C.

FndA FSt (cf. 26, *Onoclea sensibilis*) Parturition medicine. Pound 1 root, put in warm water. Make enough to drink in 1 day. Prepare fresh every day.

454. Orchid, unspecified name.

W1914A JD A remedy for bad medicine (witchcraft). Put roots of a couple of plants in 3/4 cup warm water. Drink at night, 2 tablespoons at a time.

References Cited

Ackerknecht, E.H.
1945 Primitive Medicine. *Transactions of the New York Academy of Sciences*, series II, 8:26-37.

Barber, T.X.
1961 Death by Suggestion: A Critical Note. *Psychosomatic Medicine* 23:153-155.

Beauchamp, W.M.
1888 Onondaga Indian Names of Plants. *Torrey Botanical Club Bulletin* 7:262-266.
1902 Onondaga Plant Names. *Journal of American Folklore* 15:91-103.
1922 *Iroquois Folk Lore.* Ira J. Friedman, Port Washington, New York.

Bidney, D.
1963 So-called Primitive Medicine and Religion. In *Man's Image in Medicine and Anthropology,* edited by I. Galdston, pp. 141-156. International Universities Press, New York.

Blau, H.
1963 Dream Guessing: A Comparative Analysis. *Ethnohistory* 10:233-249.
1966 Function and the False Faces: A Classification of Onondaga Masked Rituals and Themes. *Journal of American Folklore* 79:564-580.

Chafe, W.L.
1961 *Seneca Thanksgiving Rituals.* Bureau of American Ethnology Bulletin 183. Washington, D.C.
1967 *Seneca Morphology and Dictionary.* Smithsonian Press, Washington, D.C.

Clements, F.E.
1932 Primitive Concepts of Disease. *University of California Publications in American Archaeology and Ethnology* 32:185-252.

Corlett, W.T.
1935 *The Medicine Man of the American Indians and His Cultural Background.* Charles C. Thomas, Baltimore.

D'Andrade, R.G.
1975 Cultural Constructions of Reality. In *Cultural Illness and Health: Essays in Human Adaptation,* edited by L. Nader and T.W. Maretzki, pp. 115-127. American Anthropological Association, Washington, D.C.

Downs, J.F.
1971 Spirits, Power, and Man. In *Man in Adaptation: The Institutional Framework,* edited by Y.A. Cohen, pp. 187-194. Aldine Atherton, New York.

Dubos, R.
1961 *The Mirage of Health.* Doubleday, New York.

Faron, L.C.
1963 Death and Fertility Rites of the Mapuche (Araucanian) Indians of Central Chile. *Ethnology* 2:135-156.

Fegos, P.
1959 Man, Magic, and Medicine. In *Medicine and Anthropology,* edited by I. Galdston, pp 11-35. International Universities Press, New York.

Fenton, W.N.
1936 An Outline of Seneca Ceremonies at Coldspring Longhouse. *Yale University Publications in Anthropology* 9:3-23.
1940a A Further Quest for Iroquois Medicines. In *Explorations and Field-work of the Smithsonian Institution in 1939,* pp. 93-96. Washington, D.C.
1940b An Herbarium from the Allegany Senecas. In *The Historical Annals of Southwestern New York,* edited by W.J. Doty, pp. 787-796. Lewis Historical Publishing, New York.
1941a Masked Medicine Societies of the Iroquois. In *Annual Report of the Smithsonian Institution for 1940,* pp. 397-430. Washington, D.C.
1941b *Iroquois Suicide: A Study in the Stability of a Culture Pattern.* Burean of American Ethnology Bulletin 128. Washington, D.C.
1942 Contacts Between Iroquois Herbalism and Colonial Medicine. In *Annual Report of the Smithsonian Institution for 1941,* pp. 503-526. Washington, D.C.
1949 Medicinal Plant Lore of the Iroquois. *Bulletin to the Schools* 35:233-237.
1951 The Concept of Locality and the Program of Iroquois Research. In *Symposium on Local Diversity in Iroquois Culture,* edited by W.N. Fenton, pp. 3-12. Bureau of American Ethnology Bulletin 149. Washington, D.C.
1953 *The Iroquois Eagle Dance: An Offshoot of the Calumet Dance.* Bureau of American Ethnology Bulletin 156. Washington, D.C.
1962 This Island, the World on the Turtle's Back. *Journal of American Folklore* 75:283-300.
1963 The Seneca Green Corn Ceremony. *The Conservationist* 18:20-22.
1965 The Iroquois Confederacy in the Twentieth Century: A Case Study of the Theory of Lewis H. Morgan in "Ancient Society." *Ethnology* 4:251-265.
1968 *Parker on the Iroquois.* Syracuse University Press, Syracuse.
1971 The Iroquois in History. In *North American Indians in Historical Perspective,* edited by E.B. Leacock and N.O. Luire, pp 129-168. Random House, New York.
1975 The Lore of the Longhouse: Myth, Ritual, and Red Power. *Anthropological Quarterly* 48:131-146.
1979 The "Great Good Medicine." *New York State Journal of Medicine* 79:1603-1609.

Fenton, W.N. (editor)
1968 *Parker on the Iroquois. Iroquois Use of Maize and Other Food Plants; The Code of Handsome Lake, the Seneca Prophet; The Constitution of the Five Nations.* Syracuse University Press, Syracuse.

Fernald, M.L., and A.C. Kinsey
1958 *Edible Wild Plants of Eastern North America.* Harper and Row, New York.

Foster, M.K.
1974 *From the Earth to Beyond the Sky: An Ethnographic Approach to Four Longhouse Speech Events.* National Museum of Man, Mercury Series. Canadian Ethnology Service Paper 20. National Museums of Canada, Ottawa.

Frake, C.O.
1964 The Diagnosis of Disease Among the Subanum of Mindanao. In *Language in Culture and Society,* edited by D. Hymes, pp. 193-211. Harper and Row, New York.

Frank, J.D.
1961 *Persuasion and Healing.* Johns Hopkins University Press, Baltimore.

Gellhorn, E., and W.F. Kiely
1973 Autonomic Nervous System in Psychiatric Disorder. In *Biological Psychiatry,* edited by J. Mendels, pp. 235-261. John Wiley and Sons, New York.

Gleason, H.A.
1952 *The New Britton and Brown Illustrated Flora of the Northeastern United States and Adjacent Canada.* 3 vols. The New York Botanical Garden, New York.

Gonzales, N.S.
1966 Health Behavior in Cross-cultural Perspective: A Guatemalan Example. *Human Organization* 25:122-125.

Greenlee, R.F.
1944 Medicine and Curing Practices of the Modern Florida Seminoles. *American Anthropologist* 46:317-328.

Grinnell, G.B.
1935 *The Story of the Indian.* Appleton-Century, New York.

Groen, J.J.
1973 The Psychosomatic Study of Anxiety in Internal Diseases. In *Anxiety Factors in Comprehensive Patient Care: Where General Medicine and Psychiatry Meet,* edited by W.L. Rees, pp. 8-17. American Elsevier, New York.

Hallowell, A.I.
1941 The Social Function of Anxiety in a Primitive Society. *American Sociological Review* 6:869-881.

Herrick, J.W.
1976 Placebos, Psychosomatic and Psychogenic Illnesses and Psychotherapy: Their Theorized Cross-cultural Development. *The Psychological Record* 26:327-342.
1977 *Iroquois Medical Botany.* Ph.D. dissertation, State University of New York at Albany. University Microfilms International, Ann Arbor.
1978 Powerful Medicinal Plants in Traditional Iroquois Culture. *New York State Journal of Medicine* 78:979-987.
1983 The Symbolic Roots of Three Potent Iroquois Medicinal Plants. In *The Anthropology of Medicine,* edited by L. Romanucci-Ross, D.E. Moerman, and L.R. Tancredi, pp. 134-155. Praeger, New York.

Hewitt, J.N.B.
1904 Iroquois Cosmology. *Bureau of American Ethnology Annual Reports* 32:127-339.
1925-1926 Iroquois Cosmology. *Bureau of American Ethnology Annual Reports* 43:449-820.

Hewitt, J.N.B., and J. Curtin
1918 Seneca Fictions, Legends, and Myths. In *Thirty-Second Annual Report of the Bureau of American Ethnology, 1910-1911,* pp. 37-819. Washington, D.C.

Hill, A.F.
1952 *Economic Botany.* McGraw-Hill, New York.

Hogbin, H.I.
1947 Pagan Beliefs in a New Guinea Village. *Oceania* 48:120-145.

Homans, G.C.
1941 Anxiety and Ritual: The Theories of Malinowski and Radcliffe-Brown. *American Anthropologist* 43:164-172.

Horton, R.
1967 African Traditional Thought and Western Science. *Africa* 37:50-71, 155-187.

Hrdlicka, A.
1932 Disease, Medicine, and Surgery Among the American Aborigines. *Journal of the American Medical Association* 99:1661-1666.

Isaacs, H.
1976-1977 Iroquois Herbalism: The Past 100 Years. *The International Journal of Social Psychiatry* 22 (winter):272-278.

Jarcho, S.
1964 Some Observations on Disease in Prehistoric North America. *Bulletin of the History of Medicine* 38:1-19.

Kanfer, H., and A.P. Goldstein (editors)
1975 *Helping People Change: A Textbook of Methods.* Pergamon Press, New York.

King, S.H.
1955 Psychosocial Factors Associated with Rheumatoid Arthritis. *Journal of Chronic Diseases* 2:287-302.

Kluckhohn, F.R., and F.L. Strodtbeck
1961 *Variations in Value Orientations.* Row, Peterson, Elmsford, New York.

Koch, M.F., and G.D. Molnar
1974 Psychiatric Aspects of Patients with Unstable Diabetes Mellitus. *Psychosomatic Medicine* 36:57-68.

Kurath, G.
1968 Dance and Song Rituals of Six Nations Reserve, Ontario. *Folklore Series 4. National Museum of Man Bulletin* 220:3-205.

Lafitau, J.F.
1724 *Moeurs des sauvages Amériquains.* Saugrain l'Aîné, Paris.

Laughlin, W.
1963 Primitive Theory of Medicine: Empirical Knowledge. In *Man's Image in Medicine and Anthropology,* edited by I. Galdston, pp. 116-140. International Universities Press, New York.

Lefly, H.
1975 Report on Roundtable on Psychiatry and Anthropology. *Medical Anthropology Newsletter* 6:2-3.

Lehmann, H.E., and D.A. Knight
1960 Placebo-proness and Placebo-resistance of Different Psychological Functions. *Psychiatric Quarterly* 34:505-516.

Lex, B.
1973 Altered States of Consciousness in Northern Iroquoian Rituals. Unpublished MS in the author's possession.
1974 Voodoo Death: New Thoughts on an Old Explanation. *American Anthropologist* 76:818-823.

Lounsbury, F.G.
1953 *Oneida Verb Morphology.* Yale University Publications in Anthropology 48. New Haven.

Maruyama, M.
1974 Paradigmatology and Its Application to Cross-disciplinary, Cross-professional, and Cross-cultural Communication. *Cybernetica* 17:136-156, 237-281.

Mechanic, D.
1966 Response Factors in Illness: The Study of Illness Behavior. *Social Psychiatry* 1:11-20.

Michelson, G.
1973 *A Thousand Words of Mohawk*. National Museum of Man, Mercury Series. Canadian Ethnology Service Paper 5. National Museums of Canada, Ottawa.

Mitchell, R.S.
1983 *Berberidaceae through Fumariaceae of New York State*. New York State Museum Bulletin 451. Albany.
1986 *A Checklist of New York State Plants*. New York State Museum Bulletin 458. Albany.
1988 *Platanaceae through Myricaceae of New York State*. New York State Museum Bulletin 464. Albany.

Mitchell, R.S., and E.O. Beal
1979 *Magnoliaceae through Ceratophyllaceae of New York State*. New York State Museum Bulletin 435. Albany.

Mitchell, R.S., and J.K. Dean
1982 *Ranunculaceae (Crowfoot Family) of New York State*. New York State Museum Bulletin 446. Albany.

Moerman, D.E.
1979 Anthropology of Symbolic Healing. *Current Anthropology* 20(4):59-80.

Montagu, A.M.F.
1946 Primitive Medicine. *New England Journal of Medicine* 235:43-49.

Mooney, J.
1890 Sacred Formulas of the Cherokees. In *Bureau of American Ethnology 7th Annual Report*, pp. 302-397. Washington, D.C.

Mooney, J., and F.M. Olbrechts
1932 *Swimmer Manuscripts—Cherokee Sacred Formulas and Medicinal Prescriptions*. Bureau of American Ethnology Bulletin 99. Washington, D.C.

Morgan, L.H.
1962 *League of the Ho-De-No-Sau-Nee, Iroquois*. Reprinted with an introduction by W.N. Fenton. Corinth, New York. Originally published 1851.

Northrop, F.S.C.
1950 Ethics and the Integration of Natural Knowledge. In *The Stillwater Conference on the Nature of Concepts, Their Structure and Social Inter-relations*, edited by F.S.C. Northrop and H. Marenau, pp. 25-44. Foundation for Integrated Education, New York.

Ogden, E.C.
1981 *Field Guide to Northeastern Ferns*. New York State Museum Bulletin 444. Albany.

Opler, M.E.
1946 An Application of the Theory of Themes in Culture. *Journal of the Washington Academy of Sciences* 36:137-166.

Parker, A.C.
1909 Secret Medicine Societies of the Seneca. *American Anthropologist* 11:161-185. Also in Fenton 1968, pp. 113-130.
1910 *Iroquois Uses of Maize and Other Food Plants*. New York State Museum Bulletin 144. Albany. Also in Fenton 1968, pp. 5-119.
1912 Certain Iroquois Tree Myths and Symbols. *American Anthropologist* 14:608-620.
1928 *Indian Medicine and Medicine Men*. Archeological Report of the Minister of Education, Annual Reports. Toronto.

Pierce, R.V.
1875 *The People's Common Sense Medical Advisor*. World Dispensary Printing Office, Buffalo.

Polgar, S.
1962 Health and Human Behavior: Areas of Interest to the Social and Medical Sciences. *Current Anthropology* 3:159-205.
1963 Health Action in Cross-cultural Perspective. In *Handbook of Medical Sociology*, edited by H.E. Freeman, S. Levine, and

L.G. Reeder, pp. 397-419. Prentice-Hall, Englewood Cliffs, New Jersey.

Price-Williams, D.R.
1962 A Case Study of Ideas Concerning Disease Among the Tiv. *Africa* 32:123-131.

Rioux, M.
1951 Some Medical Beliefs and Practices of the Contemporary Iroquois Longhouses at Six Nations Reserve. *Journal of the Washington Academy of Sciences* 41:152-158.

Rivers, W.H.R.
1924 *Medicine, Magic, and Religion*. Harcourt, Brace, New York.

Roberts, S.L.
1944 Disease Concepts in North America. *American Anthropologist* 46:559-564.

Rubel, A.J.
1960 Concepts of Disease in Mexican-American Culture. *American Anthropologist* 62:795-814.

Sargent, C.S.
1891-1902 *The Silva of North America . . . Exclusive of Mexico*. 14 vols. Houghton, Mifflin, and Company, Boston.

Saunders, L.
1965 Healing Ways in the Spanish Southwest. In *Patients, Physicians, and Illness*, edited by E.G. Jaco, pp. 189-206. Free Press, New York.

Shimony, A.
1961 *Conservatism Among the Iroquois at the Six Nations Reserve*. Yale University Publications, New Haven.
1970 Iroquois Witchcraft. In *Systems of North American Witchcraft and Sorcery*, edited by D. Walker, Jr., pp. 239-265. University of Idaho Press, Moscow.

Skinner, B.F.
1965 *Science and Human Behavior*. Free Press, New York.

Smith, E.A.
1885 The Significance of Flora to the Iroquois. *American Association for the Advancement of Science Publications* 34:404-411.

Snyderman, G.S.
1949 The Case of Daniel P.: An Example of Seneca Healing. *Journal of the Washington Academy of Sciences* 39:217-220.

Speck, F.
1949 *Midwinter Rites of the Cayuga Longhouse*. University of Pennsylvania Press, Philadelphia.

Stone, E.
1935 Medicine Among the Iroquois. *Annals of Medical History* 6:529-539.

Tooker, E.
1960 Three Aspects of Northern Iroquoian Culture Change. *Pennsylvania Archeologist* 30:65-71.
1970 *The Iroquois Ceremonial of Midwinter*. Syracuse University Press, Syracuse.

Vogel, V.
1970 *American Indian Medicine*. University of Oklahoma Press, Norman.

Wallace, A.F.C.
1972 *The Death and Rebirth of the Seneca*. Vintage, New York.

Waugh, F.W.
1916 *Iroquis [sic] Foods and Food Preparation*. Canada Department of Mines, Geological Survey, Memoir 86. Ottawa.

Weaver, S.
1967 *Health, Culture, and Dilemma: A Study of the Non-conservative Iroquois, Six Nations Reserve, Ontario*. Ph.D. dissertation, University of Toronto, Toronto.

Wolf, S.
    1950 Effects of Suggestion and Conditioning on the Action of
          Chemical Agents in Human Subjects—The Pharmacology of
          Placebos. *Journal of Clinical Investigation* 29:100-109.

Plant Picture Credits

11. *Lycopodium digitatum:* Gleason 1952:1:7
12. *Lycopodium lucidulum:* Gleason 1952:1:3
13. *Lycopodium obscurum:* Gleason 1952:1:5
17. *Equisetum fluviatile:* Gleason 1952:1:17
19. *Botrychium virginianum:* Gleason 1952:1:21
20. *Osmunda cinnamomea:* Ogden 1981:88
21. *Osmunda claytoniana:* Ogden 1981:90
22. *Osmunda regalis:* Ogden 1981:91
23. *Polypodium virginianum:* Ogden 1981:98
24. *Adiantum pedatum:* Ogden 1981:39
26. *Onoclea sensibilis:* Ogden 1981:86
27. *Polystichum acrostichoides:* Ogden 1981:100
28. *Thelypteris palustris:* Ogden 1981:106
29. *Taxus canadensis:* Gleason 1952:1:58
30. *Abies balsamea:* Sargent 1891-1902:12:pl. 610
31. *Larix laricina:* Sargent 1891-1902:12:pl. 593
33. *Pinus rigida:* Sargent 1891-1902:11:pl. 579
34. *Pinus strobus:* Sargent 1891-1902:11:pl. 538
35. *Tsuga canadensis:* Sargent 1891-1902:12:pl. 603
36. *Juniperus communis:* Sargent 1891-1902:10:pl. 514
37. *Juniperus virginiana:* Sargent 1891-1902:10:pl. 524
38. *Thuja occidentalis:* Sargent 1891-1902:10:pl. 532
39. *Magnolia acuminata:* Mitchell and Beal 1979:3
41. *Sassafras albidum:* Mitchell and Beal 1979:12
42. *Asarum canadense:* Gleason 1952:2:61
43. *Nuphar lutea:* Mitchell and Beal 1979:28
48. *Anemone virginiana:* Gleason 1952:2:181
49. *Aquilegia canadensis:* Gleason 1952:2:183
50. *Caltha palustris:* Gleason 1952:2:168
51. *Cimicifuga racemosa:* Mitchell and Dean 1982:9
52. *Clematis virginiana:* Mitchell and Dean 1982:29
54. *Hepatica nobilis:* Gleason 1952:2:183
57. *Ranunculus abortivus:* Gleason 1952:2:173
58. *Ranunculus acris:* Gleason 1952:2:175
59. *Ranunculus bulbosus:* Gleason 1952:2:176
60. *Ranunculus hispidus:* Gleason 1952:2:177
63. *Thalictrum dioicum:* Gleason 1952:2:160
66. *Jeffersonia diphylla:* Gleason 1952:2:189
67. *Podophyllum peltatum:* Gleason 1952:2:189
68. *Sanguinaria canadensis:* Gleason 1952:2:197
73. *Celtis occidentalis:* Sargent 1891-1902:7:pl. 317
74. *Ulmus americana:* Mitchell 1988:9
75. *Ulmus rubra:* Mitchell 1988:12
76. *Cannabis sativa:* Mitchell 1988:18
77. *Laportea canadensis:* Mitchell 1988:32
79. *Urtica dioica:* Mitchell 1988:36
80. *Carya ovata:* Mitchell 1988:55
81. *Juglans cinerea:* Mitchell 1988:42
82. *Juglans nigra:* Mitchell 1988:44
84. *Fagus grandifolia:* Sargent 1891-1902:9:pl. 444

85. *Quercus alba:* Sargent 1891-1902:8:pl. 357
88. *Quercus macrocarpa:* Sargent 1891-1902:8:pl. 371
89. *Quercus rubra:* Sargent 1891-1902:8:pl. 410
92. *Betula alleghaniensis:* Sargent 1891-1902:9:pl. 449
93. *Betula lenta:* Sargent 1891-1902:9:pl. 448
94. *Betula papyrifera:* Sargent 1891-1902:9:pl. 451
95. *Betula populifolia:* Sargent 1891-1902:9:pl. 450
96. *Carpinus caroliniana:* Sargent 1891-1902:9:pl. 447
97. *Corylus americana:* Gleason 1952:2:33
98. *Corylus cornuta:* Gleason 1952:2:33
99. *Ostrya virginiana:* Sargent 1891-1902:9:pl. 445
101. *Chenopodium album:* Gleason 1952:2:89
102. *Amaranthus retroflexus:* Gleason 1952:2:104
103. *Claytonia virginica:* Gleason 1952:2:116
110. *Polygonum aviculare:* Gleason 1952:2:76
120. *Tilia americana:* Sargent 1891-1902:1:pl. 24
121. *Malva moschata:* Gleason 1952:2:526
122. *Malva neglecta:* Gleason 1952:2:526
123. *Sarracenia purpurea:* Gleason 1952:2:251
127. *Viola sagittata:* Gleason 1952:2:558
128. *Viola striata:* Gleason 1952:2:565
131. *Sicyos angulatus:* Gleason 1952:3:313
132. *Populus alba:* Gleason 1952:2:3
133. *Populus balsamifera:* Gleason 1952:2:4
135. *Populus grandidentata:* Gleason 1952:2:3
136. *Populus heterophylla:* Gleason 1952:2:4
137. *Populus tremuloides:* Gleason 1952:2:3
138. *Salix discolor:* Gleason 1952:2:23
139. *Salix nigra:* Gleason 1952:2:9
140. *Salix sericea:* Gleason 1952:2:21
142. *Armoracia rusticana:* Gleason 1952:2:227
147. *Cardamine douglassii:* Gleason 1952:2:229
149. *Rorippa sylvestris:* Gleason 1952:2:239
150. *Sisymbrium officinale:* Gleason 1952:2:245
151. *Thlaspi arvense:* Gleason 1952:2:217
154. *Kalmia latifolia:* Gleason 1952:3:15
157. *Chimaphila umbellata:* Gleason 1952:3:5
158. *Pyrola elliptica:* Gleason 1952:3:7
161. *Ribes americanum:* Gleason 1952:2:278
162. *Ribes hirtellum:* Gleason 1952:2:277
163. *Ribes rotundifolium:* Gleason 1952:2:276
165. *Mitella diphylla:* Gleason 1952:2:267
166. *Saxifraga pensylvanica:* Gleason 1952:2:265
168. *Agrimonia gryposepalla:* Gleason 1952:2:318
170. *Amelanchier arborea:* Gleason 1952:2:376
172. *Cataegus submollis:* Gleason 1952:2:371
175. *Geum aleppicum:* Gleason 1952:2:302
176. *Geum canadense:* Gleason 1952:2:301
177. *Geum rivale:* Gleason 1952:2:303
180. *Physocarpus opulifolius:* Gleason 1952:2:284

182. *Potentilla canadensis:* Gleason 1952:2:293
187. *Rosa acicularis:* Gleason 1952:2:327
189. *Rubus allegheniensis:* Gleason 1952:2:311
191. *Rubus occidentalis:* Gleason 1952:2:315
192. *Rubus odoratus:* Gleason 1952:2:317
193. *Spiraea alba:* Gleason 1952:2:286
195. *Waldsteinia fragarioides:* Gleason 1952:2:291
198. *Baptisia tinctoria:* Gleason 1952:2:393
199. *Desmodium canadense:* Gleason 1952:2:429
200. *Desmodium glutinosum:* Gleason 1952:2:427
204. *Strophostyles helvola:* Gleason 1952:2:453
205. *Trifolium hybridum:* Gleason 1952:2:401
208. *Vicia americana:* Gleason 1952:2:443
209. *Vicia sativa:* Gleason 1952:2:441
210. *Myriophyllum verticillatum:* Gleason 1952:2:601
214. *Epilobium angustifolium:* Gleason 1952:2:587
215. *Oenothera biennis:* Gleason 1952:2:591
216. *Oenothera perennis:* Gleason 1952:2:595
217. *Cornus alternifolia:* Sargent 1891-1902:5:pl. 216
219. *Cornus canadensis:* Gleason 1952:2:643
220. *Cornus drummondii:* Gleason 1952:2:645
221. *Cornus florida:* Gleason 1952:2:643
225. *Celastrus scandens:* Gleason 1952:2:503
226. *Euonymus americana:* Gleason 1952:2:503
227. *Euonymus europaea:* Gleason 1952:2:503
228. *Euonymus obovata:* Gleason 1952:2:503
231. *Chamaesyce glyptosperma:* Gleason 1952:2:483
232. *Ceanothus americanus:* Gleason 1952:2:515
236. *Parthenocissus quinquefolia:* Gleason 1952:2:522
237. *Vitis labrusca:* Gleason 1952:2:518
238. *Vitis riparia:* Gleason 1952:2:519
240. *Polygala verticillata:* Gleason 1952:2:473
242. *Aesculus hippocastanum:* Gleason 1952:2:509
243. *Acer pensylvanicum:* Sargent 1891-1902:2:pl. 85
244. *Acer rubrum:* Sargent 1891-1902:2:pl. 94
245. *Acer saccharum:* Sargent 1891-1902:2:pl. 92
246. *Acer spicatum:* Sargent 1891-1902:2:pl. 83
247. *Rhus typhina:* Sargent 1891-1902:3:pl. 103
249. *Zanthoxylum americanum:* Gleason 1952:2:467
251. *Geranium maculatum:* Gleason 1952:2:458
267. *Gentiana andrewsii:* Gleason 1952:3:63

268. *Gentianella quinquefolia:* Gleason 1952:3:61
272. *Asclepias incarnata:* Gleason 1952:3:75
273. *Asclepias syriaca:* Gleason 1952:3:77
274. *Asclepias tuberosa:* Gleason 1952:3:75
275. *Datura stramonium:* Gleason 1952:3:205
276. *Nicotiana rustica:* Gleason 1952:3:205
277. *Physalis heterophylla:* Gleason 1952:3:197
278. *Physalis pubescens:* Gleason 1952:3:195
283. *Hydrophyllum canadense:* Gleason 1952:3:107
284. *Hydrophyllum virginianum:* Gleason 1952:3:107
285. *Cynoglossum officinale:* Gleason 1952:3:121
286. *Echium vulgare:* Gleason 1952:3:115
287. *Hackelia virginiana:* Gleason 1952:3:123
291. *Verbena hastata:* Gleason 1952:3:130
294. *Collinsonia canadensis:* Gleason 1952:3:189
299. *Mentha aquatica:* Gleason 1952:3:187
301. *Mentha spicata:* Gleason 1952:3:188
303. *Nepeta cataria:* Gleason 1952:3:153
304. *Prunella vulgaris:* Gleason 1952:3:155
307. *Fraxinus americana:* Sargent 1891-1902:6:pl. 268
308. *Fraxinus nigra:* Sargent 1891-1902:6:pl. 265
310. *Chelone glabra:* Gleason 1952:3:220
311. *Linaria vulgaris:* Gleason 1952:3:229
312. *Mimulus ringens:* Gleason 1952:3:214
313. *Pedicularis canadensis:* Gleason 1952:3:251
314. *Penstemon hirsutus:* Gleason 1952:3:223
316. *Scrophularia marilandica:* Gleason 1952:3:226
317. *Verbascum thapsus:* Gleason 1952:3:218
318. *Veronica officinalis:* Gleason 1952:3:235
329. *Galium aparine:* Gleason 1952:3:284
330. *Galium triflorum:* Gleason 1952:3:285
332. *Mitchella repens:* Gleason 1952:3:279
335. *Lonicera canadensis:* Gleason 1952:3:299
336. *Lonicera dioica:* Gleason 1952:3:301
337. *Lonicera oblongifolia:* Gleason 1952:3:298
339. *Sambucus canadensis:* Gleason 1952:3:296
340. *Triosteum aurantiacum:* Gleason 1952:3:303
341. *Viburnum acerifolium:* Gleason 1952:3:292
403. *Arisaema triphyllum:* Gleason 1952:1:367
404. *Calla palustris:* Gleason 1952:1:369

Index

THE
Iroquois
AND THEIR
NEIGHBORS   Laurence M. Hauptman, *Series Editor*

This series presents a wide range of scholarship—archaeology, anthropology, history, public policy, sociology, women's studies—that focuses on the indigenous peoples of Northeastern North America. The series encourages more awareness and a broader understanding of the Iroquois Indians—the Mohawk, Oneida, Onondaga, Cayuga, Seneca, and Tuscarora—and their Native American neighbors and provides a forum for scholars to elucidate the important contributions of the first Americans from prehistory to the present day.

Selected titles in the series include: